Refereed and selected contributions from

From Parity Violation to Hadronic Structure and more

Grenoble, France

June 8–11, 2004

edited by

S. Kox, D. Lhuillier, F. Maas, and
J. Van de Wiele

Società Italiana
di Fisica

Springer

Serge Kox
Laboratoire de Physique Subatomique et de Cosmologie
53, Avenue des Martyres
38026 Grenoble Cedex, France
kox@in2p3.fr

David Lhuillier
CEA Saclay, DSM/DAPNIA/SPhN Bât. 703
91191 Gif-sur-Yvette, France
dlhuillier@cea.fr

Frank Maas
Institut für Kernphysik
A4-collaboration, Parity Violation Experiment
Johannes-Gutenberg-Universität Mainz
J.-J.-Becher-Weg 45
55099 Mainz, Germany
maas@kph.uni-mainz.de

Jacques van de Wiele
IPNO, Université Paris Sud
Bât. 100 M
91406 Orsay Cedex, France
vandewi@ipno.in2p3.fr

The articles in this book originally appeared on the internet (springeronline.com) as open access publication of the journal
The European Physical Journal A – Hadrons and Nuclei
Volume 24, Supplement 2
ISSN 1434-601X

Cataloging-in-Publication Data applied for

Bibliographic information published by Die Deutsche Bibliothek. Die Deutsche Bibliothek lists this publication in the Deutsche Nationalbibliografie; detailed bibliographic data is available in the Internet at http://dnb.ddb.de

ISBN-10 3-540-25501-X Springer Berlin Heidelberg New York
ISBN-13 978-3-540-25501-7 Springer Berlin Heidelberg New York

Springer is a part of Springer Science+Business Media
springeronline.com

Printed in Italy

Typesetting: PTP-Berlin, Berlin, Germany
Cover design: SIF Production Office, Bologna, Italy
Printing and Binding: Tipografia Compositori, Bologna, Italy

Printed on acid-free paper SPIN: 11414070 – 5 4 3 2 1 0

From Parity Violation to Hadronic Structure and more ... (Part II)

Part I :
Mainz June 5-8, 2002

Institut für Kernphysik
Universität Mainz

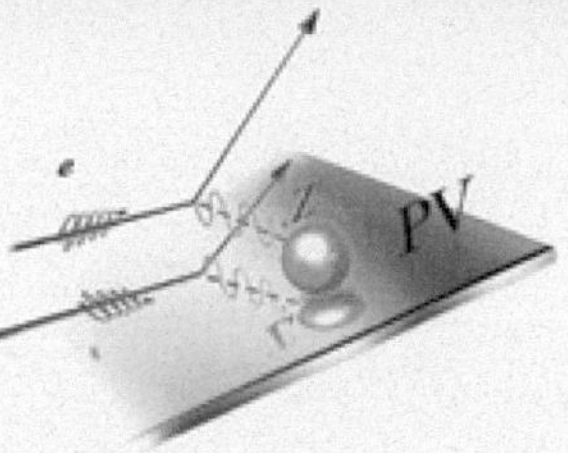

http://lpsc.in2p3.fr/congres/pavi2004/

Part II : Grenoble June 8-11, 2004

Conference secretary : Jocelyne Riffault
Laboratoire de Physique Subatomique et de Cosmologie
CNRS-IN2P3, UJF
53, avenue des Martyrs • F-38026 Grenoble cedex
e-mail : pavi2004@lpsc.in2p3.fr
phone : +33 4 76 28 40 59

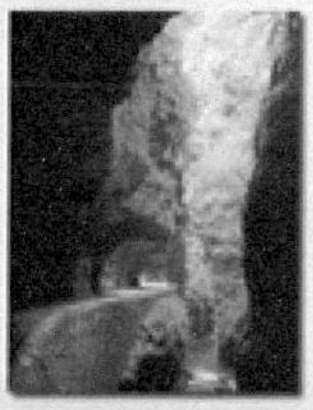

Crédits photos et logo : Fremdenverkehrsbuero der Stadt Mainz • Maison du tourisme de Grenoble • M. Caplain • J.-G. Nozé Dapnia-CEA Saclay
Réalisation poster : C. Favro LPSC-Grenoble

Organizing Committee :

S. Kox LPSC-Grenoble
D. Lhuillier DAPNIA-Saclay
F. Maas Univ. Mainz
J. Van de Wiele IPN-Orsay

International Advisory Committee :

D.H. Beck UIUC
K. de Jager TJNAF
E. De Sanctis Frascati
B. Desplanques LPSC-Grenoble
R. Frascaria IPN-Orsay
A. Lung TJNAF
A. Magnon Dapnia-Saclay
R. Milner MIT-Bates
U. Meissner Univ. of Bonn
M.J. Ramsey-Musolf Caltech/Univ. Connecticut
D. Ryckbosch Gent
P.A. Souder Univ. Syracuse
A. Thomas Adélaide
W. Van Oers Manitoba
M. Vanderhaeghen Univ. Mainz
D. Von Harrach Univ. Mainz

Foreword

Almost 50 years after the proposal of Lee and Young in 1956 to test the hypothesis of parity violation in weak interactions and the subsequent experimental verification of parity violation by C.S. Wu, parity violation has today become a useful property of weak interactions. This is due to the fact that the focus nowadays has changed: parity violation in weak interactions is no more a topic of investigation but is used as a tool in many different fields ranging from nuclear physics to the search for the hidden extra dimensions requested by string theory.

For our first workshop which took place June 5–8, 2002, at the Institut für Kernphysik of the Johannes Gutenberg-Universität Mainz, we concentrated on the investigation of the strangeness contribution in the nucleon. This book contains the refereed and selected papers of the second workshop "From Parity Violation to Hadron Structure and more (Part II)", which took place June 8–11, in the Laboratoire de Physique Subatomique et de Cosmologie, in Grenoble. These papers appear in EPJAdirect, the electronic-only part of EPJA, and they are accessible without restrictions. They will also appear in printed form and can be ordered through Springer. The excellent presentations show the dramatic and steady progress in the accuracy of measured parity violating asymmetries over the last few years. The neutral weak current appears as a powerful new probe of matter driving experimentalists and theorists towards new frontiers in the structure of the nucleon and in the exploration of the electroweak interaction and the physics beyond.

Our thanks for making the workshop such a success go to Madame J. Riffault, who as the conference secretary organized all the nonscientific aspects of the workshop. Without her tireless effort the workshop would not have become such a great success of scientific exchange.

The conference also received financial support from scientific and local organizations as well as sponsors. This has allowed the event to take place and the refereed and selected papers to be published. In addition the International Advisory Committee helped us to establish the scientific program and scope of the workshop. Thanks to all of them.

There are new experiments going on; plenty of data have been taken and are now being analyzed. We are very much looking forward to the exciting future where parity violation will bring many new insights into nature. We are currently planning part III of the workshop, which will be held on Milos Island in Greece in May 2006, where we hope to celebrate with all the participants the 50th anniversary of the discovery of parity violation.

S. Kox
D. Lhuillier
F. Maas
J. Van de Wiele
Guest Editors

Thomas Walcher
Editor-in-Chief

Eur Phys J A (2005) **24**, s2, I–XVII
DOI: 10.1140/epjad/s2005-04-100-5

The European Physical Journal A Direct

Volume 24 · Supplement 2 · 2005

■ V Hadronic structure ... and more

□ V-1/Test of the SM at low energy

□ V-2/PV in nuclear systems

□ V-3/Neutrino beam

■ VI Concluding talks

List of Participants

ARVIEUX Jacques
IPNO, Université Paris Sud
Bât 100 M
F-91406 Orsay Cedex
arvieux@ipno.in2p3.fr

BATIGNE Guillaume
INFN Torino
via Giuria, 1
I-10125 Torino
batigne@to.infn.it

BAUNACK Sebastian
Institut für Kernphysik
Johannes Gutenberg Universität
Johann-Joachim-Becher-Weg 45
D-55099 Mainz
baunack@kph.uni-mainz.de

BECIREVIC Damir
LPT, Université Paris Sud
Bât. 210)
F-91405 Orsay Cedex
Damir.Becirevic@th.u-psud.fr

BECK Douglas
UIUC, Department of Physics
1110 West Green Str.
Urbana, IL 61801, USA
dhbeck@uiuc.edu

BEISE Elizabeth
University of Maryland
Department of Physics
College Park, MD 20742-4111, USA
beise@physics.umd.edu

BIMBOT Louis
IPNO, Université Paris Sud
Bât 100 M
F-91406 Orsay Cedex
bimbotl@ipno.in2p3.fr

BLONDEL Alain
DPNC, Université de Genève
Quai Ansermet
CH-1211 Genève 4
alain.blondel@cern.ch

BREUER Herbert
University of Maryland
Department of Physics
College Park, MD 20740, USA
Breuer@enp.umd.edu

CAPOZZA Luigi
Institut für Kernphysik
Johannes Gutenberg Universität
Johann-Joachim-Becher-Weg 45
D-55099 Mainz
capozza@kph.uni-mainz.de

CARBONELL Jaume
LPSC, 53 avenue des Martyrs
F-38026 Grenoble Cedex
carbonell@lpsc.in2p3.fr

CATES Gordon
University of Virginia
Department of Physics
382 McCormick Rd.
PO Box 400714
Charlottesville, VA 22904-4714, USA
gdc4k@Virginia.edu

CHUDAKOV Eugène
Jefferson Laboratory
12000 Jefferson Avenue,
Newport News, VA 23606, USA
gen@jlab.org

COTANCH Stephen
North Carolina State University
Department of Physics
Raleigh, NC 27695-8202, USA
cotanch@ncsu.edu

DE JAGER Kees
Jefferson Laboratory
12000 Jefferson Avenue
Newport News, VA 23606, USA
kees@jlab.org

DELOUDI Sofia
ETH Zurich
Laboratory of Physical Chemistry
Hönggerberg HCI E243
CH-8093 Zürich
deloudi@phys.chem.ethz.ch

DESPLANQUES Bertrand
LPSC, 53 avenue des Martyrs
F-38026 Grenoble Cedex
desplanques@lpsc.in2p3.fr

DIEFENBACH Jürgen
Institut für Kernphysik
Johannes Gutenberg Universität
Johann-Joachim-Becher-Weg 45
D-55099 Mainz
defi@kph.uni-mainz.de

ELLIS Jonathan R.
CERN
European Organization for Nuclear Research
CH-1211 Genve 23
John.Ellis@cern.ch

FURGET Christophe
LPSC, 53 avenue des Martyrs
F-38026 Grenoble
furget@lpsc.in2p3.fr

GLAESER Boris
Institut für Kernphysk
Johannes Gutenberg Universität
Johann-Joachim-Becher-Weg 45
D-55099 Mainz
glaeser@kph.uni-mainz.de

GORSHTEYN Mikhail
Genoa University
Dipartimento di Fisica
Via Dodecaneso 33
I-16146 Genova
gorshtey@ge.infn.it

GRIMM Klaus
College of William and Mary
Department of Physics
PO Box 8795
Williamsburg, VA 23187-8795, USA
grimm@jlab.org

GUICHON Pierre
CEA Saclay
SPhN / DAPNIA
F-91191 Gif sur Yvette Cedex
pguichon@cea.fr

GUILLON Benoit
LPSC, 53 avenue des Martyrs
F-38026 Grenoble Cedex
guillon@lpsc.in2p3.fr

HENLEY Ernest
University of Washington, Physics
Box 351560
Seattle, WA 98195-1560, USA
henley@phys.washington.edu

HOLMES Richard
Syracuse University
201 Physics Building
Syracuse, NY 13244, USA
rsholmes@phy.syr.edu

HOROWITZ Charles
Indiana University
Department of Physics
Swain Hall West,
Bloomington, IN 47405, USA
horowit@indiana.edu

HYUN Chang Ho
Sungkyunkwan University
Department of Physics
440-746 Korea
hch@meson.skku.ac.kr

IMAI Yoshio
Institut für Kernphysik
Johannes Gutenberg Universität
Johann-Joachim-Becher-Weg 45
D-55099 Mainz
imai@kph.uni-mainz.de

KING Paul
Jefferson Laboratory
12000 Jefferson Avenue
MS 16B Room 96
Newport News, VA 23606 USA
pking@jlab.org

KOX Serge
LPSC, 53 avenue des Martyrs
F-38026 Grenoble Cedex
kox@in2p3.fr

KUBIS Bastian
ITP, University of Bern
Sidlerstrasse 5
CH-3012 Bern
kubis@itp.unibe.ch

KUMAR Krishna
University of Massachusetts
Department of Physics
Amherst, MA 01003, USA
kkumar@physics.umass.edu

Le GOFF Jean-Marc
CEA Saclay,
DAPNIA-SPhN
F-91191 Gif sur Yvette Cedex
jmlegoff@cea.fr

LEE Hee-Jung
Universitat de Valencia
Departamente de Fisica Teorica
E-46100 Burjassot (Valencia)
Heejung.Lee@uv.es

LEE Jeong Hang
Institut für Kernphysik
Johannes Gutenberg Universität
Johann-Joachim-Becher-Weg 45
D-55099 Mainz
lee@kph.uni-mainz.de

LEE Larry
University Manitoba / TRIUMF
4004 Wesbrook Mall,
Vancouver BC V6T2A3, Canada
lrylee@triumf.ca

LEINWEBER Derek
CSSM, University of Adelaide
SA 5005 Australia
dleinweb@physics.adelaide.edu.au

LENOBLE Jason
IPNO, Université Paris Sud
Bât.100 M
F-91406 Orsay Cedex
lenoble@ipno.in2p3.fr

LONDERGAN John Timothy
Indiana University
1331 East 10th St.
Bloomington, IN 47408, USA
tlonderg@indiana.edu

MAAS Frank E.
Institut für Kernphysik
Johannes Gutenberg Universität
Johann-Joachim-Becher-Weg 45
D-55099 Mainz
maas@kph.uni-mainz.de

MARMONIER Carole
PHOTONIS
Av. Roger Roncier, BP 520
F-19106 Brive
c.marmonier@photonis.com

McFARLAND Kevin
University of Rochester
Department of Physics and Astronomy
RC Box 270171
Rochester, NY 14627-0171, USA
Kevin.McFarland@rochester.edu

McKEOWN Robert
Caltech, Kellogg Radiation Laboratory
106-38 Caltech
Pasadena, CA 91125, USA
bmck@krl.caltech.edu

MERCHEZ Fernand
LPSC, 53 avenue des Martyrs
F-38026 Grenoble Cedex
fernand.merchez@wanadoo.fr

NAKAHARA Kazutaka
UIUC, Loomis Laboratory of Physics
1110 West Green Street,
Urbana, IL 61801, USA
nakahara@jlab.org

ONG Saro
IPNO, Université Paris Sud
Bât.100 M
F-91406 Orsay Cedex
ong@ipno.in2p3.fr

OPPER Allena
Ohio University, Physics and Astronomy
Edwards Accel. Lab.
Athens, OH 45701, USA
opper@ohiou.edu

PASQUINI Barbara
University of Pavia
Via Bassi 6
I-27100 Pavia
pasquini@pv.infn.it

PATE Stephen
New Mexico State University
Physics Department, Box 3D
Las Cruces, NM 88003, USA
pate@nmsu.edu

QUEMENER Gilles
LPSC, 53 avenue des Martyrs
F-38026 Grenoble Cedex
quemener@lpsc.in2p3.fr

RAMSEY-MUSOLF Michael
Caltech, Kellogg Radiation Laboratory
106-38 Caltech,
Pasadena, CA 91125, USA
mjrm@krl.caltech.edu

REAL Jean-Sébastien
LPSC, 53 avenue des Martyrs
F-38026 Grenoble Cedex
real@lpsc.in2p3.fr

RICHARD Jean-Marc
LPSC, 53 avenue des Martyrs
F-38026 Grenoble Cedex
richard@lpsc.in2p3.fr

ROCHE Julie
Jefferson Laboratory
12000 Jefferson Avenue
12H/C 108
Newport News, VA 23606, USA
jroche@jlab.org

ROOS Philipp
University of Maryland, Physics
College Park, MD 20742, USA
Roos@physics.umd.edu

SAINIO Mikko
Helsinki Institute of Physics
PO Box 64
FIN-00014 Helsinki
mikko.sainio@helsinki.fi

SILVA Antonio
CFC-FCT Universidade de Coimbra
CFC Faculdade de Ciencias e tecnologia
Rua Larga
P-3004-516 Coimbra
ajose@teor.fis.uc.pt

SILVESTRE-BRAC Bernard
LPSC, 53 avenue des Martyrs
F-38026 Grenoble cedex
silvestre-brac@lpsc.in2p3.fr

SMITH Gregory
Jefferson Laboratory
12000 Jefferson Avenue
12H/C 108
Newport News, VA 23606, USA
smithg@jlab.org

SPAYDE Damon
UIUC, Loomis Laboratory of Physics
1110 West Green Street
Urbana, IL 61801, USA
SPAYDED@grinnell.edu

STILIARIS Efstathios
University of Athens, IASA
University Campus
GR-15771 Athens
stiliaris@phys.uoa.gr

VACHERET Antonin
CEA Saclay
SPhN / DAPNIA
F-91191 Gif sur Yvette Cedex
avacheret@cea.fr

VAN DE WIELE Jacques
IPNO, Université Paris Sud
Bât 100 M
F-91406 Orsay Cedex
vandewi@ipno.in2p3.fr

WEINRICH Christoph
Institut für Kernphysik
Johannes Gutenberg Universität
Johann-Joachim-Becher-Weg 45
D-55099 Mainz
weinrich@kph.uni-mainz.de

WILLIAMSON Steven
University of Illinois
1110 West Green Street
Urbana, ILL 61801, USA
sew@uiuc.edu

Sponsors

L'Institut National de Physique Nucléaire et de Physique des Particules (IN2P3),
le Centre National de la Recherche Scientifique (CNRS), l'Université Joseph Fourier de Grenoble,
le Laboratoire de Physique Subatomique et de Cosmologie de Grenoble (LPSC),
le Service de Physique Nucléaire du Commissariat à l'Énergie Atomique (SPhN-CEA),
l'Institut für Kernphysik et la Johannes Gutenberg-Universität de Mainz,
la société CAEN, la société Photonis,
la Région Rhône-Alpes, le Conseil Général de l'Isère,
la Ville de Grenoble, l'Office du Tourisme de Grenoble,
le réseau européen ESOP

I Introductory talks

Eur Phys J A (2005) **24**, s2, 3–10
DOI: 10.1140/epjad/s2005-04-001-7

EPJ A direct
electronic only

Today's view on strangeness*

John Ellis

CERN, CH-1211 Geneva 23, Switzerland

Received: 1 December 2004 / Published Online: 8 February 2005

Abstract. There are several different experimental indications, such as the pion-nucleon Σ term and polarized deep-inelastic scattering, which suggest that the nucleon wave function contains a hidden $s\bar{s}$ component. This is expected in chiral soliton models, which also predicted the existence of new exotic baryons that may recently have been observed. Another hint of hidden strangeness in the nucleon is provided by copious ϕ production in various $N\bar{N}$ annihilation channels, which may be due to evasions of the Okubo-Zweig-Iizuka rule. One way to probe the possible polarization of hidden $s\bar{s}$ pairs in the nucleon may be via Λ polarization in deep-inelastic scattering.

PACS. 12.39.Dc Skyrmions – 13.75.Cs Nucleon-nucleon interactions – 14.20.-c Baryons

1 How strange is the nucleon?

Some people might argue that this is a "strange" question: why should the nucleon be strange at all - after all, is it not just made out of three up and down quarks? We should not jump to such a naïve conclusion. For a start, even the vacuum is strange: chiral symmetry for π, K mesons tells us that [1]

$$< 0|\bar{s}s|0 >= (0.8 \pm 0.1) < 0|\bar{q}q|0 > .$$

This hidden strangeness cannot be expected to disappear when one inserts a set of three quark 'coloured test charges' into the vacuum. Moreover, hidden strangeness will be generated in perturbative QCD:

$$\text{quark} \to \text{gluon} \to \bar{s}s\,\text{pair}.$$

There are also non-perturbative mechanisms for generating $\bar{s}s$ pairs in the nucleon, such as instanton effects [2].

Another objection to this 'strange' question is the fact that (at least some) experiments do not see very much strangeness in the nucleon. For example, CCFR measures a strange momentum fraction: $P_s = 4\%$ at $Q^2 = 20\text{GeV}^2$ [3], the HAPPEX measurement of a combination of strange electric and magnetic form factors gives a small value: $G_E + 0.39G_M = 0.025 \pm 0.020 \pm 0.014$ at $Q^2 = 0.48\text{GeV}^2$ [4], SAMPLE finds a small strange contribution to the nucleon magnetic moment: $-0.1 \pm 5.1\%$ [5], and the A4 Collaboration finds small strange contribution to another combination of form factors: $G_E + 0.225G_M = 0.039 \pm 0.034$ [6].

On the other hand, a few experiments indicate quite large matrix elements for some hidden-strangeness operators. One prominent example is the π-nucleon Σ term, whose value is related to the strange scalar density:

$$y = \frac{2 < p|\bar{s}s|p >}{< p|\bar{u}u|p > + < p|\bar{d}d|p >}.$$

Two recent determinations of the π-nucleon Σ term have found large values [7] :

$$\Sigma = 64 \pm 8,\ (79 \pm 7)\,\text{MeV}$$

corresponding to large values of $y = 1 - \sigma_0/\Sigma$, where octet baryon mass differences give $\sigma_0 = 36 \pm 7$ MeV [8] and hence $y \sim 0.5$. Another example is the strange spin of the nucleon: a naïve interpretation of measurements of polarized deep-inelastic structure functions would yield [9]:

$$\Delta s = dx[s_\uparrow(x) - s_\downarrow(x) + \bar{s}_\uparrow(x) - \bar{s}_\downarrow(x)] = -0.10 \pm 0.02.$$

On the other hand, HERMES measurements of single-particle inclusive particle production have been interpreted as indicating that [10]

$$\Delta s = dx[s_\uparrow(x) - s_\downarrow(x) + \bar{s}_\uparrow(x) - \bar{s}_\downarrow(x)] = 0.03 \pm 0.03 \pm 0.01.$$

However, this estimate has been questioned on the grounds that corrections to independent fragmentation may be large [11]. The overall picture is that hidden-strangeness matrix elements in the nucleon may be small or large, depending on the J^{PC} quantum numbers carried by the $\bar{s}s$ pair, which is quite compatible with theoretical ideas [12].

Even if one accepts the first estimate of Δs, there has been an argument about its interpretation, based on the observation that in one regularization scheme Δs gets a large contribution from gluons Δg: $\tilde{\Delta s} = \Delta s - (\alpha_s/\pi)\Delta g$: perhaps the 'bare' Δs vanishes, and Δg is large and positive [13]? Since the Δg correction is scheme-dependent,

Send offprint requests to: johne@mail.cern.ch
* CERN-PH-TH/2004-231 hep-ph/0411369

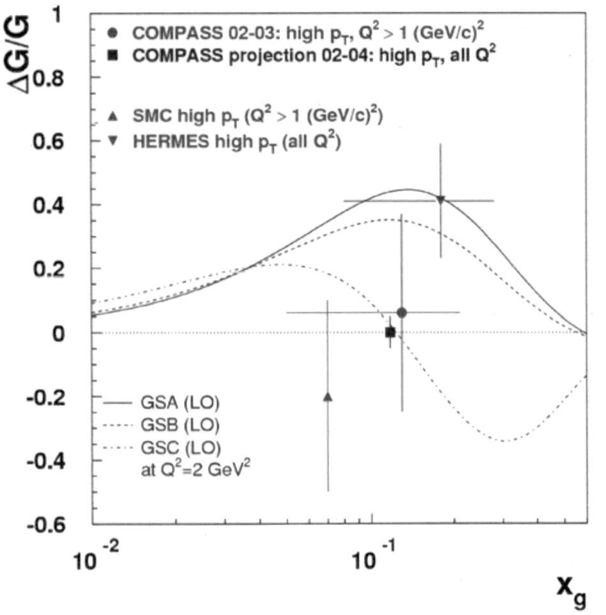

Fig. 1. Comparison of recent determinations of ΔG by HERMES [16], SMC [17] and COMPASS [18]

one may wonder how well defined it is [14]. However, this suggestion has at least raised the profile of the interesting question how large Δg may be. A first measurement of the gluon polarization was reported by FNAL experiment E581/704 [15], measuring π^0 production at high p_T, and Fig. 1 shows recent measurements of ΔG. HERMES find [16]

$$< \Delta G/G >= 0.41 \pm 0.18\,(\text{stat}) \pm 0.03\,(\text{syst})$$

for $0.06 < x_G < 0.28$, and the SMC finds [17]

$$\Delta G/G = -0.20 \pm 0.28 \pm 0.10$$

for an average longitudinal momentum fraction $< \eta \equiv x_G(1 + \hat{s}/Q^2) >= 0.07$ using the asymmetry in hadron-pair production at high p_T. Most recently, COMPASS has announced a new determination, again via the asymmetry in hadron-pair production at high p_T [18]:

$$\Delta G/G = 0.06 \pm 0.31 \pm 0.06 \qquad (1)$$

at an average $< x_G >= 0.13$, and PHENIX is preparing a new determination via the double helicity asymmetry in $pp \to \pi^0$ at high p_T. However, all these measurements have large uncertainties, both systematic and statistical. For the time being, there is no strong indication that ΔG is large, and even its sign must still be regarded as an open question.

2 Models of the nucleon

In the last millennium, the naïve quark model (NQM) [19] held pride of place. It envisages the nucleon as composed of three constituent quarks Q, each with mass $M_Q \sim 300$ MeV, in a non-relativistic wave function. Any additional $q\bar{q}$ pairs are thought to be generated perturbatively, and few of them are expected to be $s\bar{s}$ pairs. Baryons sit in the usual non-exotic SU(3) multiplets, and the combination of a UUD or UDD wave function with meagre pair creation explains the Okubo-Zweig-Iizuka (OZI) rule [20]. The proton spin is simply the algebraic sum of valence constituent-quark spins, which add up to 1/2.

Chiral soliton models [21] provide an alternative viewpoint for the new millennium. They are based on the observation that the intrinsic masses of the (current) quarks defined at short distances are much smaller: $m_{u,d} \sim$ few MeV, $m_s \sim 100$ MeV. Hence the quarks should be treated relativistically, and there are many intrinsic $q\bar{q}$ pairs in the nucleon wave function, which are treated as clouds of meson fields. In this picture, low-lying exotic SU(3) multiplets are predicted [22], as are evasions of the OZI rule due to the copious $s\bar{s}$ pairs in the nucleon. Moreover, the nucleon spin is obtained from orbital angular momentum, the sum of the quark spins vanishes in the limit of vanishing quark masses and a large number of colours, and the $s\bar{s}$ pairs are polarized [23].

In the chiral soliton model, baryons are constructed as clouds of π, K, and η_8 mesons, and the presence of the latter is one way to understand the copious $s\bar{s}$ pairs in the baryon wave function. Exotic baryons are expected as excitations of the meson cloud with non-trivial SU(3) transformation properties, which can also be interpreted as excitations of the $q\bar{q}$ sea in the baryon. The baryon spin is due to the coherent rotation of this meson cloud, motivating the interpretation of the baryon spin as orbital angular momentum, and requiring the $s\bar{s}$ pairs to be polarized.

Specifically, in the limit of massless quarks and a large number of colours, the meson cloud contains no SU(3) flavour-singlet η_0 mesons, and nor do the π, K, and η_8 mesons present have any coupling to the η_0. Since axial-current matrix elements are related in the chiral limit to pseudoscalar-meson couplings, the absence of the η_0 implies that the SU(3)-singlet axial-current matrix element between baryons also vanishes. Classically, this matrix element is in turn related to the sum of the quark spins in the baryon, which therefore vanishes. Since the sum of the u and d quark contributions to the proton spin is positive and does not vanish, there must be a negative, non-zero strange contribution that cancels them [23].

The presence of a non-trivial $q\bar{q}$ sea in the nucleon suggests that there may exist baryons with 'exotic' quantum numbers that cannot be explained in terms of naïve three-quark wave functions. It is surely too naïve to imagine that, if one places three quarks in a vacuum containing many $q\bar{q}$ pairs, there will never be any rearrangement of the $q\bar{q}$ quantum numbers. If the $q\bar{q}$ quantum numbers do not cancel each other out exactly, the resulting baryonic state will have 'exotic' quantum numbers. In the chiral-soliton language, these can be thought of as excitations of the meson cloud.

This line of argument led theorists working on chiral solitons to predict the existence of a relatively light antidecuplet of exotic baryons [22], resembling 'pentaquark'

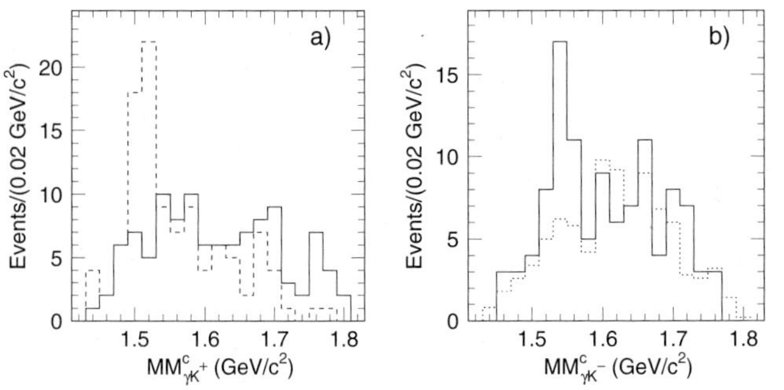

Fig. 2. The *left panel* shows that the $\Lambda(1520)$ signal can be isolated with suitable cuts. The signal for the exotic baryon Θ^+ is the *solid histogram* in the *right panel*, and the *dashed histogram* is a control sample [25]

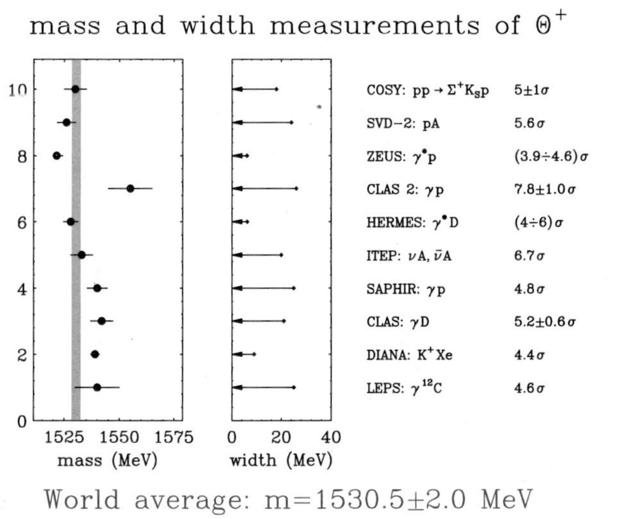

Fig. 3. Compilation of measurements of the Θ mass and decay width [28]

states in the NQM, of which the lowest-lying member would have $udud\bar{s}$ quantum numbers and weigh about 1530 MeV [24]. On the other hand, the Particle Data Group quoted in 1987 "...the prejudice against baryons not made of three quarks ...", and ceased to consider their existence.

This changed with the report by the LEPS Collaboration at Spring-8 [25] of a candidate exotic baryon with $udud\bar{s}$ quantum numbers and weighing about 1540 MeV, shown in Fig. 2. This was soon followed by an avalanche of corroborating evidence from other experiments [26], which stimulated considerable theoretical enthusiasm. However, these observations are somewhat problematic. The masses vary outside the quoted statistical and systematic errors, as seen in Fig. 3, with peaks in nK^+ final states tending to be heavier than those in pK^0 final states [27]. Moreover, KN partial-wave analyses require the decay width to be < 1 MeV [29]: this is surprisingly narrow, and some experiments have reported widths close to their experimental mass resolutions, as also seen in Fig. 3.

Nevertheless, the striking evidence in favour of the early chiral-soliton predictions motivated revisiting

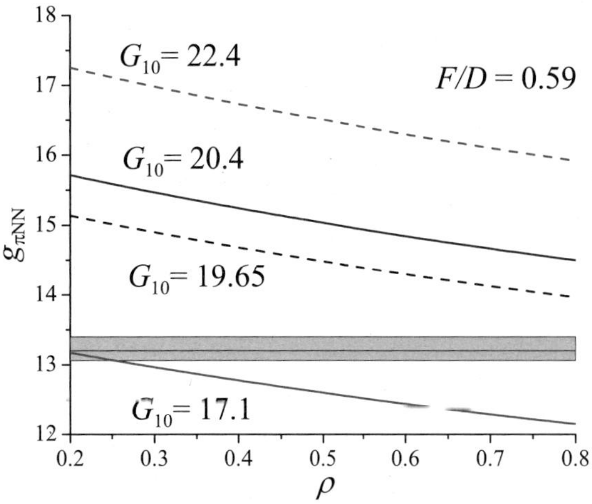

Fig. 4. Chiral-soliton calculations of the π-nucleon coupling depend on the model parameters G_{10} and ρ, but may be consistent with experiment (*shaded*) [30].

them [30]. It was soon realized that the accuracy of the mass prediction [24] was somewhat fortuitous, as it was based on a debatable assignment of another member of the baryon antidecuplet, and the mass splittings within this multiplet were calculated using an outdated value for the π-nucleon Σ term. Using plausible ranges for the chiral soliton moments of inertia that control the mean excitation energy of antidecuplet baryons, and the more modern value of the Σ term discussed above, there is an uncertainty in the θ^+ baryon mass of at least 100 MeV. As for the decay width, although it vanishes in the limit of a large number of colours [31], leading-order calculations with plausible baryon couplings had difficulty in pushing the decay width below about 10 MeV. We made a detailed study of SU(3)-symmetry breaking effects on baryon-meson couplings in chiral soliton models, finding values of the π-Nucleon and $\pi - \Delta$ couplings that are consistent with experiment, as seen in Fig. 4. These effects tend to reduce further the θ^+ decay rate, as seen in Fig. 5, though a width < 1 MeV still seems unlikely [30]. One of the key predictions of chiral soliton models is the existence of other, more 'exotic' baryon multiplets, such as a 27 and a 35 of SU(3) that are slightly heavier than the antidecuplet. In particular, there should be a θ^{++} state weighing < 100 MeV more than the θ^+, as seen in Fig. 6. It is difficult to understand how such a state could have escaped observation in many experiments, but CLAS data may hint at the existence of such a state [32].

What would be the implications of 'exotic' baryons for our understanding of strangeness in the nucleon? As has already been mentioned, the exotic baryon spectrum is sensitive to the value of the π-N Σ term. Using the chiral soliton mass formula

$$\frac{m_s}{m}\,\Sigma = \underbrace{3(4M_\Sigma - 3M_\Lambda - M_N)}_{octet} + \underbrace{4(M_\Omega - M_\Delta)}_{decuplet} - \underbrace{-4(M_{\Xi 3/2} - M_{\Theta^+})}_{antidecuplet} .$$

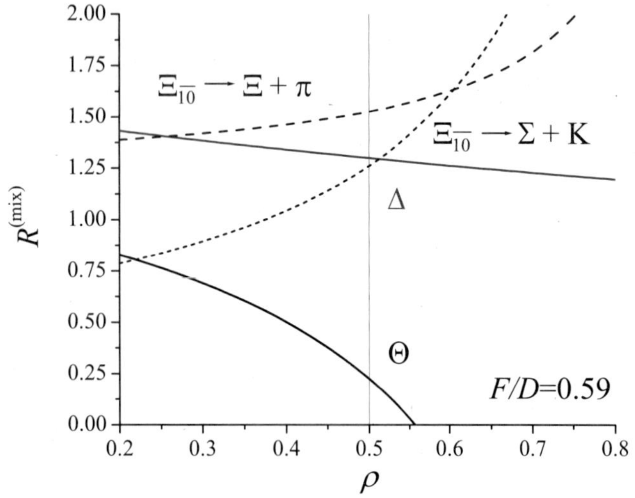

Fig. 5. Corrections to the Θ coupling tend to reduce its decay width, whereas the decays of other antidecuplet baryons are not strongly suppressed [30]

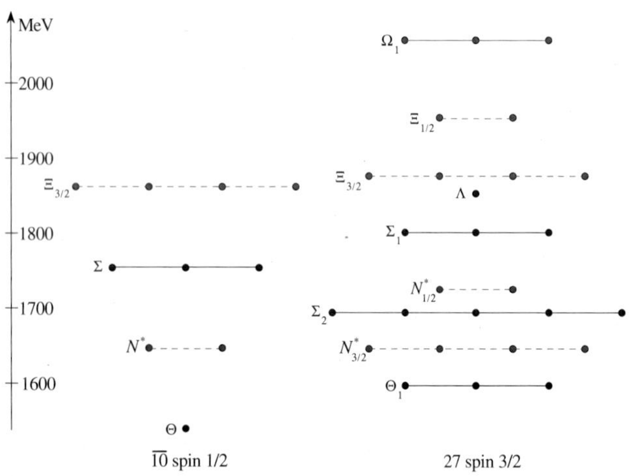

Fig. 6. Spectroscopy of the lowest-lying exotic baryons predicted in chiral-soliton models [30]

the observation of the Ξ^{--} baryon reported by NA49 [33], if confirmed, would correspond to

$$y = 2 \cdot \frac{< N|s\bar{s}|N >}{< N|u\bar{u} + d\bar{d}|N >} \approx 0.6.$$

This is quite consistent with the direct measurements of the π-N Σ term discussed earlier [7], perhaps lending some credence to the whole chiral soliton scheme.

3 OZI violation or evasion?

The Okubo-Zweig-Iizuka (OZI) rule is based on the idea that processes with disconnected quark lines are suppressed. As a corollary, it is not possible to produce $s\bar{s}$ mesons in the interactions of non-strange particles. Hence,

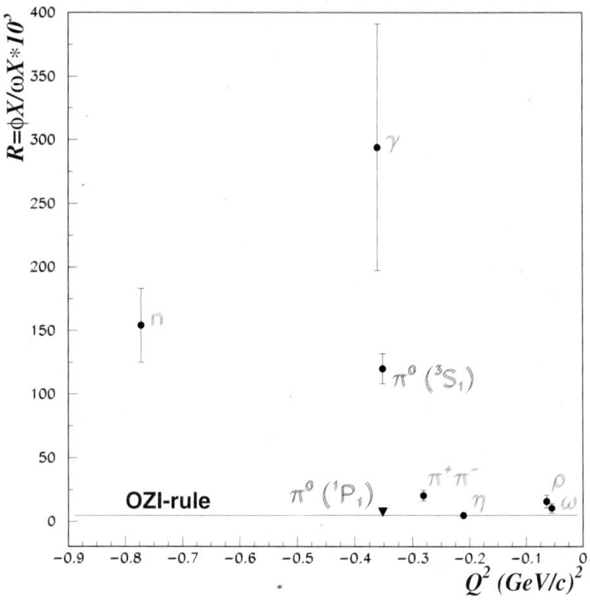

Fig. 7. The ratios of ϕ and ω production in association with various other particles in $N\bar{N}$ annihilation, as measured at LEAR [34], often exceed predictions based on the naïve OZI rule

ϕ meson production should be due only to the admixture of light quarks in the ϕ wave function, which is small, since the ϕ and ω mesons are almost ideally mixed. Generically, one would expect a production ratio

$$R(\phi/\omega) = \tan^2(\theta - \theta_I) = 4.2 \times 10^{-3}.$$

This is not very different from the weighted averages of experimental data from πN collisions:

$$R(\phi/\omega) = (3.30 \pm 0.34) \times 10^{-3}.$$

The corresponding ratios in NN collisions

$$R(\phi/\omega) = (12.78 \pm 0.34) \times 10^{-3}$$

and $\bar{N}N$ collisions

$$R(\phi/\omega) = (14.55 \pm 1.92) \times 10^{-3}$$

are somewhat larger, but not dramatically big.

On the other hand, there are large deviations from the naïve OZI relation in data from LEAR experiments on $p\bar{p}$ annihilations, particularly in the following reactions: $p\bar{p} \to \gamma\phi$, $p\bar{p} \to \pi\phi$ from the $^3S_1$ state, and $\bar{p}d \to \phi n$, as seen in Fig. 7 [34]. Moreover, the ϕ/ω ratio depends strongly on the initial-state spins of the nucleons and antinucleons, on their orbital angular momenta, on the momentum transfer and on the isospin. For example, the partial-wave dependence of annihilations into $\phi\pi$ is shown in Fig. 8, where we see that s-wave annihilations dominate. Another example of a large ϕ/ω ratio is in the Pontecorvo reaction $\bar{p}d \to \phi n$ shown in Fig. 9, where it is compared with the annihilation process $\bar{p}d \to \pi^0 n$.

The OZI rule could be evaded if there are $s\bar{s}$ pairs in the nucleon wave function, since new classes of connected

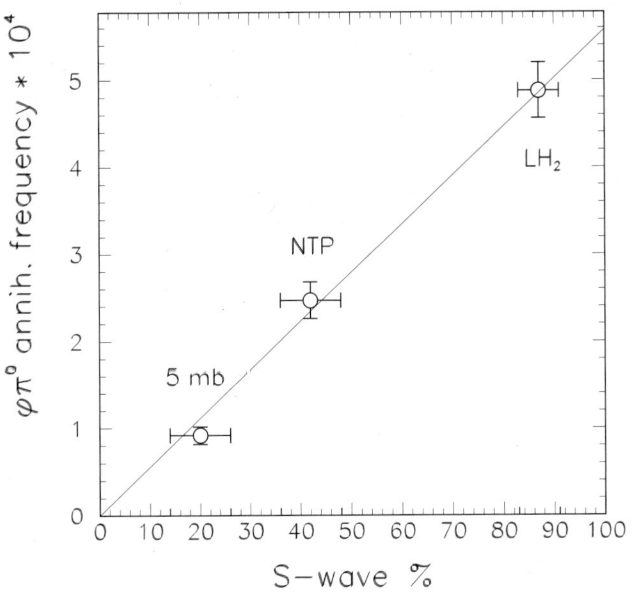

Fig. 8. The annihilation $p\bar{p} \to \phi\pi^0$ proceeds predominantly via the s wave [34]

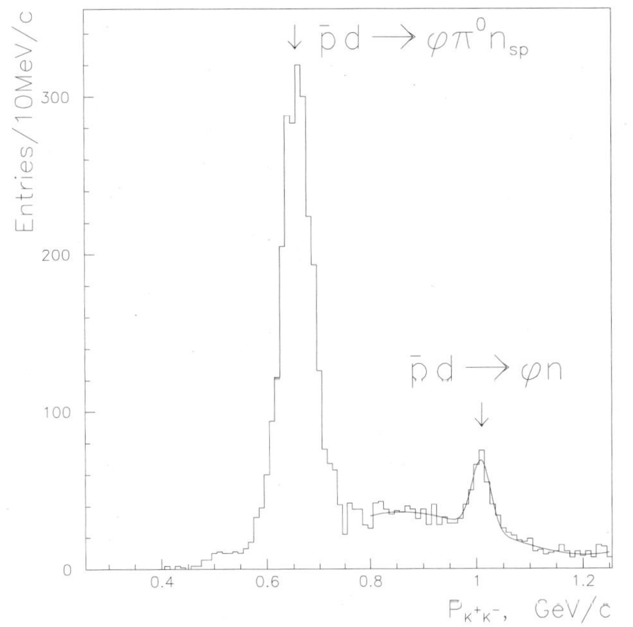

Fig. 9. Signal for the Pontecorvo reaction $\bar{p}d \to n\phi$ [34]

quark diagrams could be drawn for the production of the ϕ and other $s\bar{s}$ mesons. Motivated by the data on polarized deep-inelastic scattering [35], we have formulated a polarized intrinsic strange- ness model [36,37], in which the $s\bar{s}$ pairs in the nucleon are assumed to have negative polarization, and to be in a relative 0^{++} state, not a 1^{--} state as in the naïve ϕ wave function. The production of strangeonium states may occur via rearrangement of the s and $\bar{s}$ in different nucleons, and not via shake-out from an individual nucleon. Thus, *both* nucleons participate in the production mechanism, and their relative polarization and orbital angular momentum states are important. In particular, one would expect the ϕ and the $f_2'(1525)$ mesons to be produced more copiously from spin-triplet initial states than from spin-singlet initial states, the ϕ meson to be produced preferentially from $L = 0$ states, and the $f_2'(1525)$ to be produced preferentially from $L = 1$ states.

This model has led to several correct predictions [34]. In nucleon-antinucleon annihilations, the $p\bar{p} \to \pi^0\phi$ rates from the $^3S_1$ and $^1P_1$ initial states are in a ratio $\sim 15:1$, in agreement with the prediction of $L = 0$ dominance. On the other hand, the $p\bar{p} \to f_2'\pi^0$ rates are in a ratio $\sim 1:10$, in agreement with the prediction of $L = 1$ dominance. Moreover, there is evidence that the mechanisms for ϕ and ω production are different: the $^1P_1$ fractions in $\pi\phi^0$ and $\omega\pi^0$ are $< 7\%$ and $\sim 37\%$, respectively, and ϕ and ω production have different energy dependences in $n\bar{p}$ annihilations. Also, it has been observed that the initial states in $p\bar{p} \to \phi\phi$ are dominated by $J^{PC} = 2^{++}$, consistent with S-wave annihilations in a spin-triplet state. Additionally, spin-singlet initial states are strongly suppressed in $p\bar{p} \to \Lambda\bar{\Lambda}$: the singlet fraction $F_s = (0.1 \pm 7.3) \times 10^{-3}$. The polarized-strangeness model is also consistent with the available data on the Pontecorvo reaction $\bar{p}d \to \phi n$ and on selection rules in $p\bar{p} \to K^*K^*$. Other successful predictions include 'OZI violation' in nucleon-nucleon scattering, where $p\bar{p} \to p\bar{p}\phi$ is about 14 times more copious than $p\bar{p} \to p\bar{p}\omega$ near threshold and the ϕ and ω angular distributions are different, the 'violation' of the naïve OZI rule by a factor ~ 20 in $pd \to^3 H_e\phi(\omega)$, and the negative longitudinal polarization of Λ baryons measured in deep-inelastic neutrino scattering [38], discussed below.

However, there are also some serious problems for the polarized-strangeness model. For example, the strong OZI 'violation' in $\bar{p}p \to \gamma\phi$ takes place from a $^1S_0$ initial state, and the spin transfer D_{nn} in $p\bar{p} \to \Lambda\bar{\Lambda}$ is small, whereas $K_{nn} > 0$, indicating that the spin of the proton is transferred to the $\bar{\Lambda}$, not to the Λ [39]. Moreover, CLAS data on the reaction $\mathbf{e}p \to e'K^+\Lambda$ indicate that the spins of the s and $\bar{s}$ are anti-aligned [40]. Also serious is the problem that $\bar{p}p \to \pi^0\phi$ is not possible from a $^3S_1$ initial state without either flipping the spin of the s-quark or positive polarization of the strange quarks in the proton [41].

Many of these problems would be resolved if there are two components of polarized strangeness, one with $S_z = -1$ and one with $S_z = 0$ [42]. This would permit $p\bar{p} \to \gamma\phi$ and $p\bar{p} \to \pi^0\phi$ via rearrangement diagrams, and the CLAS data that require the spins of the s and $\bar{s}$ to be anti-aligned could be accommodated by shake-out of the $S_z = 0$ component. However, even this model does not fit all the data, as seen in the Table. One promising possibility is to assume the dominance of a spin-singlet us diquark configuration, as indicated in the last column of the Table [42]. Understanding the strange polarization of the proton is still a work in progress.

4 Probing strangeness via Λ polarization

Since the polarization of the Λ is measurable in its decays, and since the Λ polarization is inherited, at least in the

Table 1. Score card for various models of polarized strangeness in the nucleon wave function

	$0^{++}: S_z = -1$	$0^{++}: S_z = -1, 0$	$0^{-+}: (us), (s\bar{s})$
$\phi\pi/\omega\pi$: large from $^3S_1$	+	-	+
$\phi\pi$: spin state	-	+	+
$\phi\gamma/\omega\gamma$: large from $^1S_1$	-	+	glueball
$\phi\eta/\omega\eta$: small from $^3S_1$	small Q^2	small Q^2	small Q^2
$\phi\rho/\omega\rho$: small from $^1S_1$	+	-	+
f_2'/f_2: large from p-wave	+	-	+
$\phi n/\omega n$: large	large Q^2	large Q^2	large Q^2
$P(\Lambda) < 0$ in DIS	+	+	+
$ep \to \Lambda K e : P(\Lambda)$	-	+	+
$\bar{p}p \to \Lambda\bar{\Lambda} : D_{nn}$	-	+	+/-
$\bar{p}p \to \Lambda\bar{\Lambda} : K_{nn}$	-	-	+
$pp \to pp\phi$: large from $^3S_1$	+	-	+

naïve quark model, from its constituent s quark, Λ polarization is potentially a powerful way of probing polarized strangeness. Particularly interesting from this point of view is the measurement of Λ polarization in leptoproduction, where two options are available: measurements in the fragmentation region of the struck quark or in that of the target. The struck quark has net polarization, but is usually a u, so there is no interesting spin transfer to the Λ baryon. However, in the target fragmentation region the 'wounded nucleon' left behind by the polarized struck quark is itself polarized in general. A priori, it is a diquark system with the possibility of a polarized $s\bar{s}$ 'sea' attached to it. Memory of this polarization may be carried by the s and $\bar{s}$ in the wounded nucleon wave function and transferred to Λ and $\bar{\Lambda}$ baryons produced in the target fragmentation region [43].

We have modelled this idea using the Lund string fragmentation model incorporated in LEPTO 6.5.1 and JETSET 7.4, and have considered various combinations of two extreme cases in which the Λ baryon is produced by fragmentation of either the struck quark or the remnant diquark [44]. We then fix free parameters of the model by demanding consistency with data from NOMAD in deep-inelastic ν scattering [38]. In addition to providing a good fit to NOMAD data, as seen in Fig. 10, this procedure can be used directly to make predictions for electroproduction data from HERMES, and agrees very well. We have then gone on to make predictions for the COMPASS muon scattering experiment. COMPASS was originally conceived to measure the polarization of the gluons in the proton, but it may also be able to cast light on the polarization of the strange quarks!

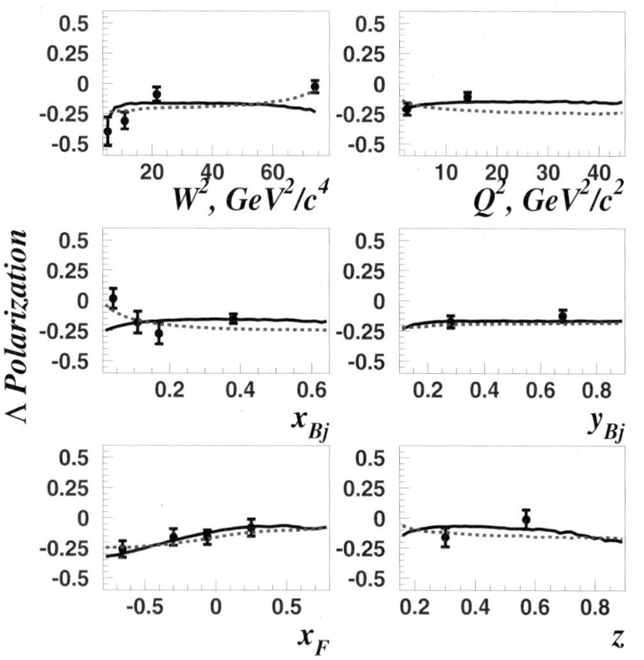

Fig. 10. Predictions for longitudinal Λ polarization in deep-inelastic scattering [44], compared with data from NOMAD [38]

5 Summary

As we have seen in this review, there are many pieces of experimental evidence for a significant amount of hidden strangeness in the proton wave function, notably the π-nucleon Σ term, polarized deep-inelastic scattering and large deviations from the naïve OZI rule. These observations may cast light on complementary models of nucleon structure, namely the 'naïve' quark model and chiral soliton models. The latter were recently boosted by reports of exotic baryons, whose existence was predicted years ago in the soliton model. Their spectroscopy depends, in particular, on the magnitude of the π-nucleon Σ term, and the tentative indication from the difference between the masses of the θ^+ and Ξ^{--}-baryons is that this should be large, in agreement with the latest direct determinations of this quantity. The situation with phenomenological models of OZI 'evasion' due to polarized $s\bar{s}$ pairs in the nucleon wave function is unclear: the data from LEAR and other low-energy experiments suggests that there must be many $s\bar{s}$ pairs, but their polarization states remain obscure. One thing is, however, clear: we may expect many more twists in the strange story of the nucleon!

Acknowledgements. It is a pleasure to thank my collaborators on the subjects discussed here, namely S. J. Brodsky, E. Gabathuler, M. Karliner, D. E. Kharzeev, A. Kotzinian, D. V. Naumov and M. G. Sapozhnikov.

References

1. M.A. Shifman, A.I. Vainshtein, V.I. Zakharov: Nucl. Phys. B **147**, 448, 519 (1979); B.L. Ioffe: Nucl. Phys. B **188**, 317 (1981), [Erratum-ibid. B **191**, 591] (1981); L.J. Reinders, H. Rubinstein, S. Yazaki: Phys. Rept. **127**, 1 (1985)
2. G. 't Hooft: Phys. Rev. D **14**, 3432 (1976), [Erratum-ibid. D **18**, 2199] (1978)
3. A.O. Bazarko et al., [CCFR Collaboration]: Z. Phys. C **65**, 189 (1995), [arXiv:hep-ex/9406007]
4. K.A. Aniol et al., [HAPPEX Collaboration]: Phys. Lett. B **509**, 211 (2001), [arXiv:nucl-ex/0006002]
5. D.T. Spayde et al., [SAMPLE Collaboration]: Phys. Lett. B **583**, 79 (2004), [arXiv:nucl-ex/0312016]
6. F.E. Maas et al., [A4 Collaboration]: Phys. Rev. Lett. **93**, 022002 (2004), [arXiv:nucl-ex/0401019]
7. These numbers are from M.M. Pavan, I.I. Strakovsky, R.L. Workman, R.A. Arndt: PiN Newslett. **16**, 110 (2002), [arXiv:hep-ph/0111066]; For other recent estimates, see: T. Inoue, V.E. Lyubovitskij, T. Gutsche, A. Faessler: arXiv:hep-ph/0311275 and references therein
8. J. Gasser, H. Leutwyler, M.E. Sainio: Phys. Lett. B **253**, 252 (1991); M. Knecht: hep-ph/9912443
9. G.K. Mallot: *Proc. of the 19th Intl. Symp. on Photon and Lepton Interactions at High Energy LP99* ed. J.A. Jaros, M.E. Peskin: Int. J. Mod. Phys. A **15S1**, 521 (2000), [eConf **C990809**, 521 (2000)], [arXiv:hep-ex/9912040]
10. K. Ackerstaff et al., [HERMES Collaboration]: Phys. Lett. B **464**, 123 (1999), [arXiv:hep-ex/9906035]
11. A. Kotzinian: Phys. Lett. B **552**, 172 (2003), [arXiv:hep-ph/0211162]
12. B.L. Ioffe, M. Karliner: Phys. Lett. B **247**, 387 (1990)
13. G. Altarelli, G.G. Ross: Phys. Lett. B **212**, 391 (1988); R.D. Carlitz, J.C. Collins, A.H. Mueller: Phys. Lett. B **214**, 229 (1988); A.V. Efremov, O.V. Teryaev: JINR-E2-88-287 (1988)
14. R.L. Jaffe, A. Manohar: Nucl. Phys. B **337**, 509 (1990)
15. D.L. Adams et al., [FNAL E581/704 Collaboration]: Phys. Lett. B **336**, 269 (1994)
16. A. Airapetian et al., [HERMES Collaboration]: Phys. Rev. Lett. **84**, 2584 (2000), [arXiv:hep-ex/9907020]
17. B. Adeva et al., [Spin Muon Collaboration (SMC)]: Phys. Rev. D **70**, 012002 (2004), [arXiv:hep-ex/0402010]
18. M. Leberig, for the COMPASS Collaboration: CERN PH Seminar, October 25th 2004, available from `http://pccosrv1.cern.ch/compass/publications/talks/`
19. J.J.J. Kokkedee: *The Quark Model* (Benjamin, New York, 1969)
20. S. Okubo: Phys. Lett. **5**, 165 (1963); G. Zweig: CERN preprint TH-412 (1964); J. Iizuka: Prog. Theor. Phys. Supp. **37**, 21 (1966)
21. T.H.R. Skyrme: Proc. Roy. Soc. Lond. A **260**, 127 (1961); Nucl. Phys. **31**, 556 (1962); see also E. Witten: Nucl. Phys. B **160**, 57 (1979) and **223**, 422, 433 (1983); G.S. Adkins, C.R. Nappi, E. Witten: Nucl. Phys. B **228**, 552 (1983); G.S. Adkins, C.R. Nappi: Nucl. Phys. B **233**, 109 (1984)
22. E. Guadagnini: Nucl. Phys. B **236**, 35 (1984); P.O. Mazur, M.A. Nowak, M. Praszałowicz: Phys. Lett. B **147**, 137 (1984); A.V. Manohar; Nucl. Phys. B **248**, 19 (1984); M. Chemtob: Nucl. Phys. B **256**, 600 (1985); S. Jain, S.R. Wadia: Nucl. Phys. B **258**, 713 (1985); M.P. Mattis, M. Karliner: Phys. Rev. D **31**, 2833 (1985); M. Karliner, M.P. Mattis: Phys. Rev. D **34**, 1991 (1986)
23. S.J. Brodsky, J.R. Ellis, M. Karliner: Phys. Lett. B **206**, 309 (1988)
24. D. Diakonov, V. Petrov, M.V. Polyakov: Z. Phys. A **359**, 305 (1997), [arXiv:hep-ph/9703373]
25. T. Nakano et al., [LEPS Collaboration]: Phys. Rev. Lett. **91**, 012002 (2003), [arXiv:hep-ex/0301020]
26. V.V. Barmin et al., [DIANA Collaboration]: Phys. Atom. Nucl. **66**, 1715 (2003), [Yad. Fiz. **66**, 1763 (2003)], hep-ex/0304040; S. Stepanyan et al., [CLAS Collaboration]: hep-ex/0307018; J. Barth et al., [SAPHIR Collaboration]: hep-ex/0307083; V. Kubarovsky, S. Stepanyan, CLAS Collaboration: hep-ex/0307088; A.E. Asratyan, A.G. Dolgolenko, M.A. Kubantsev: hep-ex/0309042 V. Kubarovsky et al., [CLAS Collaboration]: Phys. Rev. Lett. **92**, 032001 (2004), [E: ibid. **92**, 049902 (2004), [arXiv:hep-ex/0311046]; A. Airapetian et al., [HERMES Collaboration]: Phys. Lett. B **585**, 213 (2004), [arXiv:hep-ex/0312044]; S. Chekanov et al., [ZEUS Collaboration]: Phys. Lett. B **591**, 7 (2004), [arXiv:hep-ex/0403051]; R. Togoo et al.: Proc. Mongolian Acad. Sci., **4**, 2 (2003); A. Aleev et al., [SVD Collaboration]: arXiv:hep-ex/0401024
27. Q. Zhao, F.E. Close: arXiv:hep-ph/0404075
28. This compilation was kindly provided by M. Karliner
29. R.N. Cahn, G.H. Trilling: Phys. Rev. D **69**, 011501 (2004), [arXiv:hep-ph/0311245]
30. J.R. Ellis, M. Karliner, M. Praszalowicz: JHEP **0405**, 002 (2004), [arXiv:hep-ph/0401127]
31. M. Praszalowicz: Phys. Lett. B **583**, 96 (2004), [arXiv:hep-ph/0311230]
32. M. Battaglieri, [presenting CLAS Coll. data]: talk at Pentaquark Workshop, Feb. 10-12, 2004, Trento, Italy, `http://www.tp2.ruhr-uni-bochum.de/talks/trento04/battaglieri.pdf`
33. C. Alt et al., [NA49 Collaboration]: Phys. Rev. Lett. **92**, 042003 (2004), [arXiv:hep-ex/0310014]

34. For a review, see: V.P. Nomokonov, M.G. Sapozhnikov: Phys. Part. Nucl. **34**, 94 (2003), [Fiz. Elem. Chast. Atom. Yadra **34**, 189 (2003)], [arXiv:hep-ph/0204259]
35. J.R. Ellis, E. Gabathuler, M. Karliner: Phys. Lett. B **217**, 173 (1989)
36. J.R. Ellis, M. Karliner, D.E. Kharzeev, M.G. Sapozhnikov: Phys. Lett. B **353**, 319 (1995), [arXiv:hep-ph/9412334]
37. J.R. Ellis, M. Karliner, D.E. Kharzeev, M.G. Sapozhnikov: Nucl. Phys. A **673**, 256 (2000), [arXiv:hep-ph/9909235]
38. P. Astier et al., [NOMAD Collaboration]: Nucl. Phys. B **605**, 3 (2001), [arXiv:hep-ex/0103047]
39. B. Bassalleck et al., [PS185 collaboration]: Phys. Rev. Lett. **89**, 212302 (2002), [arXiv:nucl-ex/0206005]
40. D.S. Carman et al., [CLAS Collaboration]: Phys. Rev. Lett. **90**, 131804 (2003), [arXiv:hep-ex/0212014]
41. M.P. Rekalo, J. Arvieux, E. Tomasi-Gustafsson: Z. Phys. A **357**, 133 (1997)
42. J.R. Ellis, A. Kotzinian, M.G. Sapozhnikov: unpublished
43. J.R. Ellis, D. Kharzeev, A. Kotzinian: Z. Phys. C **69**, 467 (1996), [arXiv:hep-ph/9506280]
44. J.R. Ellis, A. Kotzinian, D.V. Naumov: Eur. Phys. J. C **25**, 603 (2002), [arXiv:hep-ph/0204206]

Eur Phys J A (2005) **24**, s2, 11–14
DOI: 10.1140/epjad/s2005-04-002-6

EPJ A direct
electronic only

Strange and gluonic contributions to the nucleon spin

Jean-Marc Le Goff

CEA-Saclay, DAPNIA SPhN, France

Received: 15 October 2004 / Published Online: 8 February 2005

Abstract. Inclusive deep inelastic scattering (DIS) experiments lead to a small contribution of the quark spins to the nucleon spin and a negative contribution from strange quarks. Historically this triggered the interest in measuring strange form factors. However, the result of these inclusive experiments has to be reinterpreted taking into account the axial anomaly in QCD, which depends on the gluon contribution to the nucleon spin. The COMPASS experiment at CERN and experiments at RHIC are going to measure this gluonic contribution.

PACS. 13.60.Hb Total and inclusive cross sections (including deep-inelastic processes) – 13.60.Le Meson production

1 Introduction

In 1989 the EMC experiment [1] at CERN concluded that the contribution $\Delta\Sigma$ of the quark spins to the nucleon spin was compatible with zero and that the strange contribution Δs was significantly negative, triggering the so-called spin crisis. This meant that the strange axial matrix element of the nucleon was non-zero, which raised the issue whether the strange vector matrix elements could also be non-zero and motivated a large experimental program to study strange form factors through Parity Violation experiments. Here we present an overview of the spin of the nucleon, for a full review see [2].

2 Strange quark and total quark contributions

The spin of the nucleon can be decomposed in the contributions from its constituents as

$$\frac{1}{2} = \frac{1}{2}\Delta\Sigma + \Delta g + L_q + L_g \quad (1)$$

where Δg is the contribution from the spins of the gluons and L_q and L_g are the contributions from orbital angular momenta of quarks and gluons. What do we know about $\Delta\Sigma$ from a theoretical point of view ? On the one hand we have the quark model which provides us with a large part of our understanding of hadrons. It gives $\Delta\Sigma \approx 0.75$. On the other hand we are not able to use QCD to compute $\Delta\Sigma$ from first principles, but using results from hyperon β decay experiments and assuming a strange quark contribution $\Delta s = 0$, we get $\Delta\Sigma \approx 0.6$. We then have a qualitative agreement between Quark Model and QCD.

In inclusive deep inelastic experiments (DIS) a lepton is scattered off a nucleon and only the scattered lepton is observed. Only two Lorentz invariants enter the problem. They can be chosen as the mass of the virtual photon, $Q^2 = -q_\mu^2$, which gives the resolution of the probe, and $x_{bj} = Q^2/2M(E-E')$ which is the fraction of the nucleon momentum carried by the quark which absorbed the virtual photon. DIS corresponds to the limit of large Q^2 at fixed x_{bj}. The cross section involves structure functions which depend only on the two Lorentz invariants. However, because the lepton scatters on a quark, which is a point-like particle, the Q^2 dependence vanishes, at least to leading order in QCD, a property known as scaling. In the unpolarized case we have two structure functions F_1 and F_2. They can be expressed in terms of the parton distribution function (pdf) as $F_1(x) = \frac{1}{2}\left[\frac{4}{9}u(x) + \frac{4}{9}\bar{u}(x) + \frac{1}{9}d(x) + \frac{1}{9}\bar{d}(x) + \frac{1}{9}s(x) + \frac{1}{9}\bar{s}(x)\right]$ and $F_2(x) = 2xF_1(x)$, where $u(x)$ for instance, is the probability to find inside the nucleon a quark of flavor u and a fraction x of the nucleon momentum. In the polarized case we have in addition g_1 and g_2; $g_1(x) = \frac{1}{2}\left[\frac{4}{9}\Delta u(x) + \frac{1}{9}\Delta d(x) + \frac{1}{9}\Delta s(x)\right]$ with $\Delta u(x) = u^+(x) - u^-(x) + \bar{u}^+(x) - \bar{u}^-(x)$, where the polarized pdf $u^+(x)$ is the probability to find inside the nucleon a quark of flavor u with a fraction x of the nucleon momentum and a spin parallel to the nucleon spin. The integral $\Delta u = \int_0^1 \Delta u(x)dx$ is the total contribution of spins of quark of flavor u to the nucleon spin. We then have $\Delta\Sigma = \Delta u + \Delta d + \Delta s$, where the contribution of heavier flavors (c, b and t) is negligible.

In 1989 the EMC measured $\Gamma_1^p = \int_0^1 g_1^p(x)dx = \frac{1}{2}\left[\frac{4}{9}\Delta u + \frac{1}{9}\Delta d + \frac{1}{9}\Delta s\right]$. Using $SU_f(3)$ symmetry hyperon β decays give $a_3 = \Delta u - \Delta d$ and $a_8 = \Delta u + \Delta d - 2\Delta s$. This provided 3 equations for 3 unknowns resulting in $\Delta\Sigma = 0.12 \pm 0.17$ and $\Delta s = -0.10 \pm 0.03 < 0$! This

Send offprint requests to: jmlegoff@cea.fr

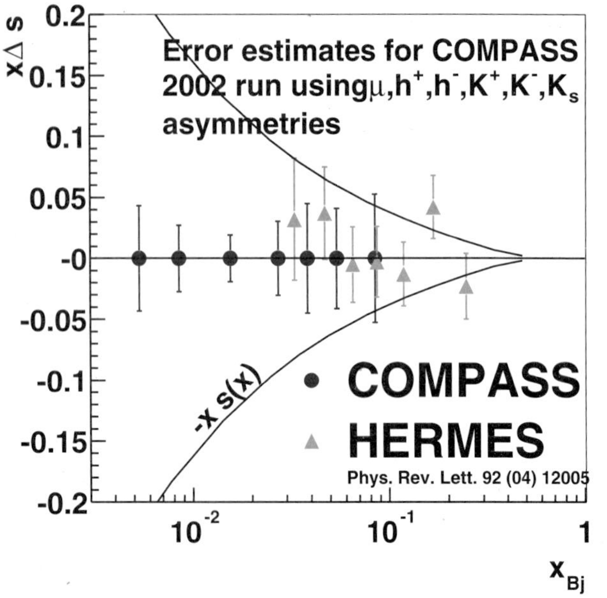

Fig. 1. Results of Hermes semi-inclusive analysis [4] and projections from COMPASS

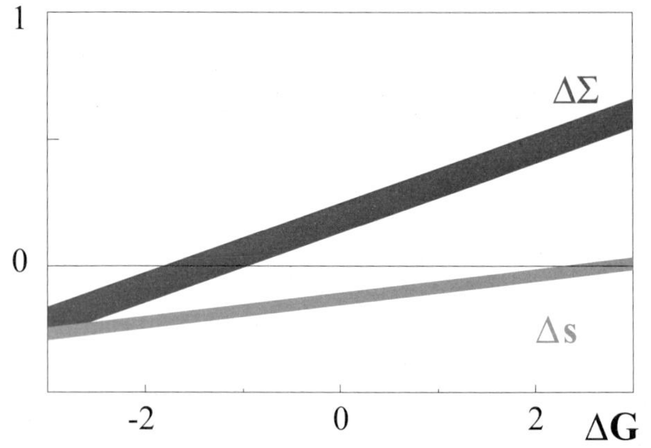

Fig. 2. $\Delta\Sigma$ and Δs as a function of the hypothesis for Δg

came as a big surprise which was advertised as the spin crisis, so that the EMC paper is one of the 3 most cited experimental papers [3]. The results were confirmed by SMC at CERN, SLAC, and Hermes at DESY. The uncertainty is now dominated by the extrapolation in the low x, unmeasured region.

In order to go further and measure the x dependence of the polarized pdf one needs to perform semi-inclusive DIS, $l + p \to l' + h + X$. The spin asymmetry for this process can be written as

$$A_1^h = \frac{\sum_q e_q^2[\Delta q(x) D_q^h(z) + \Delta\bar{q}(x) D_{\bar{q}}^h(z)]}{\sum_q e_q^2[q(x) D_q^h(z) + \bar{q}(x) D_{\bar{q}}^h(z)]} \tag{2}$$

where $z = E_h/E_{\gamma^*}$ and the fragmentation function $D_q^h(z)$ gives the probability that the fragmentation of a quark of flavor q gives a hadron h. The fact that $D_q^h(z) \neq D_{\bar{q}}^h(z)$ allows for the separation of sea from valence. Using A_1, A_1^{h+}, A_1^{h-} for proton and neutron the SMC obtained Δu_v, Δd_v, $\Delta\bar{d}$, $\Delta\bar{u}$ without using $SU_f(3)$ flavor symmetry. However, such an analysis relies on 2, which assumes that the measured hadron comes from the fragmentation of the struck quark and not from the target remnant.

In order to measure also $\Delta s(x)$ one needs to identify strange particles within the measured hadrons, either using a RICH detector, or by reconstructing the K^0 mass. Unfortunately this introduces more sensitivity to target remnants because $m_K > m_\pi$. Figure 1 presents the results obtained by the Hermes collaboration. They exhibit no indications of $\Delta s < 0$. Data down to lower x are expected from COMPASS.

Polarized pdf can also be measured through parity violating $\vec{p}\, p \to W$ at RHIC in Brookhaven [5]. The point is that $q\bar{q} \to W^{\pm}$ selects a given quark helicity. W^+ production is dominated by $u\bar{d} \to W^+$. If u comes from the polarized proton and $\bar{d}$ from the unpolarized one, we get a $\Delta u\ \bar{d}$ contribution. The opposite case gives $\Delta\bar{d}\ u$. The spin asymmetry for the process then reads

$$A_L^{W^+} = \frac{\Delta u(x_1)\bar{d}(x_2) - \Delta\bar{d}(x_1)u(x_2)}{u(x_1)\bar{d}(x_2) + \bar{d}(x_1)u(x_2)} \tag{3}$$

The same formula is obtained for W^- production with u and d exchanged. The nice thing with such an analysis is that it is independent of target fragmentation effects. It has, however, hardly any sensitivity to Δs.

We must go back to the inclusive case and note that EMC does not actually measure Δq but axial matrix elements $a_q = \langle N|\bar{q}\gamma_\mu\gamma_5 q|N\rangle$. Naively the axial matrix elements are identified with Δq. However, due to axial anomaly we rather have $a_0 = \Delta\Sigma - 3\frac{\alpha_s}{2\pi}\Delta g$ and $a_s = \Delta s - \frac{\alpha_s}{2\pi}\Delta g$. In addition $\Delta g \propto \ln Q^2$, so that the gluonic contribution to a_q does not go to zero at high Q^2 in spite of the α_s factor.

The actual value of $\Delta\Sigma$ and Δs now depends on the value of Δg as illustrated by Fig. 2. If $\Delta g \approx 0$ we are back to the spin crisis with a small $\Delta\Sigma$ and a negative Δs. If Δg is large and positive we may end up with the expected $\Delta s \approx 0$ and $\Delta\Sigma \approx 0.6$. Δg must be measured, both for itself and in order to extract the actual value of $\Delta\Sigma$ and Δs from inclusive experiments.

3 Gluon contribution

When Q^2 increases the resolution of the probe improves and what used to appear as a quark may start to appear as a quark and a gluon, or a gluon may appear as a quark anti-quark pair. In these conditions the variation $dq(x,Q^2)/d(\ln Q^2)$ tells us something about $g(x,Q^2)$. This is formalized in the Dokshitzer-Gribov-Lipatov-Altareli-Parisi (DGLAP) equations. Indeed, in the unpolarized case, performing a next-to-leading order QCD fit of F_2 data, from fixed target experiments and from the HERA collider, provides a good measurement of $g(x,Q^2)$. This is unfortunately not the case in the polarized case because there is no polarized lepton-proton collider. The Q^2 range for g_1 data, between SLAC or Hermes and SMC, is not large enough (at most a factor 10) to allow for a precise estimate of $\Delta g(x, Q^2)$. A direct measurement is needed.

It is difficult to probe gluons in lepton scattering because they have no electric charge. Photons can, however, interact with gluons through the photon gluon fusion (PGF) process, $\gamma^* g \to q\bar{q}$. This is a higher order process which has a small cross section relative to the leading order process, $\gamma^* q \to q$, so tagging is needed.

The first possibility consists in requiring the produced $q\bar{q}$ pair to be a $c\bar{c}$ pair. Since the intrinsic charm inside the nucleon is negligible, the observation of charm is a signature of the PGF process. The fragmentation of charm quarks produces a $D^0 = c\bar{u}$ meson in 60% of the cases. The easiest way to see the D^0 is through its decay to $K\pi$ which has a 4% branching ratio. Due to this low branching ratio, one requires to detect either the c or the $\bar{c}$ through this channel. The drawback is that in this case it is not possible to reconstruct the kinematics at the vertex and the momentum fraction of the gluon cannot be evaluated. One gets the mean value of $\Delta g(x)/g(x)$ averaged over the experimental acceptance.

A second possibility arises from the fact that in the leading order process, $\gamma^* q \to q$, all the produced hadrons are in the direction of the virtual photon, whereas in the PGF process the $q\bar{q}$ pair can be produced at any angle and the resulting hadrons may have a transverse momentum p_t with respect to the photon. So the idea is to search for pairs of hadrons with high p_t (or two high p_t jets at high energy). There is, however, a physical background, the so-called QCD Compton process, $\gamma^* q \to qg$, since in this process the final q and g both can produce a high p_t hadron. This background has to be evaluated by a Monte Carlo (MC) simulation, starting from the known polarized quark distribution functions.

This second method was already used by Hermes [6] and SMC [7]. Hermes used a 28 GeV electron beam, they did not measure the scattered electron and they were dominated by low Q^2, quasi real photons. The hard scale, for perturbative QCD to be valid, is then provided by p_t. The generator PYTHIA was used for background estimation. The difficulty in this case is that, in addition to QCD Compton, there is an important background of events where the photon is resolved into its partonic structure. In this analysis the dilution due to these events was taken into account but not their contribution to the asymmetry. SMC had a 190 GeV/c muon beam. They selected only events with $Q^2 > 1$ (GeV/c)^2, in which case the resolved photon contribution could be neglected and the pure DIS generator, LEPTO, could be used. The results obtained by the two collaborations are presented in Fig. 4. Error bars are still pretty large.

The COMPASS collaboration [8] at CERN is using both methods. The muon beam intensity was increased by a factor 5 relative to SMC to provide an average luminosity of $5 \cdot 10^{32}$ cm^{-2}s^{-1}. The target is made of ^{6}LiD with a dilution factor ≈ 0.5 to be compared to 0.24 for deuterated butanol, used by SMC. The spectrometer was commissioned and first data were taken in 2002. Data were taken again in 2003 and 2004 but there will be no beam in 2005, due to LHC installation.

The 2002 data gives an asymmetry for high p_t pair production, $A^{\gamma d \to hh'} = -0.065 \pm 0.036 \pm 0.010$, including all Q^2. MC studies are going on, to take into account the contribution of resolved photons to the asymmetry. Part of these events involve a gluon in the nucleon so that their contribution to the asymmetry is proportional to $\Delta g/g$. Neglecting this fact, a statistical error $\delta(\Delta g/g) = 0.17$ should be obtained. In the same conditions, all data between 2002 and 2004 should provide $\delta(\Delta g/g) \approx 0.05$, or alternatively four bins in x_g with $\delta(\Delta g/g) \approx 0.10$ in each bin. Using only data with $Q^2 > 1$ (GeV/c)^2, should provide $\delta(\Delta g/g) \approx 0.16$.

The $D^0 \to K\pi$ open charm channel suffers from an important combinatorial background. Requiring that the D^0 comes from the disintegration $D^* \to D^0\pi \to K\pi\pi$, strongly reduces the background. The first reason is the small value of $M_{D^*} - M_{D^0} - M_\pi = 6$ MeV, which leaves little space for the background. The second reason is the excellent experimental resolution in this mass difference, better than 2 MeV, to be compared to about 25 MeV for the resolution in the D^0 mass alone. Many improvements are ongoing in terms of reconstruction, in particular in the RICH detector used to identify K. An error $\delta(\Delta g/g) \approx 0.24$ is then expected out of 2002 to 2004 data.

The gluon distribution can also be probed in polarized proton-proton collisions [5]. The golden channel is the so-called direct γ production $\vec{p}\vec{p} \to \gamma$ + jet + X. At the parton level this corresponds to $qg \to q\gamma$, where quarks from one of the protons are used to probe gluons in the other proton. So the measured asymmetry is a convolution, $\Delta g \otimes \Delta q$. There is a physics background, $q\bar{q} \to g\gamma$, which also produces γ + jet + X. Its contribution to the asymmetry, which goes like $\Delta q \otimes \Delta\bar{q}$, can be computed from the measured $\Delta q(x)$ and $\Delta\bar{q}(x)$.

Other possible channels at RHIC include jet production (or high p_t leading hadron) and heavy flavor production.

RHIC, the relativistic heavy ion collider, is used part of the time as a polarized proton-proton collider at $\sqrt{s} = 50-500$ GeV. Both PHENIX and STAR collaborations are taking data in this mode. The design polarization is 70%, using siberian snakes in RHIC and partial snakes in the AGS to eliminate depolarizing resonances. The design luminosity is $\mathcal{L} = 2 \cdot 10^{32}$ cm^{-2}s^{-1}. An integrated luminosity of 7 pb^{-1} with $P \approx 50\%$ is expected from the 2004-2005 run. The first measured leading hadron asymmetry, A_{LL}^{π}, is presented in Fig. 3. This asymmetry is sensitive to the convolution $\Delta g \otimes \Delta g$. It should then be positive unless there is a node in $\Delta g(x)$.

The expected error on $\Delta g/g$ in the golden channel for an integrated luminosity of 320 pb^{-1} with $P = 70\%$ are presented in Fig. 4. We see that a wide range of x_g is covered with an excellent accuracy. Note, however, that this is only the statistical accuracy, some systematic uncertainty will arise from background subtraction and deconvolution.

In the long term there is the EIC project [10] to build a 10 GeV polarized electron linac to collide with one of the

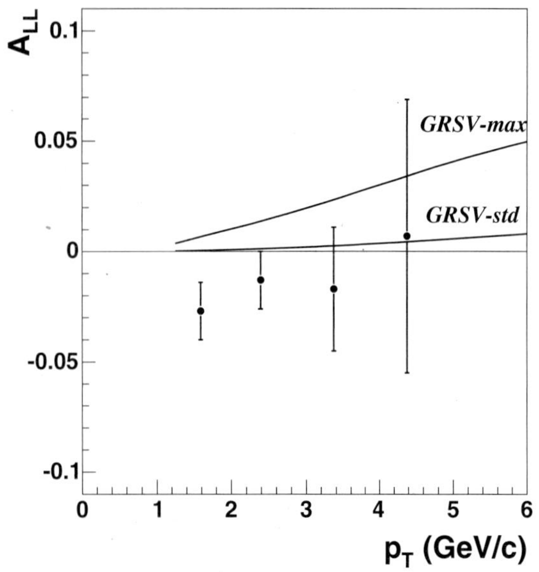

Fig. 3. Leading hadron A_{LL}^{π} asymmetry measured at RHIC by PHENIX [9], compared with prediction corresponding to the standard GRV parametrization and to a parametrization with 100% polarization at a given starting scale

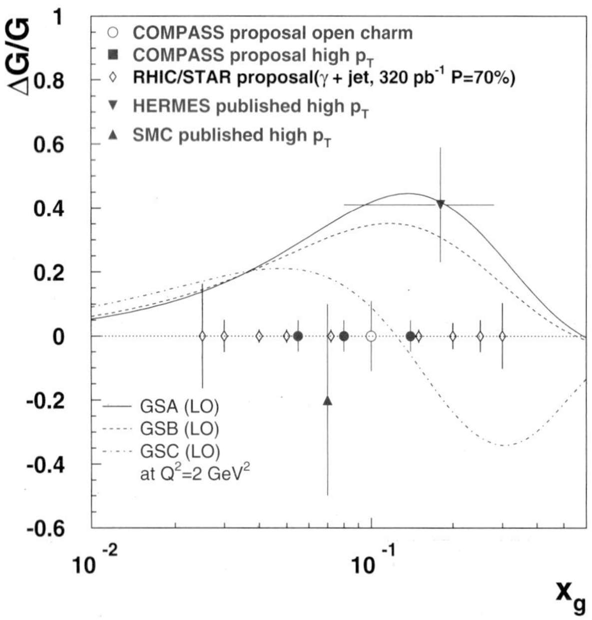

Fig. 4. Measured values of $\Delta g/g$ by Hermes and SMC, together with projections from COMPASS and RHIC

RHIC polarized proton beams with a luminosity of 10^{33} to 10^{34} $\text{cm}^{-2}\text{s}^{-1}$ and $\sqrt{s} = 100$ GeV. The beam could start between 2012 and 2014. This would allow for:

– the measurement of g_1 down to $x = 10^{-4}$ instead of $3 \cdot 10^{-3}$ now, dramatically reducing the dominant error on $\int_0^1 g_1(x)dx$ (and then on a_0) which is due to the low x extrapolation.

– a large range of Q^2 for g_1 data and then a precise estimate of Δg from QCD NLO fit.

– a direct measurement of $\Delta g(x, Q^2)$ through hadron pairs, jet pairs and charm production.

– all in all this should give an error on the integral $\Delta g = \int_0^1 \Delta g(x)dx$ on the order of 0.03 to 0.05.

4 Conclusions and perspectives

In the reconstruction of the spin puzzle,

$$\frac{1}{2} = \frac{1}{2}\Delta\Sigma + \Delta g + L_q + L_g \tag{4}$$

we know the flavor singlet axial matrix element $a_0 = \Delta\Sigma - (3\alpha_s/2\pi)\Delta g = 0.27 \pm 0.13$.

The strange axial matrix element is $a_s \approx -0.10$, but this is more sensitive to possible violations of $SU_f(3)$ symmetry than a_0. The measurement of $\Delta s(x)$ in semi-inclusive DIS is delicate because it is sensitive to target fragmentation effects. There is hardly any sensitivity to $\Delta s(x)$ in $\vec{p}p$ collisions. A possible solution would be to get the integral Δs from neutrino experiments combined with parity violation experiments [11].

The first measurements of $\Delta g(x)$ by COMPASS and RHIC will appear soon. The experimental methods and the systematic errors are completely different and in addition each experiment has several channels, so this should provide a reliable measurement of $\Delta g(x)$.

In the longer term the project of a polarized electron-proton collider at RHIC would provide a much more accurate value of a_0 and Δg and then of $\Delta\Sigma$.

The contributions from orbital momentum are very difficult to access. Generalized Parton Distributions (GPD, for a review see e.g. [12]) describe at the same time the transverse position and the longitudinal momentum, which is what is needed to compute the orbital momentum. This is formalized in the Ji sum rule which relates an integral of GPDs to $J_q = \Delta\Sigma + L_q$. GPDs can be measured in DVCS experiments, $lN \to l'N'\gamma$. First DVCS measurements were performed at JLAB and HERA and plans exist at COMPASS; due to its high luminosity an electron-proton collider at RHIC would be the ideal tool for this. There is, however, a very long way before measuring the Ji sum rule.

References

1. J. Ashman et al.: Phys. Lett. B **206**, 364 (1988), Nucl. Phys. B **328**, 1 (1989)
2. M. Anselmino, A. Efremov, E. Leader: Phys. Rep. **261**, 1 (1995); B. Lampe and E. Reya: Phys. Rep. **332**, 1 (2000)
3. `http://www.slac.stanford.edu/library/topcites/`
4. A. Airapetian et al.: Phys. Rev. Lett. **92**, 012005 (2004)
5. G. Bunce et al.: Ann. Rev. Nucl. Part. Sci. **50**, 525 (2000); hep-ph/0007218
6. A. Airapetian et al.: Phys. Rev. Lett. **84**, 2584-2588 (2000)
7. B. Adeva et al.: Phys. Rev. D **70**, 012002 (2004)
8. COMPASS, G. baum et al.: CERN-SPSLC-96-14, `http://www.compass.cern.ch/compass/proposal`
9. hep-ex/0404027
10. `www.phenix.bnl.gov/WWW/publish/abhay/Home_of_EIC/`
11. S. Pate: Don't Forget to Measure Δs, these proceedings
12. M. Diehl: Phys. Rep. **388**, 41 (2003) or hep-ph/0307382

II Form factors and two-photon effects

Eur Phys J A (2005) **24**, s2, 17–21
DOI: 10.1140/epjad/s2005-04-003-5

EPJ A direct
electronic only

Electromagnetic form factors of the nucleon

Kees de Jager

Thomas Jefferson National Accelerator Facility, Newport News, Virginia 23606, USA

Received: 15 October 2004 / Published Online: 8 February 2005

Abstract. The experimental and theoretical status of elastic electron scattering from the nucleon is reviewed. As a consequence of new experimental facilities and new theoretical insights, this subject is advancing with unprecedented precision.

PACS. 13.40.Gp Electromagnetic form factors – 29.27.Hj Polarized beams

1 Introduction

More than 40 years ago Akhiezer *et al.* [1] (followed 20 years later by Arnold *et al.* [2]) showed that the accuracy of nucleon charge form-factor measurements could be increased significantly by scattering polarized electrons off a polarized target (or equivalently by measuring the polarization of the recoiling proton). However, it took several decades before technology had sufficiently advanced to make the first of such measurements feasible and only in the past few years has a large number of new data with a significantly improved accuracy become available. For G_E^p measurements the highest figure of merit at Q^2-values larger than a few GeV^2 is obtained with a focal plane polarimeter. Here, the Jacobian focusing of the recoiling proton kinematics allows one to couple a standard magnetic spectrometer for the proton detection to a large-acceptance non-magnetic detector for the detection of the scattered electron. For studies of G_E^n one needs to use a magnetic spectrometer to detect the scattered electron in order to cleanly identify the reaction channel. As a consequence, the figure of merit of a polarized $^3\vec{\mathrm{He}}$ target is comparable to that of a neutron polarimeter.

2 Proton electric form factor

In elastic electron-proton scattering a longitudinally polarized electron will transfer its polarization to the recoil proton. In the one-photon exchange approximation the ratio of the charge and magnetic form factors is directly proportional to the ratio of the polarization components, parallel (P_l) and transverse (P_t) to the proton's momentum.

The greatest impact of the polarization-transfer technique was made by the two recent experiments [3,4] in Hall A at Jefferson Lab, which measured the ratio G_E^p/G_M^p in a Q^2-range from 0.5 to 5.6 GeV^2. The most striking feature of the data is the sharp, practically linear decline as Q^2 increases. This significant fall-off of the form-factor ratio is in clear disagreement with the results from the Rosenbluth extraction. Segel and Arrington [5] performed a high-precision Rosenbluth extraction in Hall A at Jefferson Lab, designed specifically to significantly reduce the systematic errors compared to earlier Rosenbluth measurements. The main improvement came from detecting the recoiling protons instead of the scattered electrons. One of the spectrometers was used as a luminosity monitor during an ϵ scan. Preliminary results [5] of this experiment, covering Q^2-values from 2.6 to 4.1 GeV^2, are in excellent agreement with previous Rosenbluth results. This basically rules out the possibility that the disagreement between Rosenbluth and polarization-transfer measurements of the ratio G_E^p/G_M^p is due to an underestimate of ϵ-dependent uncertainties in the Rosenbluth measurements.

2.1 Two-photon exchange

Two-(or more-)photon exchange (TPE) contributions to elastic electron scattering have been investigated both experimentally and theoretically for the past fifty years. Almost all analyses with the Rosenbluth technique have used radiative corrections that only include the infrared divergent parts of the box diagram. Thus, terms in which both photons are hard (and which depend on the hadronic structure) have been ignored.

The most stringent tests of TPE on the nucleon have been carried out by measuring the ratio of electron and positron elastic scattering off a proton. Corrections due to TPE will have a different sign in these two reactions. Unfortunately, this (e^+e^-) data set is quite limited [6], only extending (with poor statistics) up to a Q^2-value of $\sim$ 5 GeV^2, whereas at Q^2-values larger than $\sim$ 2 GeV^2 basically all data have been measured at ϵ-values larger than ~ 0.85.

Blunden *et al.* [7] carried out the first calculation of the elastic contribution from TPE effects, albeit with a simple monopole Q^2-dependence of the hadronic form factors. They obtained a practically Q^2-independent correction factor with a linear ϵ-dependence that vanishes at forward angles. However, the size of the correction only resolves about half of the discrepancy. A later calculation which used a more realistic form factor behavior, resolved up to 80% of the discrepancy. A different approach was used by Chen *et al.* [8], who related the elastic electron-nucleon scattering to the scattering off a parton in a nucleon through generalized parton distributions. TPE effects in the lepton-quark scattering process are calculated in the hard-scattering amplitudes. The results for the TPE contribution fully reconcile the Rosenbluth and the polarization-transfer data and retain agreement with positron-scattering data.

Hence, it is becoming more and more likely that TPE processes have to be taken into account in the analysis of Rosenbluth data and that they will affect polarization-transfer data only at the few percent level. Experimental confirmation of TPE effects will be difficult, but certainly should be continued. The most direct test would be a measurement of the positron-proton and electron-proton scattering cross-section ratio at small ϵ-values and Q^2-values above 2 GeV^2. A measurement in the CLAS detector at Jefferson Lab has been recently approved [9]. Additional efforts should be extended to studies of TPE effects in other longitudinal-transverse separations, such as proton knock-out and deep-inelastic scattering (DIS) experiments.

3 Neutron magnetic form factor

A significant break-through was made by measuring the ratio of quasi-elastic neutron and proton knock-out from a deuterium target. This method has little sensitivity to nuclear binding effects and to fluctuations in the luminosity and detector acceptance. A study of G_M^n at Q^2-values up to 5 GeV^2 has recently been completed in Hall B by measuring the neutron/proton quasi-elastic cross-section ratio using the CLAS detector [10]. Preliminary results [10] indicate that G_M^n is within 10% of G_D over the full Q^2-range of the experiment (0.5-4.8 GeV^2).

Inclusive quasi-elastic scattering of polarized electrons off a polarized ^{3}He target offers an alternative method to determine G_M^n through a measurement of the beam asymmetry [11]. By orienting the target polarization parallel to $\mathbf{q}$, one measure $R_{T'}$, which in quasi-elastic kinematics is dominantly sensitive to $(G_M^n)^2$. For the extraction of G_M^n corrections for the nuclear medium [12] are necessary to take into account effects of final-state interactions and meson-exchange currents.

4 Neutron electric form factor

In the past decade a series of double-polarization measurements of neutron knock-out from a polarized ^{2}H or ^{3}He target have provided accurate data on G_E^n. The ratio of the beam-target asymmetry with the target polarization perpendicular and parallel to the momentum transfer is directly proportional to the ratio of the electric and magnetic form factors. A similar result is obtained with an unpolarized deuteron target when one measures the polarization of the knocked-out neutron as a function of the angle over which the neutron spin is precessed with a dipole magnet.

At low Q^2-values corrections for nuclear medium and rescattering effects can be sizeable: 65% for ^{2}H at 0.15 GeV^2 and 50% for ^{3}He at 0.35 GeV^2. These corrections are expected to decrease significantly with increasing Q. The latest data from Hall C at Jefferson Lab, using either a polarimeter [13] or a polarized target [14], extend up to $Q^2 \approx 1.5$ GeV^2 with an overall accuracy of ~10%, in mutual agreement. From ~ 1 GeV^2 onwards G_E^n appears to exhibit a Q^2-behavior similar to that of G_E^p. Schiavilla and Sick [15] have extracted G_E^n from available data on the deuteron quadrupole form factor $F_{C2}(Q^2)$ with a much smaller sensitivity to the nucleon-nucleon potential than from inclusive (quasi-)elastic scattering.

5 Model calculations

The recent production of very accurate EMFF data, especially the surprising G_E^p data from polarization transfer, has prompted the theoretical community to intensify their investigation of nucleon structure. The first EMFF models were based on the principle of vector meson dominance (VMD), in which one assumes that the virtual photon couples to the nucleon as a vector meson. With this model Iachello *et al.* [16] predicted a linear drop of the proton form factor ratio, similar to that measured by polarization transfer, more than 20 years before the data became available. Gari and Krümpelmann [17] extended the VMD model to conform with pQCD scaling at large Q^2-values. An improved description requires the inclusion of the isovector $\pi\pi$ channel through dispersion relations [18, 19]. By adding more parameters, such as the width of the ρ-meson and the masses of heavier vector mesons [20], the VMD models succeeded in describing new EMFF data as they became available, but with little predictive power. Figure 1 confirms that Lomon's calculations provide an excellent description of all EMFF data. Bijker and Iachello [21] have extended the original calculations by also including a meson-cloud contribution in F_2. The intrinsic structure of the nucleon is estimated to have an rms radius of ~ 0.34 fm. These new calculations are in good agreement with the proton form-factor data, but do rather poorly for the neutron.

Many recent theoretical studies of the EMFFs have applied various forms of a relativistic constituent quark model (RCQM). Because the momentum transfer can be several times the nucleon mass, the constituent quarks require a relativistic quantum mechanical treatment. Although most of these calculations correctly describe the EMFF behaviour at large Q^2-values, effective degrees of freedom, such as a pion cloud and/or a finite size of the

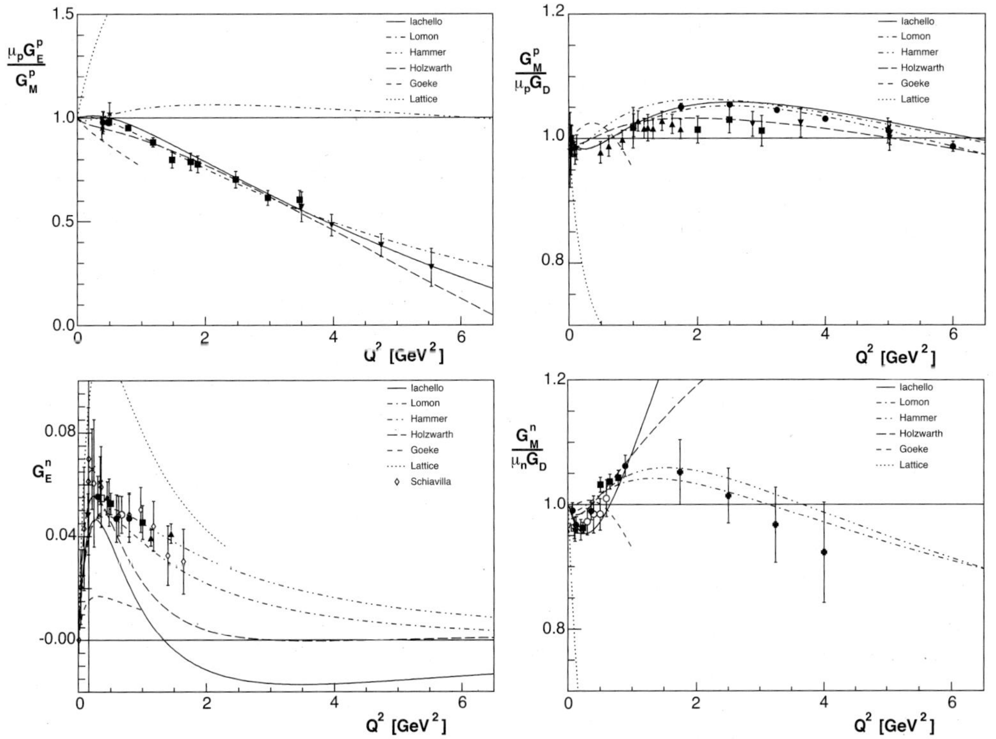

Fig. 1. Comparison of various calculations with available EMFF data. For G_E^p only polarization-transfer data are shown. For G_E^n the results of Schiavilla and Sick [15] have been added. The calculations shown are from [19,20,21,28,29,33]. Where applicable, the calculations have been normalized to the calculated values of $\mu_{p,n}$

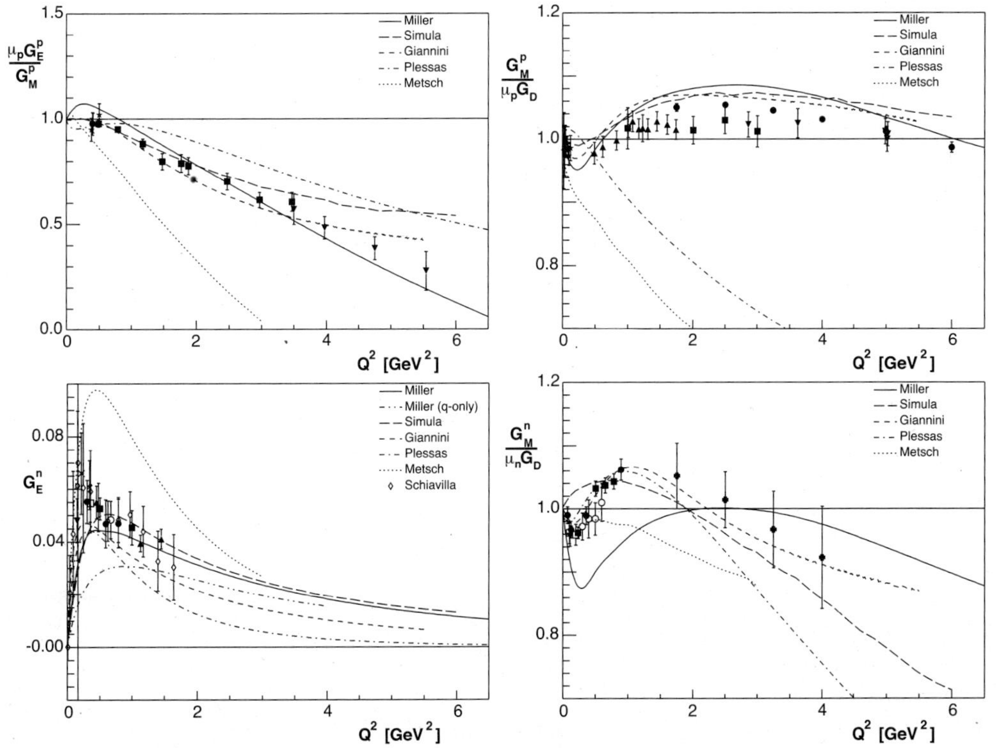

Fig. 2. Comparison of various RCQM calculations with available EMFF data, similarl to the comparison in Fig. 1. The calculations shown are from [22,24,26,25,27]. Miller (q-only) denotes a calculation by Miller [22] in which the pion cloud has been suppressed. Where applicable, the calculations have been normalized to the calculated values of $\mu_{p,n}$

constituent quarks, are introduced to correctly describe the behaviour at lower Q^2-values.

Miller [22] uses an extension of the cloudy bag model [23], three relativistically moving (in light-front kinematics) constituent quarks, surrounded by a pion cloud. Cardarelli and Simula [24] also use light-front kinematics, but they calculate the nucleon wave function by solving the three-quark Hamiltonian in the Isgur-Capstick one-gluon-exchange potential. In order to get good agreement with the EMFF data they introduce a finite size of the constituent quarks in agreement with recent DIS data. The results of Wagenbrunn *et al.* [25] are calculated in a covariant manner in the point-form spectator approximation (PFSA). In addition to a linear confinement, the quark-quark interaction is based on Goldstone-boson exchange dynamics. The PFSA current is effectively a three-body operator (in the case of the nucleon as a three-quark system) because of its relativistic nature. It is still incomplete but it leads to surprisingly good results for the electric radii and magnetic moments of the other light and strange baryon ground states beyond the nucleon. Giannini *et al.* [26] have explicitly introduced a three-quark interaction in the form of a gluon-gluon interaction in a hypercentral model, which successfully describes various static baryon properties. Relativistic effects are included by boosting the three quark states to the Breit frame and by introducing a relativistic quark current. All previously described RCQM calculations used a non-relativistic treatment of the quark dynamics, supplemented by a relativistic calculation of the electromagnetic current matrix elements. Merten *et al.* [27] have solved the Bethe-Salpeter equation with instantaneous forces, inherently respecting relativistic covariance. In addition to a linear confinement potential, they used an effective flavor-dependent two-body interaction. The results of these five calculations are compared to the EMFF data in Fig. 2. The calculations of Miller do well for all EMFFs, except for G_M^n at low Q^2-values. Those of Cardarelli and Simula, Giannini *et al.* and Wagenbrunn *et al.* are in reasonable agreement with the data, except for that of Wagenbrunn *et al.* for G_M^p, while the results of Merten *et al.* provide the poorest description of the data.

Before the Jefferson Lab polarization transfer data on G_E^p/G_M^p became available Holzwarth [28] predicted a linear drop in a chiral soliton model. In such a model the quarks are bound in a nucleon by their interaction with chiral fields. Holzwarth's model introduced one vector-meson propagator for both isospin channnels in the Lagrangian and a relativistic boost to the Breit frame. His later calculations used separate isovector and isoscalar vector-meson form factors. He obtained excellent agreement for the proton data, but only a reasonable description of the neutron data. Christov *et al.* [29] used an SU(3) Nambu-Jona-Lasinio Lagrangian, an effective theory that incorporates spontaneous chiral symmetry breaking. This procedure is comparable to the inclusion of vector mesons into the Skyrme model, but it involves many fewer free parameters (which are fitted to the masses and decay constants of pions and kaons). A constituent quark mass of 420 MeV provided a reasonable description of the EMFF data (Fig. 1).

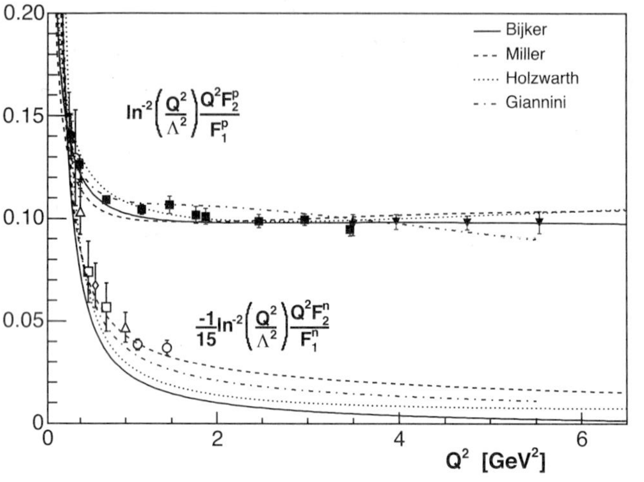

Fig. 3. The ratio $(Q^2F_2/F_1)/\ln^2(Q^2/\Lambda^2)$ as a function of Q^2 for the polarization-transfer data and the calculations of [21, 22,28,26]. The same ratio, scaled by a factor -1/15, is shown for the neutron with *open symbols*. For Λ a value of 300 MeV has been used

In the asymptotically free limit, QCD can be solved perturbatively, providing predictions for the EMFF behavior at large Q^2-values. Recently, Brodsky *et al.* [30] and Belitsky *et al.* [31] have independently revisited the pQCD domain. Belitsky *et al.* derive the following large Q^2-behavior:

$$\frac{F_2}{F_1} \propto \frac{\ln^2 Q^2/\Lambda^2}{Q^2}, \tag{1}$$

where Λ is a soft scale related to the size of the nucleon. Even though the Jefferson Lab data follow this behavior (Fig. 3), Belitsky *et al.* warn that this could very well be precocious, since pQCD is not expected to be valid at such low Q^2-values.

However, all theories described until now are at least to some extent effective (or parametrizations). They use models constructed to focus on certain selected aspects of QCD. Only lattice gauge theory can provide a truly ab initio calculation, but accurate lattice QCD results for the EMFFs are still several years away. One of the most advanced lattice calculations of EMFFs has been performed by the QCDSF collaboration [32]. The technical state of the art limits these calculations to the quenched approximation (in which sea-quark contributions are neglected), to a box size of 1.6 fm and to a pion mass of 650 MeV. Ashley *et al.* [33] have extrapolated the results of these calculations to the chiral limit, using chiral coefficients appropriate to full QCD. The agreement with the data (Fig. 1) is poorer than that of any of the other calculations, a clear indication of the technology developments required before lattice QCD calculations can provide a stringent test of experimental EMFF data.

6 Experimental review and outlook

In recent years highly accurate data on the nucleon EMFFs have become available from various facilities around the world, made possible by the development of high luminosity and novel polarization techniques. These have established some general trends in the Q^2-behavior of the four EMFFs. The two magnetic form factors G_M^p and G_M^n are close to identical, following G_D to within 10% at least up to 5 GeV^2, with a shallow minimum at ~ 0.25 GeV^2 and crossing G_D at ~ 0.7 GeV^2. Highly accurate measurements with the Rosenbluth technique have established that the discrepancy between results on G_E^p/G_M^p with the Rosenbluth techniques and with polarization transfer is not an instrumentation problem. Recent advances on two-photon exchange contributions make it highly likely that the application of TPE corrections will resolve that discrepancy. G_E^p/G_M^p drops linearly with Q^2 and G_E^n appears to drop from ~ 1 GeV^2 onwards at the same rate as G_E^p.

Measurements that extend to higher Q^2-values and offer improved accuracy at lower Q^2-values, will become available in the near future. In Hall C at Jefferson Lab Perdrisat *et al.* [34] will extend the measurements of G_E^p/G_M^p to 9 GeV^2 with a new polarimeter and large-acceptance lead-glass calorimeter. Wojtsekhowski *et al.* [35] will measure G_E^n in Hall A at Q^2-values of 2.4 and 3.4 GeV^2 using the $^3\vec{\mathrm{He}}(\mathbf{e}, e'n)$ reaction with a 100 msr electron spectrometer. The Bates Large Acceptance Spectrometer Toroid facility (BLAST, http://blast.lns.mit.edu/) at MIT with a polarized hydrogen and deuteron target internal to a storage ring will provide highly accurate data on G_E^p and G_E^n in a Q^2-range from 0.1 to 0.8 GeV^2. Thus, within a couple of years G_E^n data with an accuracy of 10% or better will be available up to a Q^2-value of 3.4 GeV^2. Once the upgrade to 12 GeV [36] has been implemented at Jefferson Lab, it will be possible to extend the data set on G_E^p and G_M^n to 14 GeV^2 and on G_E^n to 8 GeV^2.

Acknowledgements. This work was supported by DOE contract DE-AC05-84ER40150 Modification No. M175, under which the Southeastern Universities Research Association (SURA) operates the Thomas Jefferson National Accelerator Facility.

References

1. A.I. Akhiezer, L.N. Rozentsweig, I.M. Shmushkevich: Sov. Phys. JETP **6**, 588 (1958)
2. R. Arnold, C. Carlson, F. Gross: Phys. Rev. C **23**, 363 (1981)
3. V. Punjabi et al.: submitted to Phys. Rev. C; M.K. Jones et al.: Phys. Rev. Lett. **84**, 1398 (2000)
4. O. Gayou et al.: Phys. Rev. Lett. **88**, 092301 (2002)
5. R. Segel, J. Arrington: spokespersons, Jefferson Lab experiment E00-001 (2000); private communication
6. J. Arrington: Phys. Rev. C **69**, 032201R (2004) and references therein
7. P.G. Blunden, W. Melnitchouk, J.A. Tjon: Phys. Rev. Lett. **91**, 142304 (2003)
8. Y.C. Chen et al.: Phys. Rev. Lett. **93**, 122301 (2004)
9. W. Brooks et al.: spokespersons, Jefferson Lab experiment E04-116 (2004)
10. W. Brooks, M.F. Vineyard: spokespersons, Jefferson Lab experiment E94-017 (1994); private communication
11. T.W. Donnelly, A.S. Raskin: Ann. Phys. **169**, 247 (1986)
12. J. Golak et al.: Phys. Rev. C **63**, 034006 (2001)
13. R. Madey et al.: Phys. Rev. Lett. **91**, 122002 (2003)
14. G. Warren et al.: Phys. Rev. Lett. **92**, 042301 (2004); Zhu H. et al.: Phys. Rev. Lett. **87**, 081801 (2001)
15. R. Schiavilla, I. Sick: Phys. Rev. C **64**, 041002 (2001)
16. F. Iachello, A. Jackson, A. Lande: Phys. Lett. B **43**, 191 (1973)
17. M.F. Gari, W. Krümpelmann: Z. Phys. A **322**, 689 (1985); Phys. Lett. B **274**, 159 (1992)
18. G. Höhler et al.: Nucl. Phys. B **114**, 505 (1976)
19. H.W. Hammer, U.-G. Meissner, D. Drechsel: Phys. Lett. B **385**, 343 (1996); H.W. Hammer, U.-G. Meissner: Eur. Phys. Jour. A **20**, 469 (2004); P. Mergell, U.-G. Meissner, D. Drechsel: Nucl. Phys. A **596**, 367 (1996)
20. E.L. Lomon: Phys. Rev. C **64**, 035204 (2001); Phys. Rev. C **66**, 045501 (2002)
21. R. Bijker, F. Iachello: Phys. Rev. C **69**, 068201 (2004)
22. G.A. Miller: Phys. Rev. C **66**, 032001R (2002)
23. S. Théberge, A.W. Thomas, G.A. Miller: Phys. Rev. D **24**, 216 (1981)
24. F. Cardarelli, S. Simula: Phys. Rev. C **62**, 065201 (2000)
25. R.F. Wagenbrunn et al.: Phys. Lett. B **511**, 33 (2001); S. Boffi et al.: Eur. Phys. Jour. A **14**, 17 (2002)
26. M. Giannini, E. Santopinto, A. Vassallo: Prog. Part. Nucl. Phys. **50**, 263 (2003); M. De Sanctis et al.: Phys. Rev. C **62**, 025208 (2000); M. Ferraris et al.: Phys. Lett. B **364**, 231 (1995)
27. D. Merten et al.: Eur. Phys. Jour. A **14**, 477 (2002)
28. H. Holzwarth: Z. Phys. A **356**, 339 (1996); hep-ph/0201138
29. C.V. Christov et al.: Nucl. Phys. A **592**, 513 (1995); H.C. Kim et al.: Phys. Rev. D **53**, 4013 (1996)
30. S.J. Brodsky et al.: Phys. Rev. D **69**, 076001 (2004)
31. A.V. Belitsky, X. Ji, F. Yuan: Phys. Rev. Lett. **91**, 092003 (2003)
32. M. Göckeler et al.: hep-lat/0303019
33. J.D. Ashley et al.: Eur. Phys. Jour. A **19 (Suppl. 1)**, 9 (2004)
34. C.F. Perdrisat et al.: spokespersons, Jefferson Lab experiment E01-109 (2001)
35. B. Wojtsekhowski et al.: spokespersons, Jefferson Lab experiment E02-013 (2002)
36. L.S. Cardman et al., eds.: *Pre-Conceptual Design Report for the Science and Experimental Equipment for the 12 GeV Upgrade of CEBAF*, 2003, http://www.jlab.org /div_dept/physics_division/pCDR_public/pCDR_final

Eur Phys J A (2005) **24**, s2, 23–27
DOI: 10.1140/epjad/s2005-04-004-4

EPJ A direct
electronic only

Two photon effects in electron scattering

P.A.M. Guichon

SPhN/DAPNIA, CEA Saclay, F91191 Gif sur Yvette, France

Received: 15 November 2004 / Published Online: 8 February 2005

Abstract. The apparent discrepancy between the Rosenbluth and the polarization transfer method for the ratio of the electric to magnetic proton form factors can be explained by a two-photon exchange correction which does not destroy the linearity of the Rosenbluth plot. Though intrinsically small, of the order of a few percent of the cross section, this correction is kinematically enhanced in the Rosenbluth method while it is small for the polarization transfer method, at least in the range of Q^2 where it has been used until now.

PACS. 25.30.Bf Elastic electron scattering – 13.40.Gp Electromagnetic form factors – 24.85.+p Quarks, gluons, and QCD in nuclei and nuclear processes

1 Introduction

The electro-magnetic form factors are essential pieces of our knowledge of the nucleon structure and this justifies the efforts devoted to their experimental determination. They are defined by the matrix elements of the electromagnetic current $J^\mu(x)$ according to[1]:

$$< N(p')|J^\mu(0)|N(p) >= e\,\bar{u}(p')\left[G_M(Q^2)\gamma^\mu - F_2(Q^2)\frac{p+p'}{2M}\right]u(p)\ , \qquad (1)$$

where $e \simeq \sqrt{4\pi/137}$ is the proton charge and M the nucleon mass. The magnetic form factor G_M is related to the Dirac (F_1) and Pauli (F_2) form factors by $G_M = F_1 + F_2$. Here we consider only the proton case, so we have $F_1(0) = 1$, $F_2(0) = \mu_p - 1 = 1.79$. In the one photon exchange or Born approximation, elastic lepton scattering:

$$l(k) + N(p) \to l(k') + N(p') \qquad (2)$$

gives direct access to the form factors in the spacelike region ($Q^2 > 0$) where they are real. Here we adopt the usual definitions:

$$P = \frac{p+p'}{2},\ K = \frac{k+k'}{2},\ q = k - k' = p' - p, \qquad (3)$$

and choose

$$Q^2 = -q^2,\ \nu = K.P \qquad (4)$$

as the independent invariants of the scattering. In the Born approximation the elastic cross section is written:

$$d\sigma_B = C_B(Q^2,\varepsilon)\left[G_M^2(Q^2) + \frac{\varepsilon}{\tau}G_E^2(Q^2)\right] \qquad (5)$$

where the electric form factor is defined by $G_E = F_1 - \tau F_2$ with $\tau = Q^2/4M^2$ and $C_B(Q^2,\varepsilon)$ is a known phase space factor which is irrelevant in what follows. The polarization parameter of the virtual photon has the expression[2]

$$\varepsilon = \frac{\nu^2 - M^4\tau(1+\tau)}{\nu^2 + M^4\tau(1+\tau)}, \qquad (6)$$

so, at fixed Q^2, giving ε is equivalent to give ν.

For a given Q^2, (5) shows that it is sufficient to measure the cross section for two values of ε to determine the form factors G_M and G_E. In the following the determination of G_M and G_E using (5) will be referred to as the Rosenbluth method [1]. The fact that the combination $d\sigma/C_B(Q^2,\varepsilon)$ is a linear function of ε (Rosenbluth plot criterion) is generally considered as a test of the validity of the Born approximation. We shall see below that this criterion is not strong enough.

Polarized lepton beams give another way to access the form factors [2]. In the Born approximation, the polarization of the recoiling proton along its motion (P_l) is proportional to G_M while the component perpendicular to the motion (P_t) is proportional to G_E. We call this the polarization method for short. Because it is much easier to measure ratios of polarizations, it has been used mainly to determine the ratio G_E/G_M through a measurement of P_t/P_l for which one finds the expression [3]:

$$\frac{P_t}{P_l} = -\sqrt{\frac{2\varepsilon}{\tau(1+\varepsilon)}}\frac{G_E}{G_M}. \qquad (7)$$

So, in the framework of the Born approximation, one has two independent measurements of $R = G_E/G_M$. In Fig. 1 we show the corresponding results, which we call

[1] The spinors satisfy $\bar{u}(p)u(p) = 2M$ and the free states are normalized as $< N(p')|N(p) >= (2\pi)^3 2p^0\delta(\mathbf{p} - \mathbf{p}')$.

[2] This expression assumes a negligible electron mass.

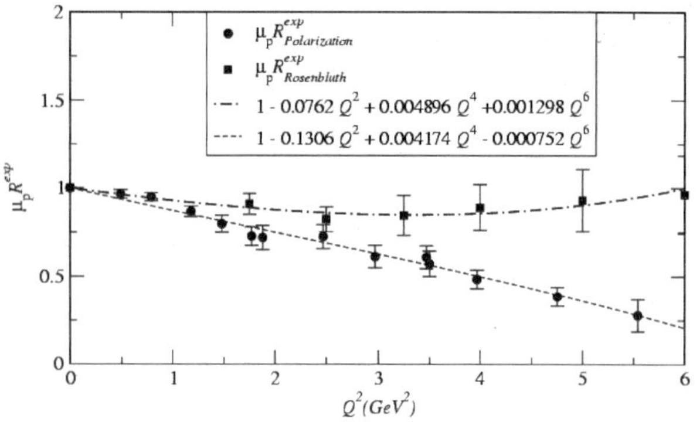

Fig. 1. Experimental values of $R^{exp}_{Rosenbluth}$ and $R^{exp}_{Polarization}$ and their polynomial fit

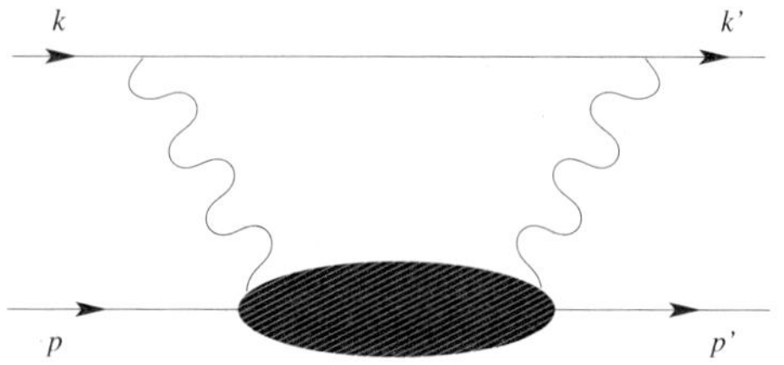

Fig. 2. The box diagram. The filled blob represents the response of the nucleon to the scattering of the virtual photon

$R^{exp}_{Rosenbluth}$ and $R^{exp}_{Polarization}$, for the range of Q^2 which is common to both methods. The data are taken from [4, 7,8]. The deviation between the two methods starts at $Q^2 = 2 - 3GeV^2$ and increases with Q^2, reaching a factor 4 at about $Q^2 = 6$ GeV^2. This discrepancy is a serious problem as it generates confusion and doubt about the whole methodology of lepton scattering experiments.

To unravel this problem we have to give up the beloved one photon exchange concept and enter the not well paved path of multi-photon physics. By this we do not mean the effect of soft (real or virtual) photons, that is the radiative corrections. The effect of the latter is well under control because their dominant (infra-red) part can be factorized in the observables and therefore does not affect the ratio G_E/G_M. Here we must consider genuine exchange of hard photons between the lepton and the hadron. Even if we restrict to the two photon exchange case, the evaluation of the box diagram 2 involves the *full* response of the nucleon to the scattering of a virtual photon and we do not know how to perform this calculation in a model independent way. Therefore we adopt a modest strategy based on the phenomenological consequences of using the full eN scattering matrix rather than its Born approximation. Though it cannot lead to a full answer it produces the following interesting results [5]:

- the 2-photon exchange amplitude needed to explain the discrepancy is actually of the expected order of magnitude, that is a few percent of the Born amplitude.
- there may be a simple explanation to the fact that the Rosenbluth plot looks linear even though it is strongly affected by the 2-photon exchange.
- the polarization method result is little affected by the 2-photon exchange, *at least in the range of Q^2 which has been studied until now.*

2 Amplitude decomposition

The simplest way to get the general form of the (e,p) scattering amplitude is to consider its helicity matrix elements $T(h'_e h'_p; h_e h_p)$ in the center of mass $\mathbf{p} + \mathbf{k} = 0$. Due to rotational invariance T depends on $E_{cm}, \cos\theta_{cm}$ which do not change under the parity and time reversal operations. Since QED is invariant with respect to these operations, one must have

$$T(h'_e h'_p; h_e h_p) = T(-h'_e - h'_p; -h_e - h_p), \tag{8}$$
$$T(h'_e h'_p; h_e h_p) = T(h_e h_p; h'_e h'_p). \tag{9}$$

Note that these equalities generally involve a phase factor which depends on the phase convention for the helicity states. For our purposes this factor is irrelevant. One can also check that charge conjugation does not bring in additional constrains. If one applies the relations (8, 9) to the $2^4 = 16$ helicity amplitudes one finds that only 6 of them are independent. Moreover 3 of them change the electron helicity which implies that they are suppressed by an electron mass factor. So one ends with

$$T\left(\frac{1}{2}\frac{1}{2};\frac{1}{2}\frac{1}{2}\right), T\left(\frac{1}{2}\frac{1}{2};\frac{1}{2}-\frac{1}{2}\right), T\left(\frac{1}{2}-\frac{1}{2};\frac{1}{2}-\frac{1}{2}\right)$$

as the only independent amplitudes. The next step is to write a covariant decomposition with 3 independent Lorentz structures. For obvious reasons we choose two of them to be the same as in the one photon approximation. For the third structure several choices are possible and we found convenient to choose

$$\bar{u}(k')\gamma.Pu(k)\ \bar{u}(p')\gamma.Ku(p).$$

so that the T matrix can be written as

$$T = \frac{e^2}{Q^2}\bar{u}(k')\gamma_\mu u(k)$$
$$\times\ \bar{u}(p')\left(\tilde{G}_M\gamma^\mu - \tilde{F}_2\frac{P^\mu}{M} + \tilde{F}_3\frac{\gamma.KP^\mu}{M^2}\right)u(p), \tag{10}$$

where $\tilde{G}_M$, $\tilde{F}_2$, $\tilde{F}_3$ are complex functions of ν and Q^2 and the factor e^2/Q^2 has been introduced for convenience. By analogy we define:

$$\tilde{G}_E = \tilde{G}_M - (1+\tau)\tilde{F}_2 \tag{11}$$

which is equal to G_E in the Born approximation. The last step is to evaluate the matrix elements of T in the helicity basis and in the CM frame and to check that the set of equations

$$T\left(\frac{1}{2}\frac{1}{2};\frac{1}{2}\frac{1}{2}\right), T\left(\frac{1}{2}\frac{1}{2};\frac{1}{2}-\frac{1}{2}\right), T\left(\frac{1}{2}-\frac{1}{2};\frac{1}{2}-\frac{1}{2}\right)$$

$$\leftrightarrow \tilde{G}_M,\ \tilde{F}_2,\ \tilde{F}_3$$

can be inverted, which is indeed the case. Note that one can also start directly from the general amplitude for elastic scattering of two spin 1/2 particles as derived by Goldberger et al. [6]. Neglecting the amplitudes which flip the helicity of the electron this amplitude is written [6]:

$$T = \bar{u}(k')\gamma.Pu(k)\,\bar{u}(p')\,(S + V\,\gamma.K)\,u(p) \\ +A\,\bar{u}(k')\gamma_5\gamma.Pu(k)\,\bar{u}(p')\gamma_5\gamma.Ku(p) \qquad (12)$$

Using Dirac equation and elementary relations among the Dirac matrices, it is then a simple exercise to write (12) in the form (10).

If one compares with the Born amplitude:

$$T_B = e^2\bar{u}(k')\gamma_\mu u(k)\frac{1}{Q^2}\bar{u}(p')\left(G_M\,\gamma^\mu - F_2\frac{P^\mu}{M}\right)u(p), \qquad (13)$$

one gets the relations:

$$\tilde{G}_M^{Born}(\nu, Q^2) = G_M(Q^2), \\ \tilde{F}_2^{Born}(\nu, Q^2) = F_2(Q^2), \\ \tilde{F}_3^{Born}(\nu, Q^2) = 0. \qquad (14)$$

Since $\tilde{F}_3$ and the phases of $\tilde{G}_M$ and $\tilde{F}_2$ vanish in the Born approximation, they must originate from processes involving at least the exchange of 2-photon. If we take care of the factor e^2 introduced in the definition (13), we see that they are at least of order e^2. This, of course, assumes that the phases of $\tilde{G}_M$ and $\tilde{F}_2$ are defined, which amounts to suppose that, in the kinematical region of interest, the modulus of $\tilde{G}_M$ and $\tilde{F}_2$ do not vanish. We take it for granted in the following but this restriction must be kept in mind if one goes in regions where some amplitude becomes very small.

3 Cross section and polarization transfer

If we define:

$$\tilde{G}_M = e^{i\phi_M}\left|\tilde{G}_M\right|, \\ \tilde{G}_E = e^{i\phi_E}\left|\tilde{G}_E\right|, \qquad (15) \\ \tilde{F}_i = e^{i\phi_i}\left|\tilde{F}_i\right|,$$

then, using standard techniques, we get the following expressions for the observables of interest:

$$\frac{d\sigma}{C_B(\varepsilon, Q^2)} = \left|\tilde{G}_M\right|^2 + \frac{\varepsilon}{\tau}\left|\tilde{G}_E\right|^2 \\ + 2\varepsilon\rho\mathcal{R}\left(\left(\tilde{G}_M + \frac{1}{\tau}\tilde{G}_E\right)\tilde{F}_3^*\right) \\ + \left(\frac{1}{\tau} + \frac{2\varepsilon}{1+\varepsilon}\right)\varepsilon\rho^2\left|\tilde{F}_3\right|^2, \qquad (16)$$

$$\frac{P_t}{P_l} = -\sqrt{\frac{2\varepsilon}{\tau(1+\varepsilon)}}\,\frac{\left|\tilde{G}_E\right|\cos\phi_{ME} + \rho\left|\tilde{F}_3\right|\cos\phi_{3M}}{\left|\tilde{G}_M\right| + \frac{2\varepsilon}{1+\varepsilon}\rho\left|\tilde{F}_3\right|\cos\phi_{3M}}, \qquad (17)$$

where:

$$\phi_{ME} = \phi_M - \phi_E,\ \phi_{3M} = \phi_3 - \phi_M,\ \rho = \frac{\nu}{M^2}. \qquad (18)$$

If one substitutes the Born approximation values of the amplitudes (14) then (16, 17) give back the familiar expressions (5, 7).

To simplify the general expressions (16, 17) we make the very reasonable assumption that only the two photons exchange needs to be considered. This amounts to keep only the terms of order e^2 with respect to the leading one in (16, 17). Using the fact that ϕ_M, ϕ_E and $\tilde{F}_3$ are of order e^2 we get the approximate expressions:

$$\frac{d\sigma}{C_B(\varepsilon, Q^2)} = \left|\tilde{G}_M\right|^2 \qquad (19) \\ \times\left\{1 + \frac{\varepsilon}{\tau}\frac{\left|\tilde{G}_E\right|^2}{\left|\tilde{G}_M\right|^2} + 2\varepsilon\left(1 + \frac{1}{\tau}\frac{\left|\tilde{G}_E\right|}{\left|\tilde{G}_M\right|}\right)Y_{2\gamma}\right\}$$

and

$$\frac{P_t}{P_l} = -\sqrt{\frac{2\varepsilon}{\tau(1+\varepsilon)}} \qquad (20) \\ \times\left\{\frac{\left|\tilde{G}_E\right|}{\left|\tilde{G}_M\right|} + \left(1 - \frac{2\varepsilon}{1+\varepsilon}\frac{\left|\tilde{G}_E\right|}{\left|\tilde{G}_M\right|}\right)Y_{2\gamma}\right\}$$

To set a scale for the size of the two photon correction we have introduced the dimensionless ratio:

$$Y_{2\gamma}(\nu, Q^2) = \mathcal{R}\left(\frac{\nu\tilde{F}_3}{M^2\left|\tilde{G}_M\right|}\right)$$

which should be a good measure of the effect since, if we neglect $\left|\tilde{G}_E\right|/\left|\tilde{G}_M\right|$ with respect to τ in (19), we see that the cross section would be of the form $\left|\tilde{G}_M\right|^2(1+Y_{2\gamma})^2$. Therefore we expect $Y_{2\gamma}$ to be of the order of $\alpha \simeq 1/137$.

The equations (19,20) already exhibit the solution to our problem. In the cross section the ratio $\left|\tilde{G}_E\right|/\left|\tilde{G}_M\right|^2$ comes with a term $2\left(\tau + \left|\tilde{G}_E\right|/\left|\tilde{G}_M\right|\right)Y_{2\gamma}$ which at large Q^2 is essentially $2\tau Y_{2\gamma} = Q^2 Y_{2\gamma}/2M^2$. This produces an amplification of the two photon effect which is not present in P_t/P_l. As a rough estimate let us take $\left|\tilde{G}_E\right|/\left|\tilde{G}_M\right| \sim G_E(0)/G_M(0) = 1/2.79$ and choose $Q^2 = 4M^2$. Then the coefficient of ε in (19), which is supposed to measure G_E/G_M in the Born approximation, is equal to $0.128 + 2.7\,Y_{2\gamma}$. So even if $Y_{2\gamma}$ is as small as 1% it produces a relative correction of 21%! By contrast if we do the same for the expression in parenthesis in (20), with $\varepsilon = 0.8$ which is a typical value used in [7,4], we get $0.36+0.68\,Y_{2\gamma}$. For $Y_{2\gamma} = 1\%$ this only produces a 2% correction. Now that the origin of the discrepancy has been identified we can try to analyze the data in a more quantitative manner.

4 Analysis

From (19,20) we see that the experimental couple $(d\sigma, P_t/P_l)$ depends on $\left|\tilde{G}_M\right|$, $\left|\tilde{G}_E\right|$ and $\mathcal{R}(\tilde{F}_3)$. In first approximation we know that

$$\left|\tilde{G}_M(\nu\, Q^2)\right| \simeq G_M(Q^2),\ \left|\tilde{G}_E(\nu\, Q^2)\right| \simeq G_E(Q^2)$$

so only $\mathcal{R}(\tilde{F}_3)$ is really a new unknown parameter. If we look at the data of [8] for $\sigma/C_B(\varepsilon, Q^2)$ as a function of ε we observe that for each value of Q^2 the set of points are pretty well aligned. We see on (19) that this can be understood if, at least in first approximation, the product $\nu\, \tilde{F}_3$ is independent of ε. We do not have a first principle explanation for this but we feel allowed to take it as an experimental evidence. To explain the linearity of the plots one should also suppose that $\left|\tilde{G}_M\right|$ and $\left|\tilde{G}_E\right|$ are independent of ε (that is ν) but since the dominant term of these amplitudes depends only on Q^2 this is a very mild assumption. We then see from (19) that what is measured using the Rosenbluth method is:

$$(R^{exp}_{Rosenbluth})^2 = \frac{\left|\tilde{G}_E\right|^2}{\left|\tilde{G}_M\right|^2} + 2\left(\tau + \frac{\left|\tilde{G}_E\right|}{\left|\tilde{G}_M\right|}\right) Y_{2\gamma}, \qquad (21)$$

with $\left|\tilde{G}_E\right| / \left|\tilde{G}_M\right|$ and $Y_{2\gamma}$ essentially independent of ε, rather than

$$(R^{exp}_{Rosenbluth})^2 = \left(\frac{G_E}{G_M}\right)^2, \qquad (22)$$

as implied by the one photon exchange approximation.

On the other hand the experimental results of the polarization method have been obtained for a narrow range of ε, typically [3] from $\varepsilon = .6$ to $.9$. So, in practice, we can neglect the ε dependence of $R^{exp}_{Polarization}$ and from (20) we see that this experimental ratio must be interpreted as:

$$R^{exp}_{Polarization} = \left|\frac{\tilde{G}_E}{\tilde{G}_M}\right| + \left(1 - \frac{2\varepsilon}{1+\varepsilon}\left|\frac{\tilde{G}_E}{\tilde{G}_M}\right|\right) Y_{2\gamma}, \qquad (23)$$

rather than

$$R^{exp}_{Polarization} = \frac{G_E}{G_M}. \qquad (24)$$

In order that (23) be consistent with our hypothesis we should find that $Y_{2\gamma}$ is small enough that the factor $2\varepsilon/(1+\varepsilon)$ introduces no noticeable ε dependence in $R^{exp}_{Polarization}$. We have now a system of (21, 23) for $\left|\tilde{G}_E\right| / \left|\tilde{G}_M\right|$ and $Y_{2\gamma}$ that we can solve for each value of Q^2 . Due to the kinematical enhancement of the two photons effect in the cross section we cannot treat it as a perturbation when solving the system of equations. Since the latter is equivalent to a quartic equation it is more efficient to solve it numerically. For this we have fitted the data by a polynomial in Q^2. The result is shown on Fig. 1 and we shall consider this fit as the experimental values. In particular we do not attempt to represent the effect of the error bars which can be postponed to a more complete re-analysis of the data.

[3] except at the lowest values of Q^2 , where there is anyway no discrepancy between the Rosenbluth and the polarization method.

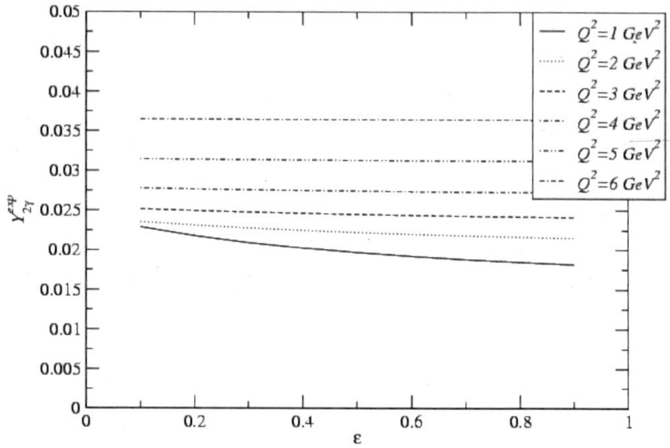

Fig. 3. The ratio $Y^{exp}_{2\gamma}$ versus ε for several values of Q^2

Using the fit we solve numerically the system (21, 23) for the couple $(Y^{exp}_{2\gamma},\ R^{exp}_{1\gamma+2\gamma} = \left|\tilde{G}_E\right| / \left|\tilde{G}_M\right|)$. The solution for the ratio $Y^{exp}_{2\gamma}$ is shown on Fig. 3 where we can see that it is actually small, of the order of a few percents. Also we observe that it is essentially flat as a function of ε which is consistent with our hypothesis. In fact it is a direct consequence of the smallness of $Y^{exp}_{2\gamma}$ which multiplies the only factor which depends on ε in (23). For the same reason $R^{exp}_{1\gamma+2\gamma}$ is also essentially independent of ε.

The above result for $Y^{exp}_{2\gamma}$ indicates that the corrections to the Born approximation are actually small in absolute value. In the Rosenbluth method their effect is accidentally amplified but there is no reason to think that this kind of accident will also occur in $\tilde{G}_E - G_E$ or $\tilde{G}_M - G_M$. So it makes sense to compare the value we get for $R^{exp}_{1\gamma+2\gamma}$ with the starting experimental ratios $R^{exp}_{Rosenbluth}$ and $R^{exp}_{Polarization}$. This is shown on Fig. 4 where we see that[4], as expected, $R^{exp}_{1\gamma+2\gamma}$ is close to $R^{exp}_{Polarization}$.

5 Conclusion

Within the hypothesis of our analysis, we come to the conclusion that the data for G_E/G_M from the Rosenbluth and the polarization method are compatible once the exchange of 2-photon is allowed in the analysis. The two photon effect is intrinsically small, as it should, but is strongly amplified in the Rosenbluth method. Assuming that the difference between G_E/G_M and $\tilde{G}_E/\tilde{G}_M$ is of the same order as $Y^{exp}_{2\gamma}$ one can consider that the value

[4] The calculation here has been done (arbitrarily) at $\varepsilon = 0.6$ but the result is essentially independent of ε.

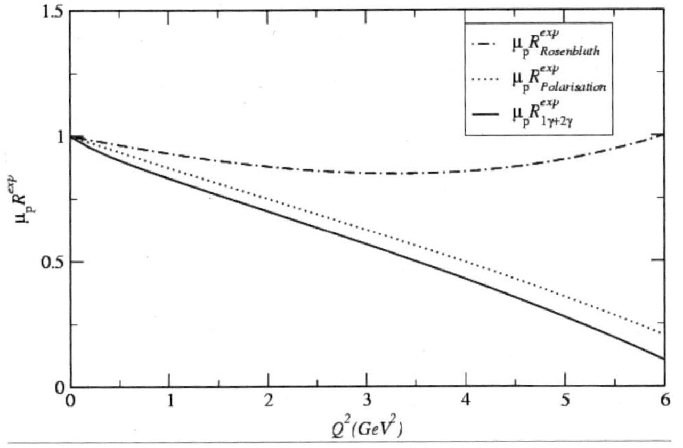

Fig. 4. Comparison of the experimental ratios $\mu_p R^{exp}_{Rosenbluth}$ and $\mu_p R^{exp}_{Polarization}$ with the value of $\mu_p R^{exp}_{1\gamma+2\gamma} = \mu_p \left|\tilde{G}_E\right| / \left|\tilde{G}_M\right|$ deduced from our analysis

we obtain for $\tilde{G}_E/\tilde{G}_M$ is essentially the value of G_E/G_M obtained in the Born approximation but corrected by the two photon effect due to $\tilde{F}_3$. Our conclusion is then that the correction which must be applied to the results of the polarization method is negligible while it is huge in the case of the Rosenbluth method.

To confirm our results one needs some model calculation of the central quantity of our analysis, namely $\tilde{F}_3$. An explicit calculation of the two photon effect has been performed in [9] and it indicates that it can explain part of the discrepancy. However this calculation limits the intermediate states to the nucleon itself, which is certainly not a realistic hypothesis. A more ambitious calculation, where the intermediate states excitations are implicitly taken into account through the use of generalized parton distributions, is now close to completion [10]. The preliminary results are in agreement with our analysis.

Another important point is the study of observables which can directly test our understanding of the two photon effects. As an example we consider the beam charge asymmetry BCA, which is defined by:

$$\frac{\sigma(positron)}{\sigma(electron)} = 1 - 2BCA.$$

In the 2-photon approximation one gets

$$BCA = 2\frac{\varepsilon G_E \mathcal{R}(\delta\tilde{G}_E) + \tau G_M \mathcal{R}(\delta\tilde{G}_M) + \varepsilon G_M (G_E + \tau G_M) Y_{2\gamma}}{\varepsilon G_E^2 + \tau G_M^2},$$

which in the limit $G_E/(\tau G_M) \to 0$ implies

$$\frac{\sigma(+)}{\sigma(-)} \simeq 1 - 4\varepsilon Y_{2\gamma} - 4\frac{\mathcal{R}(\delta\tilde{G}_M)}{G_M}.$$

According to our analysis we have $Y_{2\gamma} > 0$, while the scarce existing data [11] at large Q^2 indicate that $\sigma(+)/\sigma(-)$ is compatible with one or even a bit larger. This would imply that $\delta\tilde{G}_M/G_M$ is negative and of the order of $Y_{2\gamma}$, which is of course in line with our working hypothesis. This illustrates how dedicated beam charge asymmetry experiments could be used to strengthen our understanding of two photon effects in electron scattering.

References

1. M.N. Rosenbluth: Phys. Rev. **79**, 615 (1950)
2. A.I. Akhiezer, L.N. Rozentseig, I.M. Shmushkevich: Sov. Phys. JETP **6**, 588 (1958)
3. A.I. Akhiezer, M.P. Rekalo: Sov. Phys. Doklady **13**, 572 (1968)
4. O. Gayou et al.: Phys. Rev. Lett. **88**, 092301 (2002)
5. P.A.M. Guichon, Marc Vanderhaeghen: Phys. Rev. Lett. **91**, 142303 (2003)
6. M.L. Goldberger, Y. Nambu, R. Oehme: Ann. of Phys. **2**, 226 (1957)
7. M.K. Jones et al.: Phys. Rev. Lett. **84**, 1398 (2000)
8. L. Andivahis et al: Phys. Rev. D **50**, 5491 (1994)
9. P.G. Blunden, W. Melnitchouk, J.A. Tjon: Phys. Rev. Lett. **91**, 142304 (2003)
10. M. Vanderhaeghen: private communication
11. J. Mar et al.: Phys. Rev. Lett. **21**, 482 (1968)

Eur Phys J A (2005) **24**, s2, 29–32
DOI: 10.1140/epjad/s2005-04-005-3

EPJ A direct
electronic only

Single spin asymmetries in elastic electron–nucleon scattering

B. Pasquini[1,2] and M. Vanderhaeghen[3,4]

[1] Dipartimento di Fisica Nucleare e Teorica, Università degli Studi di Pavia and INFN, Sezione di Pavia, Pavia, Italy
[2] ECT*, Villazzano (Trento), Italy
[3] Thomas Jefferson National Accelerator Facility, Newport News, VA 23606, USA
[4] Department of Physics, College of William and Mary, Williamsburg, VA 23187, USA

Received: 15 October 2004 / Published Online: 8 February 2005

Abstract. We discuss the target and beam normal spin asymmetries in elastic electron-nucleon scattering which depend on the imaginary part of two-photon exchange processes between electron and nucleon. In particular, we estimate these transverse spin asymmetries for beam energies below 2 GeV, where the two-photon exchange process is dominated by the resonance contribution to the doubly virtual Compton scattering tensor of the nucleon.

PACS. 25.30.Bf Elastic electron scattering – 25.30.Rw Electroproduction reactions

1 Introduction

Elastic electron-nucleon scattering in the one-photon exchange approximation is a valuable tool to access information on the structure of hadrons. New experimental techniques exploiting polarization observables have made possible precision measurements of hadron structure quantities, such as its electroweak form factors, parity violating effects, $N \to \Delta$ transition form factors, and spin dependent structure functions. However, to push the precision frontier further in electron scattering, one needs a good control of the two-photon (2γ) exchange mechanism and needs to understand how they may affect different observables. The imaginary (absorptive) part of the 2γ exchange amplitude can be accessed through a single spin asymmetry (SSA) in elastic electron-nucleon scattering, when either the target or beam spin are polarized normal to the scattering plane. As time reversal invariance forces this SSA to vanish for one-photon exchange, it is of order $\alpha = e^2/(4\pi) \simeq 1/137$. Furthermore, to polarize an ultra-relativistic particle in the direction normal to its momentum involves a suppression factor m/E (with m the mass and E the energy of the particle), which typically is of order $10^{-4} - 10^{-3}$ when the electron beam energy is in the 1 GeV range. Therefore, the resulting target normal SSA can be expected to be of order 10^{-2}, whereas the beam normal SSA is of order $10^{-6} - 10^{-5}$. In the case of a polarized lepton beam, asymmetries of the order ppm are currently accessible in parity violation (PV) elastic electron-nucleon scattering experiments. While the PV asymmetry measurements involve a beam spin polarized along its momentum, the SSA for an electron beam spin normal to the scattering plane can be accessed using the same experimental apparatus. Results from first measurements of this beam normal SSA have been presented in this conference.

Model calculations for such observables have been recently performed in different kinematical regimes [1,2,3, 4]. Here we report on a study of the imaginary part of the 2γ exchange entering in the normal SSA's at low and intermediate beam energies [5]. Using unitarity, one can relate the imaginary part of the 2γ amplitude to the electroabsorption amplitudes on a nucleon. Below or around two-pion production threshold, one is in a regime where these electroproduction amplitudes are relatively well known using pion electroproduction experiments as input. Therefore the aim is to gain a good knowledge of the imaginary part of the two-photon exchange amplitude, and then to use such information as input for dispersion relations which will allow to quantify the contribution of the real part of the 2γ exchange processes. In addition, observables such as normal SSA's are sensitive to the electroproduction amplitudes on the nucleon for a wide range of photon virtualities. This may provide information on resonance transition form factors complementary to the information obtained from pion electroproduction experiments.

2 Single spin asymmetries in elastic electron-nucleon scattering

The target spin asymmetries in elastic electron-nucleon scattering is defined by

$$A_n = \frac{\sigma_\uparrow - \sigma_\downarrow}{\sigma_\uparrow + \sigma_\downarrow}, \tag{1}$$

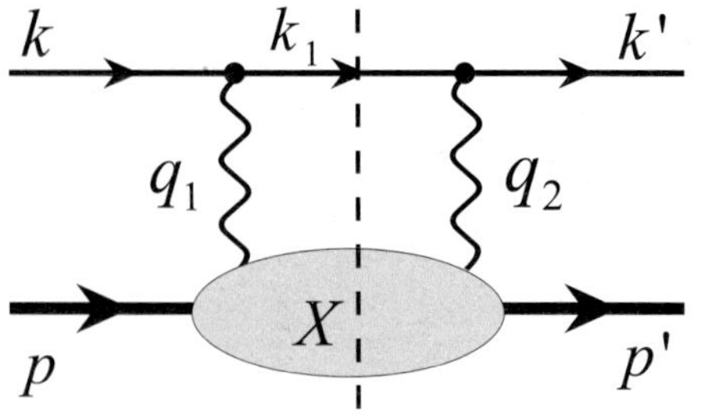

Fig. 1. The 2γ exchange diagram. The blob represents the response of the nucleon to the scattering of the virtual photon

where $\sigma_\uparrow$ ($\sigma_\downarrow$) denotes the cross section for an unpolarized beam and for a nucleon spin parallel (anti-parallel) to the normal polarization vector $\mathbf{S}_n = (\mathbf{k}\times\mathbf{k}')/|\mathbf{k}\times\mathbf{k}'|$ (with $\mathbf{k}$ and $\mathbf{k}'$ the three-momenta of the initial and final electron, respectively). Analogous expression as in (1) holds for the beam spin asymmetry (B_n) when we interpret $\sigma_\uparrow$ ($\sigma_\downarrow$) as the cross section for an unpolarized target and for an electron beam spin parallel (antiparallel) to the normal polarization vector. As has been shown by de Rujula *et al.* [6], the target and beam normal spin asymmetry are related to the absorptive part of the elastic eN scattering amplitude. Since the one-photon exchange amplitude is purely real, the leading contribution to SSA's is of order $O(e^2)$, and is due to an interference between one- and two-photon exchange amplitudes, i.e.

$$\mathrm{SSA} = \frac{2\,\mathrm{Im}(\sum_{spins} T^*_{1\gamma}\cdot \mathrm{Abs}\,T_{2\gamma})}{\sum_{spins}|T_{1\gamma}|^2}, \qquad (2)$$

where $T_{1\gamma}$ is the one-photon exchange amplitude, and $\mathrm{Abs}\,T_{2\gamma}$ is the absorptive part of the doubly virtual Compton scattering tensor on the nucleon, as shown in Fig. 1. Equation (2) can be expressed in terms of a 3-dimensional phase-space integral as

$$\mathrm{SSA} = -\frac{1}{(2\pi)^3}\frac{e^2Q^2}{D(s,Q^2)}\int_{M^2}^{(\sqrt{s}-m_e)^2} dW^2\,\frac{|\mathbf{k}_1|}{4\sqrt{s}} \times \int d\Omega_{k_1}\frac{1}{Q_1^2\,Q_2^2}\,\mathrm{Im}\left\{L_{\alpha\mu\nu}\,H^{\alpha\mu\nu}\right\}, \qquad (3)$$

where $W^2 = p_X^2$ is the squared invariant mass of the intermediate state X, and $s=(p+k)^2$. In (3), the momenta are defined as in Fig. 1, $Q_1^2 \equiv -q_1^2$ and $Q_2^2 \equiv -q_2^2$ correspond with the virtualities of the two spacelike photons, $D(s,Q^2) \equiv Q^4/e^4\sum_{spins}|T_{1\gamma}|^2$, and $L_{\alpha\mu\nu}$ and $H^{\alpha\mu\nu}$ are the leptonic and hadronic tensors, respectively. Furthermore, (3) reduces to the target or beam asymmetry once we specify the helicities for the polarized particles and take the sum over the helicities of the unpolarized particles. The explicit expression for the tensor $H^{\alpha\mu\nu}$ is given by :

$$H^{\alpha\mu\nu} = W^{\mu\nu}\cdot\left[\bar{u}(p',\lambda'_N)\Gamma^\alpha(p',p)u(p,\lambda_N)\right]^*, \qquad (4)$$

where $\Gamma^\alpha(p',p)$ is the elastic photon-nucleon vertex and $W^{\mu\nu}$ corresponds to the absorptive part of the doubly virtual Compton scattering tensor with two *spacelike* photons. The latter is given by

$$W^{\mu\nu}(p',\lambda'_N;p,\lambda_N) = \sum_X (2\pi)^4\,\delta^4(p+q_1-p_X) \times <p'\,\lambda'_N|J^{\dagger\mu}(0)|X><X|J^\nu(0)|p\,\lambda_N>, \qquad (5)$$

where the sum goes over all possible *on-shell* intermediate hadronic states X. The number of intermediate states X which one considers in the calculation sets a limit on how high in energy one can reliably calculate the hadronic tensor in (5). In addition to the elastic contribution ($X=N$) which is exactly calculable in terms of on-shell nucleon electromagnetic form factors, we approximate the remaining inelastic part of $W^{\mu\nu}$ with a sum over all πN intermediate states (i.e. $X=\pi N$ in the blob of Fig. 1). The calculation is performed by using the unitarity relation which allows to express $W^{\mu\nu}$ in terms of electroabsorption amplitudes $\gamma^* N \to X$ at different photon virtualities. To estimate the pion electroproduction amplitudes we use the phenomenological MAID analysis (version 2000) [7], which contains both resonant and non-resonant pion production mechanisms. This same strategy has been used before in the description of real and virtual Compton scattering in the resonance region, and checked against data in [8].

3 Results and discussion

In this section we show our results for both beam and target normal spin asymmetries for elastic electron-proton scattering. Our calculation covers the whole resonance region, and addresses measurements performed or in progress at MIT-Bates [9], MAMI [10], JLab [11,12], and SLAC [13]. In Fig. 2, we show the beam normal spin asymmetry B_n for elastic $e^{-\uparrow}p \to e^-p$ scattering at a low beam energy of $E_e = 0.2$ GeV. At this energy, the elastic contribution is sizeable. The inelastic contribution is dominated by the region of threshold pion production, as is shown in Fig. 3, where we display the integrand of the W-integration for B_n. When integrating the full curve in Fig. 3 over W, one obtains the total inelastic contribution to B_n (i.e. dashed-dotted curve in Fig. 2). The present calculation (MAID) of the threshold pion electroproduction is consistent with chiral symmetry predictions, and is therefore largely model independent. One notices that at backward *c.m.* angles (i.e. with increasing Q^2) the π^+n and π^0p intermediate states contribute with opposite sign. The peaked structure at the maximum possible value of the integration range in W, i.e. $W_{max} = \sqrt{s}-m_e$, is due to the near singularity (in the electron mass) corresponding with quasi-real Compton scattering (RCS), in which both photons in the 2γ exchange process become quasi-real. This contribution at large W mainly drives the results for the inelastic part of the beam asymmetry. Furthermore, it is seen from Fig. 2 that the inelastic and elastic contributions at a low energy of 0.2 GeV have opposite sign, resulting in quite a small asymmetry. It is somewhat puzzling that the only experimental data point at this energy

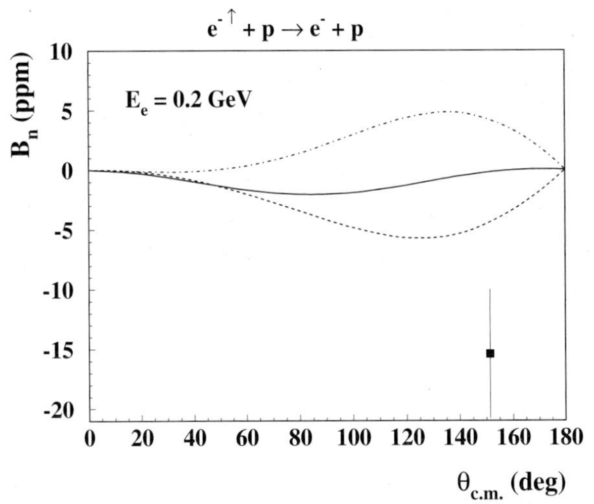

Fig. 2. Beam normal spin asymmetry B_n for $e^{-\uparrow}p \to e^-p$ at a beam energy $E_e = 0.2$ GeV as function of the *c.m.* scattering angle, for different hadronic intermediate states (X) in the blob of Fig. 1: N (*dashed curve*), πN (*dashed-dotted curve*), sum of the N and πN (*solid curve*). The data point is from the SAMPLE Collaboration (MIT-Bates) [9]

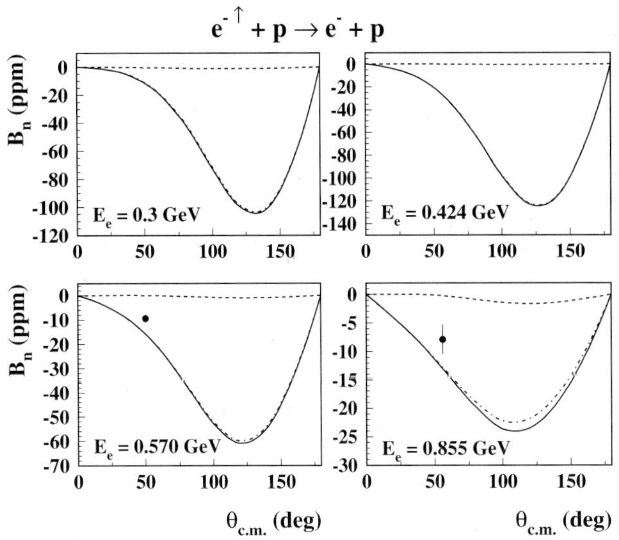

Fig. 4. Beam normal spin asymmetry B_n for $e^{-\uparrow}p \to e^-p$ as function of the *c.m.* scattering angle at different beam energies, as indicated on the figure. The meaning of the different lines is the same as in Fig. 2. The data points are from the A4 Collaboration (MAMI) [10]

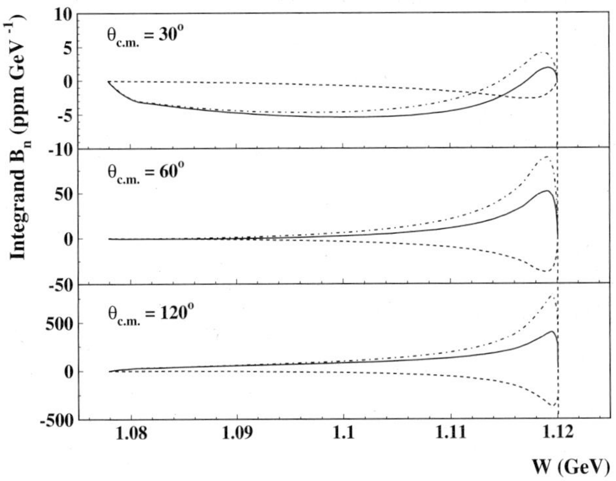

Fig. 3. Integrand in W of the beam normal spin asymmetry B_n for $e^{-\uparrow}p \to e^-p$ at a beam energy of $E_e = 0.2$ GeV and at different *c.m.* scattering angles as indicated on the figure. The *dashed curves* are the contribution from the $\pi^0 p$ channel, the *dashed-dotted curves* show the contribution from the $\pi^+ n$ channel, and the *solid curves* are the sum of the contributions from the $\pi^+ n$ and $\pi^0 p$ channels. The *vertical dashed line* indicates the upper limit of the W integration, i.e. $W_{max} = \sqrt{s} - m_e$

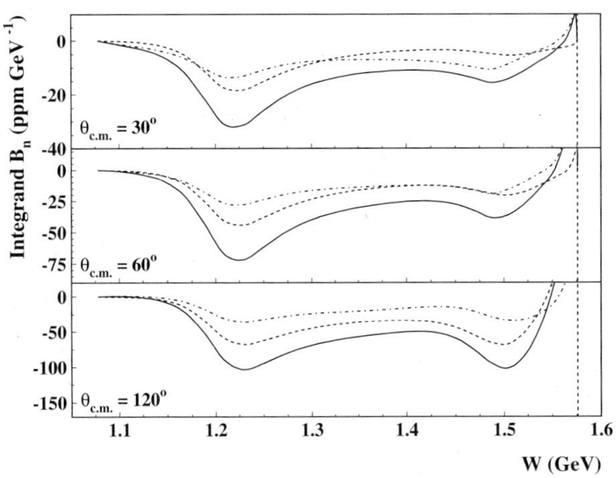

Fig. 5. Integrand in W of the beam normal spin asymmetry B_n for $e^{-\uparrow}p \to e^-p$ at a beam energy of $E_e = 0.855$ GeV and at different *c.m.* scattering angles as indicated on the figure. The meaning of the different lines is the same as in Fig. 3

indicates a larger negative value at backward angles, although with quite large error bar.

In Fig. 4, we show B_n at different beam energies below $E_e = 1$ GeV. It is clearly seen that at energies $E_e = 0.3$ GeV and higher the elastic contribution yields only a very small relative contribution. Therefore B_n is a direct measure of the inelastic part which gives rise to sizeable large asymmetries, of the order of several tens of ppm in the backward angular range. At forward angles, the size of the predicted asymmetries is compatible with the first high precision measurements performed at MAMI. It will be worthwhile to investigate if the slight overprediction (in absolute value) of B_n, in particular at $E_e = 0.57$ GeV, is also seen in a backward angle measurement, which is planned in the near future at MAMI.

To gain a better understanding of how the inelastic contribution to B_n arises, we show in Fig. 5 the integrand of B_n at $E_e = 0.855$ GeV and at different scattering angles. The resonance structure is clearly reflected in the integrands for both $\pi^+ n$ and $\pi^0 p$ channels. At forward

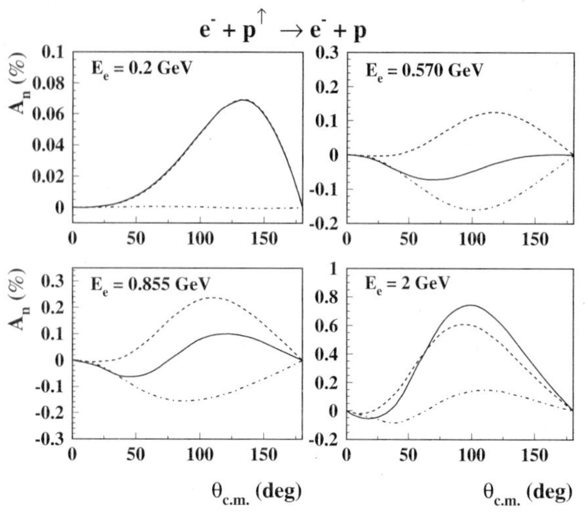

Fig. 6. Target normal spin asymmetry A_n for $e^- p^\uparrow \to e^- p$ as function of the $c.m.$ scattering angle at different beam energies, as indicated on the figure. The meaning of the different lines is the same as in Fig. 2

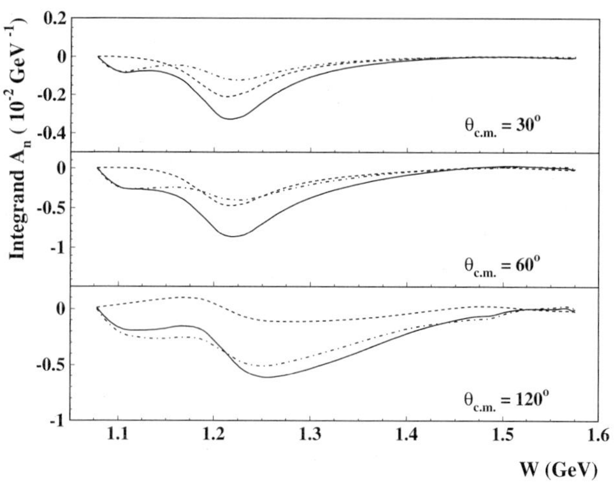

Fig. 7. Integrand in W of the target normal spin asymmetry A_n for $e^- p^\uparrow \to e^- p$ for a beam energy of $E_e = 0.855$ GeV and at different $c.m.$ scattering angles as indicated on the figure. The meaning of the different lines is the same as in Fig. 3

angles, the quasi-RCS at the endpoint $W = W_{max}$ only yields a very small contribution, which grows larger when going to backward angles. This quasi-RCS contribution is of opposite sign as the remainder of the integrand, and therefore determines the position of the maximum (absolute) value of B_n when going to backward angles.

We next discuss the target normal spin asymmetry A_n. In Fig. 6, we show the results for both elastic and inelastic contributions to A_n at different beam energies. At a low beam energy of $E_e = 0.2$ GeV, A_n is completely dominated by the elastic contribution. Going to higher beam energies, the inelastic contribution becomes of comparable magnitude as the elastic one. This is in contrast with the situation for B_n where the elastic contribution already becomes negligible for beam energies around $E_e = 0.3$ GeV. The integrand of the inelastic contribution at a beam energy of $E_e = 0.855$ GeV is shown in Fig. 7. The total inelastic result displays a $\pi^+ n$ threshold region contribution and a peak at the $\Delta(1232)$ resonance. Notice that the higher resonance region is suppressed in comparison with the corresponding integrand for B_n. Also the quasi-RCS peak around the maximum W value is absent. As a result, the elastic contribution to A_n can be of comparable magnitude as the inelastic one. Due to the partial cancelation between elastic and inelastic contributions, A_n is significantly reduced for the proton, taking on values around or below 0.1 % for beam energies below 1 GeV.

4 Conclusions

In this contribution, we presented calculations for beam and target normal SSAs in the kinematics where several experiments are performed or in progress.

Besides providing estimates for ongoing experiments, this work can be considered as a first step in the construction of a dispersion formalism for elastic electron-nucleon scattering amplitudes. In such a formalism, one needs a precise knowledge of the imaginary part as input in order to construct the real part as a dispersion integral over this imaginary part. The real part of the two-photon exchange amplitudes may yield corrections to elastic electron-nucleon scattering observables, such as the unpolarized cross sections or double polarization observables. Therefore it is of primary importance to quantify this piece of information, in order to increase the precision in the extraction of hadron structure quantities such as the nucleon form factors.

References

1. A.V. Afanasev et al.: hep-ph/0208260
2. A.V. Afanasev, N.P. Merenkov: Phys. Lett. B **599**, 48 (2004)
3. M. Gorchtein et al.: Nucl. Phys. A **741**, 234 (2004)
4. L. Diaconescu, M.J. Ramsey-Musolf: nucl-th/0405044
5. B. Pasquini, M. Vanderhaeghen: hep-ph/0405303
6. A. De Rujula et al.: Nucl. Phys. B **35**, 365 (1971)
7. D. Drechsel et al.: Nucl. Phys. A **645**, 145 (1999)
8. D. Drechsel et al.: Phys. Rep. **378**, 99 (2003)
9. S.P. Wells et al., (SAMPLE Coll.): Phys. Rev. C **63**, 064001 (2001)
10. F. Maas et al., (MAMI/A4 Coll.): nucl-ex/0410013
11. JLab HAPPEX-2 experiment (E-99-115), spokespersons G. Cates, K. Kumar, D. Lhuillier
12. JLab G0 experiment (E-00-006, E-01-116), spokesperson D. Beck
13. SLAC E158 experiment, contact person K. Kumar

Eur Phys J A (2005) **24**, s2, 33–34
DOI: 10.1140/epjad/s2005-04-006-2

EPJ A direct
electronic only

Transverse single spin asymmetry in elastic electron–proton scattering

M. Gorchtein[1], P.A.M. Guichon[2], and M. Vanderhaeghen[3]

[1] Università di Genova and Sezione INFN di Genova, 16146 Genova, Italy
[2] SPhN/DAPNIA, CEA Saclay, F-91191 Gif sur Yvette, France
[3] College of William and Mary & Jefferson Laboratory, Newport News, VA 23606, USA

Received: 15 October 2004 / Published Online: 8 February 2005

Abstract. We discuss the two-photon exchange contribution to observables which involve lepton helicity flip in elastic lepton-nucleon scattering. This contribution is accessed through the single spin asymmetry for a lepton beam polarized normal to the scattering plane. We estimate this beam normal spin asymmetry at large momentum transfer using a parton model and we express the corresponding amplitude in terms of generalized parton distributions.

PACS. 25.30.Bf Elastic electron scattering – 12.38.Bx Perturbative calculations – 13.40Gp Electromagnetic form factors

1 Introduction

Elastic electron-nucleon scattering in the one-photon exchange approximation gives direct access to the electromagnetic form factors of the nucleon, an essential piece of information about its structure. In recent years, the ratio G_{E_p}/G_{M_p} of the proton's electric to magnetic form factors has been measured up to large momentum transfer Q^2 in precision experiments [1,2] using the polarization transfer method. It came as a surprise that these experiments for Q^2 up to 5.6 GeV^2 extracted a ratio of G_{E_p}/G_{M_p} which is incompatible with unpolarized experiments [3,4,5] using the Rosenbluth separation technique. In [6], it was pointed out that this discrepancy may be resolved by a precise account of the two-photon exchange effects which enter the radiative corrections to elastic form factors.

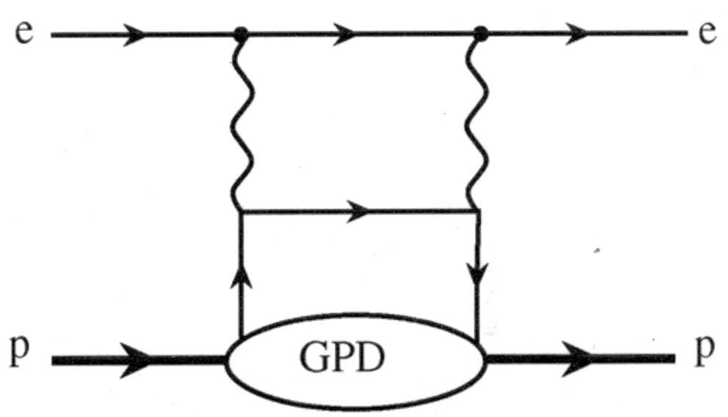

Fig. 1. Handbag contribution to the 2γ-exchange amplitude for elastic ep scattering

2 Elastic lepton-nucleon scattering amplitude beyond the one-photon exchange

In this work, we consider the elastic lepton-nucleon scattering process $l(k)+N(p) \to l(k')+N(p')$. The general amplitude for elastic scattering of two spin-1/2 particles can be parameterized in terms of six independent amplitudes. Three of them describe such scattering without helicity flip on the lepton side [6], and the other three amplitudes do flip the lepton helicity [7]. We note that in the one-photon exchange (Born) approximation, only two of them survive, the well known electromagnetic form factors G_E and G_M which are real functions of the momentum transfer Q^2 only. In this work, we study the single spin asymmetry, $B_n = \frac{\sigma_\uparrow - \sigma_\downarrow}{\sigma_\uparrow + \sigma_\downarrow}$, where $\sigma_{\uparrow(\downarrow)}$ denotes the cross section for an unpolarized target and a lepton beam spin parallel (anti-parallel) to the normal polarization vector defined as $S_n^\mu = (0, \mathbf{S}_n)$, $S_n^2 = -1$ with $\mathbf{S}_n \equiv [\mathbf{k} \times \mathbf{k}']/|[\mathbf{k} \times \mathbf{k}']|$. Its leading non-vanishing contribution is linear in the lepton mass. Furthermore, B_n vanishes in the Born approximation, and is therefore of relative order e^2. Keeping only the leading term in e^2, B_n arises from an interference between the one-photon exchange (Born) amplitude and the imaginary part of the two-photon exchange amplitude [8]. To calculate the latter, we proceed with the model used in [7,9] where the partonic (handbag) model (cf. Fig. 1) has been adopted.

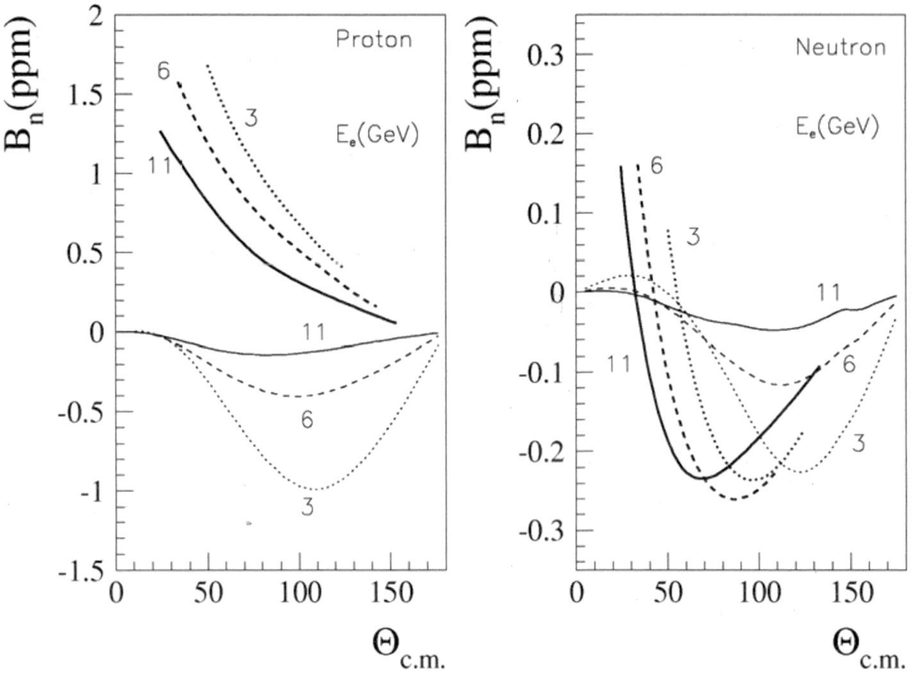

Fig. 2. Beam normal spin asymmetry for elastic $e^- N$ scattering as function of c.m. scattering angle at different values of beam energy as indicated on the figure for the proton target (*left panel*), and the neutron target (*right panel*). The *thick curves* are the GPD calculations for the kinematical range where $s, -u > M^2$. For comparison, the nucleon pole contribution is also displayed (*thin curves*)

3 Results

In this section we present our results for the asymmetry B_n on the proton and neutron targets. In our calculation, we used the most recent parameterizations of the GPDs [10,11] for the evaluation of the lower blob of Fig. 1, while we calculate exactly the loop integral appearing in the upper part of Fig. 1, for details see [7,9]. As one can see from Fig. 2, the handbag mechanism predicts the effect of ~ 1.5 ppm for the proton, and ~ -0.2 ppm for the neutron. For comparison, the contribution of the nucleon intermediate state (instead of the lower blob with GPD in Fig. 1) is shown for the same kinematics. This latter can be calculated exactly since it only contains the on-shell elastic form factors of the nucleon. For the proton, the forward kinematics ($30° < \Theta_{cm} < 90°$) looks promising for disentangling the inelastic contribution from the elastic one on the experiment. For the neutron, the effect is quite small due to partial cancellation of competing contributions. Further investigations of this observable at high momentum transfers is necessary to obtain a valuable cross-check for the real part of the 2γ-exchange amplitude in order to resolve the present experimental situation with the elastic form factors. At present, there exist experimental data on B_n [12,13], while the further several experiments are planned, aiming to measure this beam normal spin asymmetry in different kinematics [14,15,16].

References

1. M.K. Jones et al.: Phys. Rev. Lett. **84**, 1398 (2000)
2. O. Gayou et al.: Phys. Rev. Lett. **88**, 092301 (2002)
3. L. Andivahis et al.: Phys. Rev. D **84**, 5491 (1994)
4. M.E. Christy et al.: nucl-ex/0401030
5. J. Arrington (JLab E01-001 Collaboration): nucl-ex/0312017
6. P.A.M. Guichon, M. Vanderhaeghen: Phys. Rev. Lett. **91**, 142303 (2003)
7. M. Gorchtein, P.A.M. Guichon, M. Vanderhaeghen: Nucl. Phys. A **741**, 234 (2004)
8. A. de Rujula, J.M. Kaplan, E. de Rafael: Nucl. Phys. B **35**, 365 (1971)
9. Y.C. Chen, A. Afanasev, S.J. Brodsky, C.E. Carlson, M. Vanderhaeghen: hep-ph/0403058
10. A.D. Martin, R.G. Roberts, W.J. Stirling, R.S. Thorne: Phys. Lett. B **531**, 216 (2002); Feng Yuan, Phys. Rev. D **69**, 051501 (2004)
11. A.V. Radyushkin: Phys. Rev. D **58**, 114008 (1998)
12. S.P. Wells et al. (SAMPLE Collaboration): Phys. Rev. C **63**, 064001 (2001)
13. F. Maas et al. (MAMI/A4 Collaboration): in preparation
14. SLAC E158 Experiment: contact person K. Kumar
15. G. Cates, K. Kumar, D. Lhuillier, spokesperson(s): HAPPEX-2 Experiment (JLab E-99-115)
16. D. Beck, spokesperson(s): JLab/G0 Experiment (JLab E-0-006, E-01-116)

Eur Phys J A (2005) **24**, s2, 35–38
DOI: 10.1140/epjad/s2005-04-007-1

EPJ A direct
electronic only

Transverse spin asymmetry at the A4 experiment

Experimental results

Sebastian Baunack[a], for the A4-collaboration

Johannes Gutenberg Universität Mainz, Institut für Kernphysik, J.J. Becherweg 45, 55299 Mainz, Germany

Received: 15 October 2004 / Published Online: 8 February 2005

Abstract. The A4 collaboration at the MAMI accelerator has measured the transverse spin asymmetry in the cross section of elastic scattering of transversely polarized electrons off unpolarized protons. An azimuthal dependence of the asymmetry has been observed, the amplitudes have been determined as $A_\perp(Q^2 = 0.106\ (\mathrm{GeV/c})^2) = (-8.59 \pm 0.89_{stat} \pm 0.75_{syst}) \cdot 10^{-6}$ and $A_\perp(Q^2 = 0.230\ (\mathrm{GeV/c})^2) = (-8.52 \pm 2.31_{stat} \pm 0.87_{syst}) \cdot 10^{-6}$. $A_\perp$ arises from the imaginary part of the 2γ-exchange amplitude. Our experimentally determined values of $A_\perp$ show that in the intermediate hadronic state not only the ground state of the proton, but also excited states contribute to the asymmetry.

PACS. 13.40.Gp Electromagnetic form factors – 11.30.Er Charge conjugation, parity, time reversal, and other discrete – 13.40.-f Electromagnetic processes and properties – 14.20.Dh Properties of protons and neutrons

1 Introduction

The A4 collaboration was founded to investigate the asymmetry in the elastic scattering of longitudinally polarized electrons off unpolarized protons. In recent years, theoretical effort has been spent to calculate the asymmetry for transversely polarized electrons [1,2]. The asymmetry arises from the interference between 1γ and 2γ exchange (Fig. 1). The calculations take into account the intermediate hadronic state of the 2γ exchange and might explain the discrepancy between the Rosenbluth separation technique and the polarization transfer method for the dermination of the ratio G_E^p/G_M^p of the electromagnetic form factors of the proton [3].

The A4 segmented lead fluoride (PbF_2) calorimeter which covers the full 2π azimuthal range is an appropiate apparatus to reveal the sinusoidal dependence of the asymmetry on the angle between the electron spin and the scattering plane. We have measured $A_\perp$ at two beam energies 569.31 MeV and 855.15 MeV at scattering angles between $30° \leq \theta \leq 40°$ corresponding to momentum transfers 0.106 $(\mathrm{GeV/c})^2$ and 0.230 $(\mathrm{GeV/c})^2$. In contrast to the parity violation measurements, the transverse electron spin causes physical asymmetries of nonnegligible order in our luminosity monitors. An extensive analysis of these asymmetries, which come from the 2γ exchange in Møller scattering, has been made in order to understand and control this effect. Our experimentally determined values of $A_\perp$ show that in the intermediate hadronic state not only the ground state of the proton, but also excited states contribute to the physical asymmetry [2].

[a] comprises part of PhD thesis

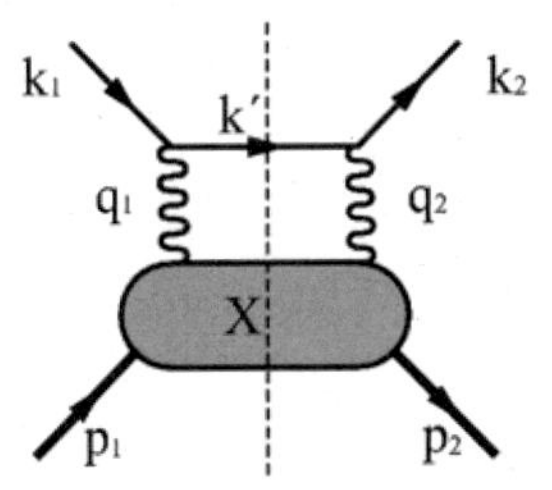

Fig. 1. 2γ exchange

2 Experimental setup

The measurement principle is quite simple (Fig. 2): the polarised electron beam hits a hydrogen target and is scattered into the detector which counts the number $N^\pm$ of the elastic scattered particles for two opposite spin directions. The asymmetry is then $A = (N^+ - N^-)/(N^+ + N^-)$.

The measurement took place in the MAMI accelerator facility using the A4-experiment setup, which has been described in detail in [4]. We used an electron beam with an intensity of 20 μA. Thanks to the polarized electron source using a strained layer GaAs crystal the averaged beam polarization P_e was about 80 %. The spin of the electrons was reversed every 20 ms following a randomly selected pattern, (+ - - +) or (- + + -). The polarization degree of the beam was measured by a Møller polarimeter situated in another experimental hall. The spin of the

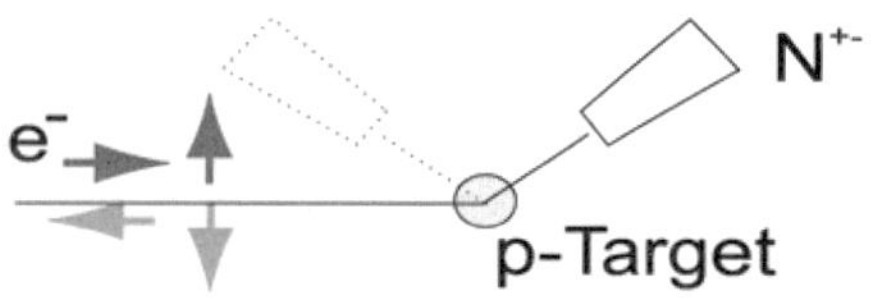

Fig. 2. Measurement principle: the polarized electrons hit the liquid hydrogen target. The detector counts the number $N^{\pm}$ of elastic scattered electrons for two opposite spin directions. The scattering angle has been so far $\theta = (35 \pm 5)^{\circ}$. The future measurements will include backward scattering with $\theta = (145 \pm 5)^{\circ}$

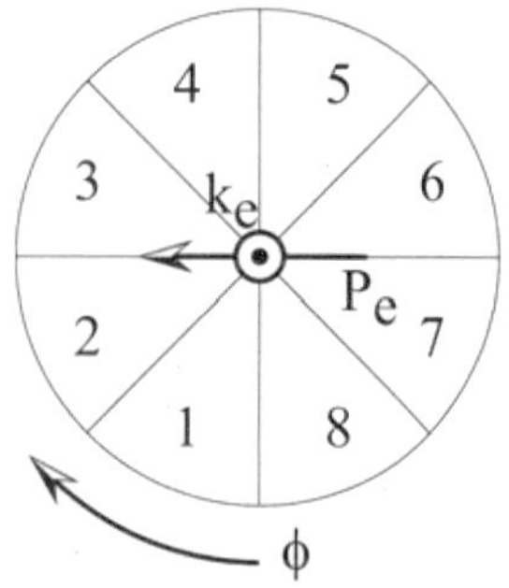

Fig. 3. Schematic view of the PbF_2 calorimeter from behind. The incoming electron's momentum vector $\mathbf{k}_e$ is pointing out of the paper plane. The momentum vector $\mathbf{k}_{out}$ of the outgoing electron can take all possible ϕ values. Both together define the coordinate system. The direction of the electron polarization vector $\mathbf{P}_e$ for the '+' helicity state is indicated by the arrow. ϕ is counted as indicated

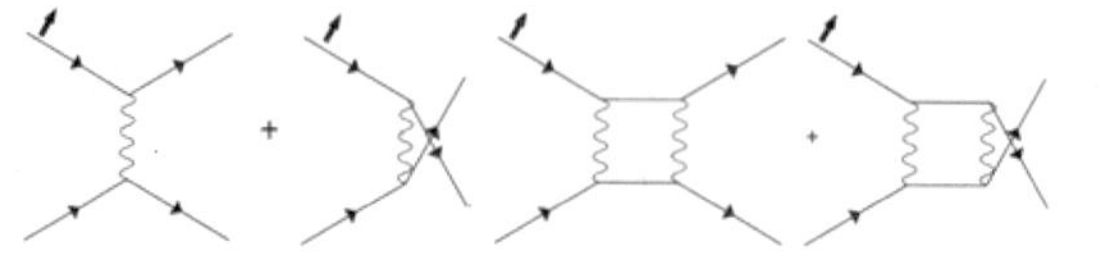

Fig. 4. Feynman graphs for Møller scattering: one loop diagrams (*left*) and box diagrams (*right*). The *arrow* indicates the transverse polarisation of the incoming electron. The asymmetry arises from the interference term of the 1γ and the 2γ-exchange

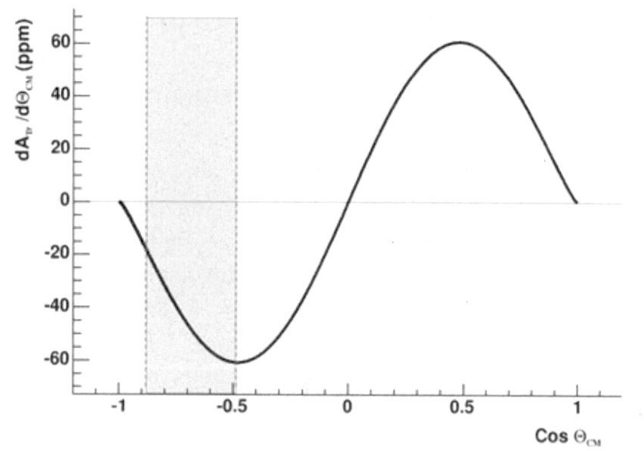

Fig. 5. Leading order asymmetry $A_{\perp}^{\text{Møller}}$ in Møller scattering for a beam energy of E=569.31 MeV as a function of the scattering angle Θ_{CM} in the center of mass system. The two *dashed lines* indicate the polar acceptance of the luminosity monitors

electrons was put into transverse direction with the help of a Wien filter located between the 100 keV electron source and the injector linac of the accelerator. The exact spin angle was determined by a Møller-Mott polarimeter located at the beam dump. Several monitor and stabilization systems have been installed along the accelerator to minimize beam fluctuations that introduce false asymmetries, such as differences in position, angle and energy for the two helicity states. The electrons were scattered on a 10 cm long liquid hydrogen target. The resulting luminosity is $L \approx 0.5 \cdot 10^{38}\ cm^{-2}s^{-1}$. The luminosity was monitored by eight water-cerenkov monitors put under small scattering angles $\theta = 4.4^{\circ}...8^{\circ}$ covering the full azimuthal range $\phi = 0...2\pi$.

The scattered electrons are detected in a total absorbing calorimeter that consists of 1022 individual lead fluoride (PbF_2) crystals placed in 7 rings and 146 rows. It covers scattering angles from $\theta = 30^{\circ}...40^{\circ}$, $\phi = 0...2\,\pi$ and a solid angle $\Delta\,\Omega = 0.64\ sr$. To see the azimuthal angular dependence of the asymmetry, we divide our calorimeter in the analysis into eight sectors, each sector covering $\Delta\Phi \approx 45^{\circ}$. Figure 3 shows our convention of chosen angles and directions.

3 Physical asymmetry in Møller scattering

In order to calculate the experimental asymmetry A_{exp}, target density fluctuations have to be taken into account. The elastic counts $N^{\pm}$ are normalised to the target density $\rho^{\pm} = L^{\pm}/I^{\pm}$, the ratio of luminosity L and beam current I:

$$A_{exp} = \frac{\frac{N^+}{\rho^+} - \frac{N^-}{\rho^-}}{\frac{N^+}{\rho^+} + \frac{N^-}{\rho^-}} \approx \frac{N^+ - N^-}{N^+ + N^-} - \frac{L^+ - L^-}{L^+ + L^-} + \frac{I^+ - I^-}{I^+ + I^-} \qquad (1)$$

The approximation is valid up to large asymmetries and certainly for the asymmetries in the A4 experiment. One sees that the asymmetry measured in the luminosity monitors enters linearly into the experimental asymmetry A_{exp}. While it is desired to eliminate asymmetries coming from real target density fluctuations, a physical asymmetry in the process of the luminosity measurement would enter false asymmetries into A_{exp}. The eight luminosity monitors covering the same ϕ-ranges as indicated for the calorimeter sectors in Fig. 3 detect mainly Møller scattered electrons. It is known for many years [5] that a ϕ-depending asymmetry occurs in the Møller scattering of transversely polarized electrons. Most recently, new calculations have been done for the E-158 experiment [6]. Figure 4 shows the one-loop diagrams and the box diagrams for Møller scattering. The physical asymmetry arises from the interference term between these two diagrams. The leading order asymmetry $A_{\perp}^{\text{Møller}}$ is plotted in Fig. 5 for a beam energy of 569.31 MeV as a function of the scattering angle Θ_{CM} in the center of mass system. The polar acceptance of the luminosity monitors are indicated by the grey box. Since the acceptance is non-symmetric with respect to Θ_{CM}, a nonvanishing physical asymmetry is observable.

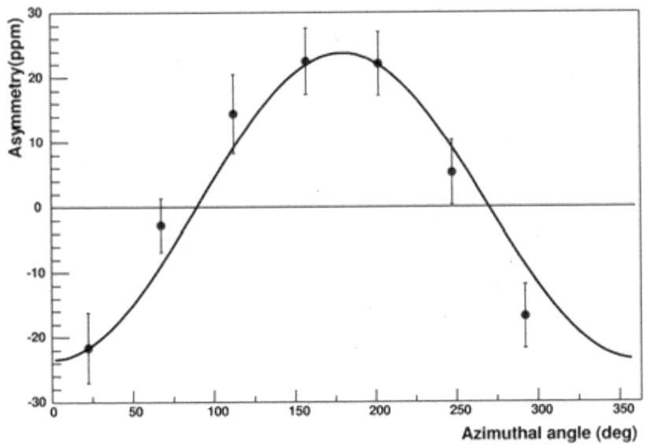

Fig. 6. Extracted physical asymmetry in the luminosity monitors as a function of the azimuthal scattering angle ϕ at Q^2=0.106 $(\text{GeV/c})^2$. The electron spin was almost transverse with $\Phi_s = 85.1°$. The fit gives $A^{Lumi} = (-24.3 \pm 2.9) \cdot 10^{-6}$

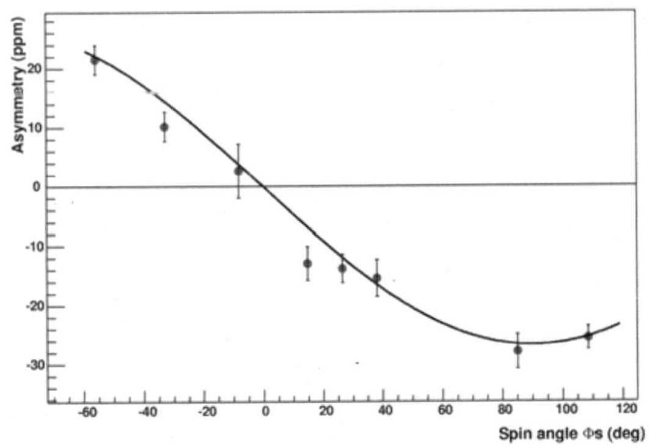

Fig. 7. Extracted asymmetry A^{Lumi} of the luminosity monitor as a function of the angle of the electron's spin angle Φ_s for Q^2 =0.106 $(\text{GeV/c})^2$. The fit to the data gives $A_\perp^{\text{Lumi}} = (-26.7 \pm 1.3) \cdot 10^{-6}$

An experimental verification and systematic check of the asymmetries in the luminosity monitors is possible during a spin rotation measurement. The Wien filter rotates the spin of the electrons from longitudinal to transverse direction in discrete steps by applying various currents. The corresponding spin angles are determined by the dump polarimeter. According to the spin angle a change in size and sign of the azimuthal dependence of the asymmetries in the luminosity monitors should be observable. Figure 6 shows the observed asymmetries for an almost transverse spin angle Φ_s = 85.1° and Q^2 =0.106 $(\text{GeV/c})^2$. The sinusoidal behaviour is obvious. The amplitude is $A^{\text{Lumi}} = (-24.3 \pm 2.9) \cdot 10^{-6}$. Measurements at various spin angles Φ_s of $A^{\text{Lumi}}(\Phi_s)$ make a fit possible to determine $A_\perp^{\text{Lumi}}$. We performed such an analysis for both of our so far applied momentum transfers. Figure 7 shows the result for our low Q^2 value. Due to target density fluctuations and shorter measurement times, the result for the higher Q^2 value has a larger uncertainty.

Table 1 gives the results of the NLO calculations for $A_\perp^{\text{Møller}}$ and our observed asymmetries $A_\perp^{\text{Lumi}}$ in the luminosity monitors. For a comparison of these values one has to keep in mind that the luminosity monitors do not only see Møller electrons, but a certain 'dilution' of elastic scattered electrons. We understand qualitatively the occuring asymmetries. For a normalisation of the elastic counts, the luminosity signals have to added Φ-symmetrically, since the physical asymmetry in Møller scattering then averages out.

Table 1. NLO calculation for the transverse spin asymmetry $A_\perp^{\text{Møller}}$ in Møller scattering and comparison with the observed asymmetries $A_\perp^{\text{Lumi}}$ in the luminosity monitors

Q^2	$A_\perp^{\text{Møller}}$ NLO calc.	$A_\perp^{\text{Lumi}}$ exp.
0.106 $(\text{GeV/c})^2$	$-29.6 \cdot 10^{-6}$	$(-26.7 \pm 1.3) \cdot 10^{-6}$
0.230 $(\text{GeV/c})^2$	$-16.6 \cdot 10^{-6}$	$(-23.3 \pm 4.3) \cdot 10^{-6}$

Table 2. Beam parameters for momentum transfer Q^2=0.230 $(\text{GeV/c})^2$ achieved within 47 hours and Q^2=0.106 $(\text{GeV/c})^2$ßachieved within 54 hours of pure data taking

Parameter	0.23 $(GeV/c)^2$	0.11 $(GeV/c)^2$
A_I	(2.5 ± 0.6) ppm	(0.4 ± 1.4) ppm
Δx	(15.1 ± 3.4) nm	(157.0 ± 150.0) nm
Δy	(-126.7 ± 18.0) nm	(-504.3 ± 43.7) nm
$\Delta x'$	(5.4 ± 0.8) nrad	(20.5 ± 12.9) nrad
$\Delta y'$	(21.5 ± 3.0) nrad	(45.4 ± 3.9) nrad
ΔE	(0.2 ± 0.3) eV	(41.4 ± 2.6) eV

4 Determination of transverse asymmetry $A_\perp$

For the determination of the physical asymmetry we use the same technique that we already used in the parity violating case [4]. The experimental measured asymmetry $A_{exp}(\Phi_i)$ of each sector i can be written as the sum of the physical asymmetry $A_{phys}(\Phi_i)$ and the helicity correlated false asymmetries A_j, $j = 1..6$:

$$A_{exp}(\Phi_i) = P \cdot A_{phys}(\Phi_i) + \sum_{i=1}^{6} a_i X_i \quad (2)$$

with P the beam polarisation, X_1 the asymmetry in beam current, X_2, X_3 the differences in horizontal and vertical position, X_3, X_4 differences in horizontal and vertical angle and X_6 difference in beam energy. We determine the coefficients a_i via a multiple linear regression out of the asymmetry data itself. Thanks to the stabilisation systems, the quantities X_i could be kept small. Table 2 shows the achieved values for the two momentum transfers. We have 47 hours beam data for E=569.31 MeV and 54 hours for E=855.15 MeV, corresponding to a number of elastic events of $N_{tot} = 3.9 \cdot 10^{12}$ and $N_{tot} = 6.3 \cdot 10^{11}$ respectively. Figure 8 show the physical asymmetries as a function of the azimuthal scattering angle Φ. At the time of the

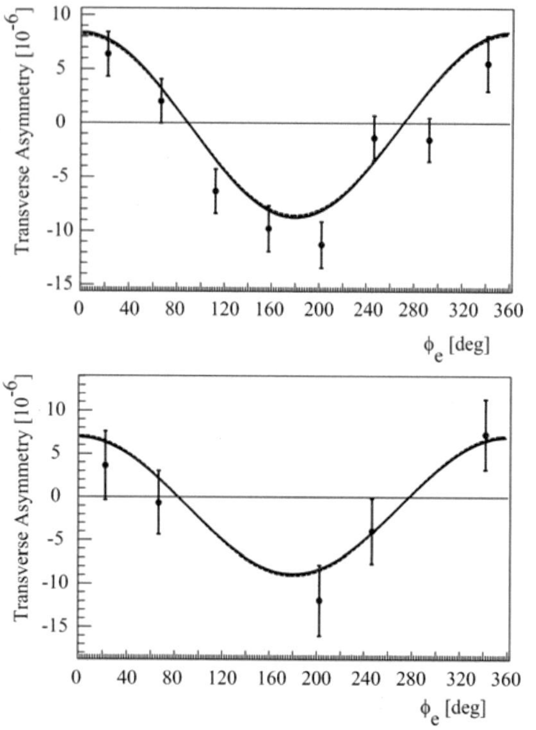

Fig. 8. Extracted physical asymmetries for the eight sectors of the calorimeter for Q^2=0.106 (GeV/c)^2 (*upper plot*) and Q^2=0.230 (GeV/c)^2

0.230 (GeV/c)^2-measurement, only sectors 1, 2, 5, 6 and part of sector 8 have been equipped with detectors. We fit to the data an averaged function $A_\perp^m$ which takes into account that each sector covers about 45° in the azimuthal range:

$$A_\perp^m(\phi) = \int_{\phi-\pi/8}^{\phi+\pi/8} A_\perp \cos\phi' d\phi' = 0.765 A_\perp \cos\phi \quad (3)$$

Applying dead time corrections and corrections for aluminum dilution from the target entry and exit windows, we obtain:

$$A_\perp(Q^2 = 0.11) = (-8.59 \pm 0.89_{stat} \pm 0.75_{syst}) \cdot 10^{-6}$$
$$A_\perp(Q^2 = 0.23) = (-8.52 \pm 2.31_{stat} \pm 0.87_{syst}) \cdot 10^{-6}$$

The first error represents the statistical uncertainty and the second error the systematic error.

5 Conclusion and outlook

A comparison of our measured values of $A_\perp$ with model calculations from [2] shows that the size of the observed asymmetries can not be explained if one takes into account only the elastic contribution to the intermediate nucleon state (Fig. 9). Even at our low Q^2 nucleon resonances play an important role. For the future we plan an extensive measurement program at various energies under forward and backward angles on hydrogen as well as on deuterium (Table 3 and Table 4).

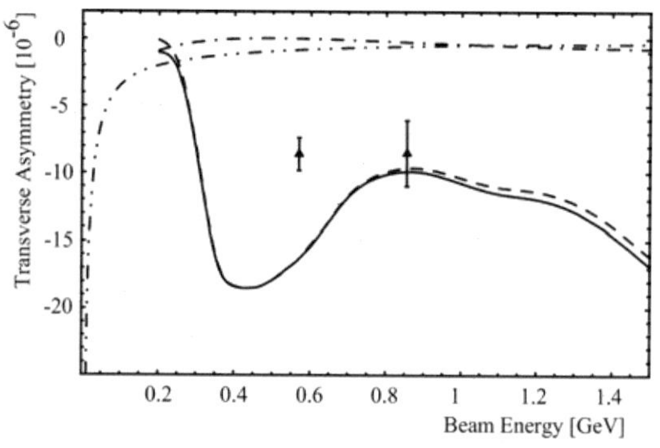

Fig. 9. Our extracted asymmetries $A_\perp$ compared with model calculations from [2] as a function of the beam energy. For the *dashed dotted line* the intermediate nucleon state has been the ground state only. The *dashed line* represents the contribution of all possible πN intermediate states. The *solid line* is the sum of both contributions

Table 3. Projected measurements for transverse single spin asymmetries $A_\perp$ with hydrogen and deuterium under forward angles ($\Theta = (35 \pm 5)°$)

E_e [*MeV*]	$\delta A_\perp^p$ [*ppm*]	hours proton	$\delta A_\perp^D$ [*ppm*]	hours deuteron	$\delta A_\perp^n$ [*ppm*]
300	0.5	20	0.5	20	20
420	0.5	40	0.5	35	11
570	0.5	90	0.5	70	7
854	0.5	300	0.5	220	4
1200	1	260	1	180	6
1500	2	180	2	120	11

Table 4. Projected measurements for transverse single spin asymmetries $A_\perp$ with hydrogen and deuterium under backward angles ($\Theta = (145 \pm 5)°$)

E_e [*MeV*]	$\delta A_\perp^p$ [*ppm*]	hours proton	$\delta A_\perp^D$ [*ppm*]	hours deuteron	$\delta A_\perp^n$ [ppm]
300	3	90	2	130	10
420	3	230	2	320	10
570	4	370	3	390	13
854	8	490	7	380	28

References

1. P.A.M. Guichon, M. Vanderhaeghen: Phys. Rev. Lett. **91**, 142303 (2003)
2. B. Pasquini, M. Vanderhaeghen: hep-ph/0405303 (2003)
3. J. Arrington: nucl-ex/0408020
4. F.E. Maas et al.: Phys. Rev. Lett. **93**, 022002 (2004)
5. A.O. Barut, C. Fronsdal: Phys. Rev. **120**, 1871 (1960)
6. L. Dixon, M. Schreiber: Phys. Rev. D **69**, 113001 (2004)

Eur Phys J A (2005) **24**, s2, 39–40
DOI: 10.1140/epjad/s2005-04-008-0

EPJ A direct
electronic only

Normal beam spin asymmetries during the G^0 forward angle measurement

P.M. King[1], for the G^0 Collaboration[2a]

[1] University of Illinois at Urbana-Champaign, Urbana, IL, USA
[2] Experiment E-00-006, Jefferson Laboratory, Newport News, VA, USA

Received: 1 November 2004 / Published Online: 8 February 2005

Abstract. The vector analyzing power measured in elastic scattering of transversely polarized electrons from an unpolarized nucleon is directly proportional to the imaginary part of the two photon exchange amplitude. There has been recent interest in explorations of the two photon exchange amplitude, as the real part has been proposed as a possible resolution of the discrepancy between Rosenbluth separation and polarization observable measurements of the ratio of the electric to magnetic proton form factor. The vector analyzing power appears in the experiment as an azimuthal asymmetry. It has been measured previously in the SAMPLE and A4 experiment with different kinematics than those achievable with the G^0 apparatus. As part of the systematic checks for the G^0 forward angle measurement at TJNAF, the normal beam spin asymmetry in the G^0 detector array was measured with a 3 GeV beam incident upon a liquid hydrogen target. The experimental configuration was identical to the standard G^0 forward angle running except that the beam was transversely polarized in the plane of the accelerator. The data collected cover a range in center of mass angle from 19° to 37°, with an eight-fold azimuthal symmetry. About 30 hours of data were taken in this configuration, resulting in an extracted vector analyzing power with a precision of a few ppm, which may already be able to provide some constraint on model predictions.

PACS. 13.60.-r Photon and charged-lepton interactions with hadrons – 14.20.Dh Protons and neutrons – 29.27.Hj Polarized beams in accelerators

1 Introduction

Elastic scattering of transversely polarized electrons from an unpolarized nucleon results in an asymmetry due to two-photon exchange. The two-photon exchange produces a cross section dependence on the angle between the polarization vector, $\mathbf{P}$, and the scattering plane [1]:

$$\sigma = \sigma_0 \; [1 + A_n \mathbf{P} \cdot \hat{\mathbf{n}}] \tag{1}$$

where the vector analyzing power, A_n, is proportional to the imaginary terms of the two-photon exchange amplitude. In a parity violating electron scattering measurement, in the general case, the normal beam spin asymmetry will contribute to the measured asymmetry,

$$\begin{aligned} A(\theta,\phi) = & P\cos(\theta_{spin}) A_{PV}(\theta) \\ & + P\sin(\theta_{spin})\sin(\phi-\phi_{spin}) A_n(\theta) \end{aligned} \tag{2}$$

where A_{PV} is the parity violating asymmetry, θ and ϕ are the electron scattering angles, and θ_{spin} and ϕ_{spin} are the angles of the electron polarization vector.

The G^0 experiment in Hall C of Jefferson Laboratory is primarily designed to measure the parity violating electron scattering asymmetry from the proton. In the forward angle measurement, the recoil protons scattered by a 3 GeV polarized electron beam incident on a 20 cm long liquid hydrogen target are momentum analyzed by a superconducting magnet, and are detected in 8 azimuthally symmetric detector packages with 16 detectors in each octant. The parity violating asymmetry in each detector varies from about 2 ppm to about 20 ppm through the Q^2 range. To minimize the effect on the parity violating asymmetry measurement due to the normal beam spin asymmetry, we set the polarization to be longitudinal for the G^0 measurements, and conducted the measurements of A_n discussed in this paper.

2 Experiment

The transverse data collection took place from 22 March through 26 March, 2004. At Jefferson Lab, the beam spin can be oriented in the horizontal plane (the plane of the ac-

[a] The G^0 Experiment is supported by the U.S. National Science Foundation (NSF) under grant PHY94-10768, the U.S. Department of Energy (DOE), the Natural Sciences and Engineering Research Council (NSERC) of Canada, and the Centre National de la Recherche Scientifique (CNRS) of France through the Institut National de Physique Nucléaire et de Physique des Particules (IN2P3).

Table 1. Elastic proton kinematic coverage for the G^0 forward angle measurements. The estimated error corresponds to the statistical error on the measurements only

Detectors	$\langle Q^2 \rangle$	$\langle \theta_{CM} \rangle$	Est. Error
1-4	0.13	19.03	1.3 ppm
5-8	0.17	21.65	1.3 ppm
9-12	0.25	26.12	1.3 ppm
13-14	0.38	32.35	2.4 ppm
15	0.6	37.4	2.9 ppm

celerator) by adjusting the Wien filter in the injector. The Hall C Møller polarimeter is only sensitive to the longitudinal component of the beam polarization. To determine the Wien filter setting which would produce transversely polarized beam in Hall C, a set of polarization measurements were made with Wien filter settings corresponding to spin angles of -95° to +100°. The data were fit to a cosine to find the setting which gave a zero of the longitudinal polarization; this was at a Wien angle of -85°. A set of three points at ±5° from the zero crossing were also taken at the beginning and end of the transverse running period, to check for drift of the zero-crossing; the zero crossing was found to have drifted by 3° during the four days of the measurement. The total spin precession through the accelerator was about 23π.

The detector, target, and spectrometer magnet configurations were identical to those used in the standard G^0 running. Data are simultaneously collected for elastically scattered protons, pions, and inelastic protons, which are separated by time-of-flight from the target. The analysis is primarily concerned with the elastic protons, and will apply corrections for the contributions to the background under the proton peak which come from the pions and inelastic protons.

The kinematic coverage and statistical error for the G^0 detectors are shown in Table 1. The systematic error estimates are not shown. The detectors cover a range in center of mass scattering angle from 19° to 38°. The statistical errors should be compared to the results of a calculation [2], shown in Fig. 1, for the G^0 beam energy which included intermediate states of the proton and π-N states with $W < 2$ GeV.

3 Future plans

The G^0 collaboration had submitted a proposal to the Jefferson Lab PAC26 to carry out transverse asymmetry measurements in conjunction with the three beam energy running periods of the G^0 backward angle measurement. The goal of this proposal was to make a 3 ppm measurement at a center of mass angle of about 130° at the three beam energies of 0.424, 0.585, and 0.799 GeV. At the highest beam energy, this would be a 10% measurement of the predictions of [2], as shown in Fig. 2.

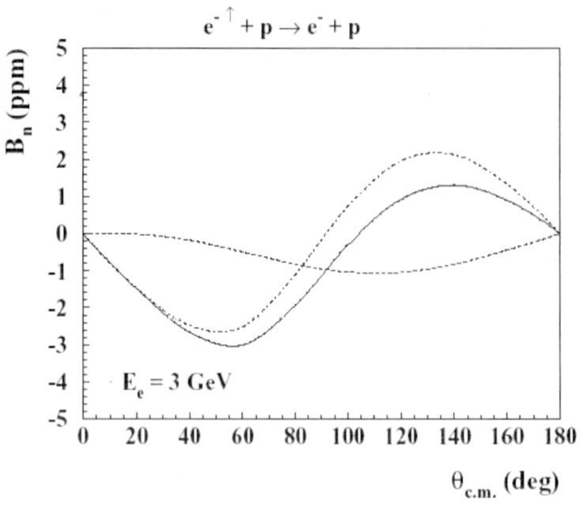

Fig. 1. Theory prediction for G^0 forward angle beam energy from [2]. The *dashed line* is the nucleon intermediate state, the *dash-dot line* is the π-N intermediate states, and the *solid line* is the total

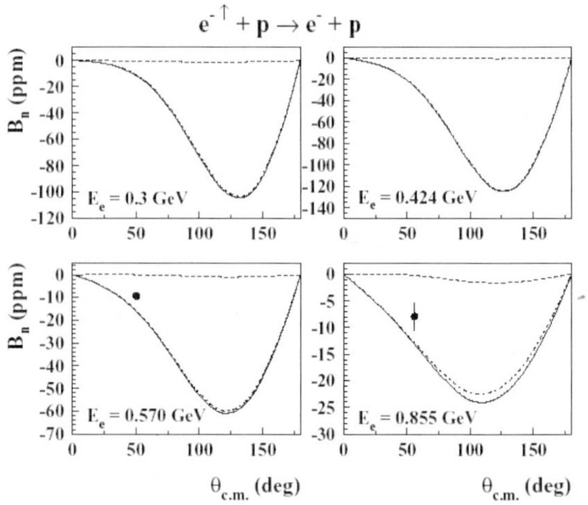

Fig. 2. Theory prediction for beam energies comparable to the G^0 backward angle measurements from [2]. The *line* types have the same meaning as in Fig. 1. The *points* are from the A4 measurements

4 Conclusion

We have made measurements of the beam normal asymmetry for elastic e-p scattering at a beam energy of 3 GeV for center of mass angles of 19-38 degrees, with an estimated statistical precision of a few ppm. We plan to make measurements at backward angles, with center of mass angles of about 130 degrees, at beam energies of 799, 585, and 424 MeV. The projected statistical error, for a dedicated experiment, is 3 ppm.

References

1. S.P. Wells et al. (SAMPLE Collaboration): Phys. Rev. C **63**, 064001 (2001)
2. B. Pasquini, M. Vanderhaeghen: hep-ph/0405303

III Weak form factors of the nucleon

III-1 PV experiment

Eur Phys J A (2005) **24**, s2, 43–46
DOI: 10.1140/epjad/s2005-04-009-y

EPJ A direct
electronic only

The axial form factor of the nucleon

Elizabeth Beise

University of Maryland, College Park, MD 20742, USA

Received: 1 November 2004 / Published Online: 8 February 2005

Abstract. The parity violation programs at MIT-Bates, Jefferson Lab and Mainz are presently focused on developing a better understanding of the sea-quark contributions to the vector matrix elements of nucleon structure. The success of these programs will allow precise semi-leptonic tests of the Standard Model such as that planned by the QWeak collaboration. In order to determine the vector matrix elements, a good understanding of the nucleon's axial vector form factor as seen by an electron, G_A^e, is also required. While the vector electroweak form factors provide information about the nucleon's charge and magnetism, the axial form factor is related to the nucleon's spin. Its $Q^2 = 0$ value at leading order, g_A, is well known from nucleon and nuclear beta decay, and its precise determination is of interest for tests of CKM unitarity. Most information about its Q^2 dependence comes from quasielastic neutrino scattering and from pion electroproduction, and a recent reanalysis of the neutrino data have brought these two types of measurements into excellent agreement. However, these experiments are not sensitive to additional higher order corrections, such as nucleon anapole contributions, that are present in parity-violating electron scattering. In this talk I will attempt to review what is presently known about the axial form factor and its various pieces including the higher order contributions, discuss the various experimental sectors, and give an update on its determination through PV electron scattering.

PACS. 12.15.Lk Neutral currents – 11.30.Er Charge conjugation, parity, time reversal, and other discrete symmetries – 13.60.-r Photon and charged-lepton interactions with hadrons – 13.15.+g Neutrino interactions – 14.20.Dh Protons and neutrons

1 Introduction

The neutral weak interaction between leptons and nucleons can be described by a set of three form factors that contain information about nucleon structure. The goal of present-day experiments in parity-violating electron scattering has been to determine the two vector weak form factors, G_E^Z and G_M^Z. These two form factors can be used along with the nucleon's electromagnetic form factors to disentangle the contributions of up, down and strange quarks to the nucleon's charge and magnetization distributions. However, in order to carry this out one also needs to know the third weak form factor coming from the nucleon's axial current, G_A^Z. The axial form factor has been determined at low momentum transfer in both quasielastic neutrino scattering and in pion electroproduction, but very little information on G_A^Z is available at momentum transfers greater than 1 $(\mathrm{GeV/c})^2$. In addition, the axial form factor as seen by an electron is substantially modified by electroweak radiative corrections that cannot yet be computed with high precision, including the term related to parity-violating coupling of a photon to the nucleon, known as the anapole coupling. Two experimental directions, quasielastic neutrino scattering and parity-violating quasielastic electron scattering from deuterium, can improve our knowledge of both G_A^Z and of the anapole contribution.

2 Lepton–nucleon scattering

The nucleon electromagnetic current associated with lepton-nucleon scattering can be written as

$$\langle N'|J_\mu^\gamma|N\rangle = \overline{u}_N \Big[F_1^\gamma(q^2)\gamma_\mu + \frac{i\sigma_{\mu\nu}q^\nu}{2M_N}F_2^\gamma(q^2) + \frac{G_F}{M_N^2}F_A^\gamma(q^2)\left(q^2\gamma_\mu - q^\nu\gamma_\nu q_\mu\right)\gamma_5 - \frac{i\sigma_{\mu\nu}q^\nu\gamma_5}{2M_N}F_E(q^2)\Big] u_N \,,$$

where F_1^γ and F_2^γ are the well-known Pauli and Dirac electromagnetic form factors, $q^2 = -Q^2$ is the four-momentum transfered from the lepton to the nucleon. The term containing the anapole form factor $F_A^\gamma(q^2)$ violates parity, and $F_E^\gamma(q^2)$ is a form factor that would arise with time-reversal violation. The anapole form factor has been computed by several authors [1,2,3] and is expected to be small at $Q^2 = 0$, but its computation is complicated by strong interaction effects in the nucleon, and its momentum transfer dependence is unknown. It is negligible in electron scattering cross section measurements but it enters the asymmetry in parity-violating electron scattering at the same order as the weak nucleon axial form factor. The nucleon's neutral weak current is

$$\langle N'|J_\mu^Z + J_{\mu 5}^Z|N\rangle = \overline{u}_N \left[F_1^Z(q^2)\gamma_\mu + \frac{i\sigma_{\mu\nu}q^\nu}{2M_N}F_2^Z(q^2) \right. \\ \left. + \gamma_\mu\gamma_5 G_A^Z(q^2)\right] u_N \, .$$

The vector form factors F_1^Z and F_2^Z are of primary interest in determining s-quark effects in PV electron scattering, and the axial form factor G_A^Z contains information about the nucleon spin. At leading order, G_A^Z can further be explicitly deconstructed using SU(3) symmetry into isovector and isospin singlet components to separate out the contribution of s-quarks to nucleon spin

$$G_A^Z(Q^2) = -\tau_3 G_A(Q^2) + G_A^s(Q^2)\,,$$

where $\tau_3 = +1(-1)$ for $p(n)$, $G_A(0) = -(g_A/g_V) = 1.2670 \pm 0.0035$ [4] as determined in nucleon β decay, and $G_A^s(0) = \Delta s$, the strange quark spin content of the nucleon. The Q^2 dependence of G_A has generally been characterized by a dipole form, $1/(1+Q^2/M_A^2)^2$, which can then be linked to a determination of an axial radius in a low momentum expansion of G_A with Q^2:

$$\langle r_A^2\rangle = -\frac{6}{g_a}\frac{dG_A}{dQ^2}|_{Q^2=0} = \frac{12}{M_A}\,.$$

3 Available data

Two methods have been used to determine this lowest order Q^2 behavior of the axial form factor. The most direct method is to use quasielastic neutrino-nucleon scattering. Very little neutral current scattering data is available, so cross section data from the charged current process $\nu_\mu + n \rightarrow \mu^- + p$ has typically been use to extract M_A. Recently, a new global fit to neutrino data was carried out by Budd, *et al.* [5], which improved over earlier fits both by using the most recent determination of (g_A/g_V) along with new results for nucleon electromagnetic form factors. The improved fit gives $M_A = 1.001 \pm 0.020$ GeV. In pion electroproduction, M_A can also be extracted from the transverse component of the near-threshold electroproduction cross section by associating it with the electric dipole transition amplitude $E_{0+}^{(-)}$ through the low energy theorem of Nambu, Lurie and Shrauner [6] under the assumption that $m_\pi = 0$. A measurement was recently carried out at the Mainz Microtron [7], resulting in $M_A = 1.068 \pm 0.015$ GeV. In a recent topical review, Bernard *et al.* [8] used chiral perturbation theory to compute a finite mass correction to this extraction, which is substantial and results in a corrected M_A of 1.013 ± 0.015 GeV, bringing it into agreement with the neutrino data. Therefore, it appears that M_A is reasonably well determined and the low Q^2 behavior of G_A can at least be described phenomenologically. This does not, however, give a first principles theoretical description of $G_A(Q^2)$, nor does it provide an adequate description at momentum transfers above 1 $(\text{GeV/c})^2$. In addition, better modeling of neutrino scattering, guided by improved data, will be required for upcoming neutrino oscillation experiments. A new experiment, Minerνa [9], has been proposed that would consist of a high granularity neutrino detector located at the NUMI beam line at Fermilab. This experiment would be able to provide a precise determination of G_A, including possible departures from the nominal dipole behavior, at $Q^2 < 2$ $(\text{GeV/c})^2$ and a first determination of G_A for $Q^2 > 2$ $(\text{GeV/c})^2$. Planning for another potential experiment is underway at Jefferson Lab, using the reaction $\mathbf{e} + p \rightarrow \nu + n$, covering the range $Q^2 \sim 1 - 3$ $(\text{GeV/c})^2$ [10]. This very challenging experiment would require detection of the recoiling neutrons at very forward angles, and the parity-violating asymmetry in the detected neutrons would be measured in order to constrain backgrounds.

While the above measurements would be able to better constrain $G_A(Q^2)$ and provide improved models of cross section data for neutrino oscillation experiments, it is the axial form factor as seen by an electron that is relevant to the parity violation program that is the topic of this workshop. The axial form factor seen in PV electron scattering can be written, going beyond first order, as

$$G_A^e(Q^2) = -\tau_3(1 + R_A^{T=1})G_A(Q^2) + R_A^{T=0}G_A^8(Q^2) + G_A^s(Q^2)$$

where $R_A^{T=0,1}$ are electroweak radiative corrections arising from higher order diagrams [1]. The SU(3) octet form factor G_A^8 is not present at tree-level, but appears once radiative corrections are included. Its $Q^2 = 0$ value can be estimated from the ratio of axial vector to vector couplings in hyperon β decay which, assuming SU(3) flavor symmetry, can be related to the octet axial charge a_8 and to the hyperon F and D coefficients [4],

$$G_A^8(0) = \frac{(3F-D)}{2\sqrt{3}} = \frac{1}{2}a_8 = 0.217 \pm 0.043\,.$$

Its Q^2 behavior has also not been measured, but it is usually assumed to have the same dipole form as the isovector form factor $G_A(Q^2)$ with the same mass parameter M_A.

It should be noted that while a decade of measurements related to the "spin crisis" have indirectly determined $G_A^s(0) = \Delta s$ from polarized deep-inelastic scattering, its Q^2 behavior is also unknown. There has been one determination of $G_A^s(0)$ from quasielastic neutrino scattering [12], which is in reasonable agreement with the polarized DIS data. An improved analysis of these data was carried out by Garvey *et al.* [13] who included possible effects of nonzero strange vector form factors. A recent further improved analysis was carried out by S. Pate [14], who combined the neutrino data with results from HAPPEX to perform a global fit to the three strange form factors (so far with only two constraints) to extract G_A^s at the mean of the two experiments, $Q^2 = 0.5$ $(\text{GeV/c})^2$, rather than extrapolating to $Q^2 = 0$. A new direct measurement of G_A^s at low momentum transfer has been proposed using

[1] I am here following the notation in [11].

Table 1. Electroweak radiative corrections, computed in the $\overline{MS}$ scheme, for the axial form factor measured in PV electron scattering. The values are taken from [1]

Source	$R_A^{T=1}$	$R_A^{T=0}$
1-quark	-0.18	0.07
anapole	-0.06 ± 0.24	0.01 ± 0.14
total	-0.24 ± 0.24	0.08 ± 0.14

the ratio of neutral current to charged current neutrino scattering at low momentum transfer, FiNeSSE, using a highly segmented detector with wavelength shifting optical fibers embedded in mineral oil to identify tracks left by the recoiling protons [15]. This experiment would potentially improve the determination of Δs by about a factor of two over the DIS data, and with less theoretical uncertainty.

Of more direct interest to the parity violation program is the radiative correction to the isovector $G_A(Q^2)$, of which the nucleon's anapole form factor F_A^γ is one component. The dominant contributions to $R_A^{T=0}$ and $R_A^{T=1}$ come from 1-quark terms such as γ-Z mixing and vertex corrections, which have been computed by several authors [1,4]. Multi-quark or anapole contributions were also computed [1], who modeled them in terms of hadronic parity-violating NN couplings cast within a heavy baryon chiral perturbation theory framework. The results are shown in Table 1. While the 1-quark contributions dominate the correction, the anapole contributions dominate the uncertainty.

The axial form factor G_A^e, or at least its isovector piece $G_A^{e\,(T=1)}$, can be determined from the PV asymmetry in quasielastic scattering from deuterium, where the strange quark effects in the neutron and proton tend to cancel. Nuclear effects, including both parity conserving [17] and parity-violating [18,19] contributions, have been shown to be small. The first measurement of $G_A^{e\,(T=1)}$ was carried out by the SAMPLE collaboration. The measured asymmetry at two momentum transfers are shown in Fig. 1. They agree fairly well with the calculation, which was carried out at $Q^2 = 0$, indicating that there is no anomalously large Q^2 dependence to the anapole term or to the correction at these very low momentum transfers. However, very little else is known about its behavior away from $Q^2 = 0$. Two model calculations [3,2] of $F_A^\gamma(Q^2)$ have been carried out, and they indicate a much softer behavior with Q^2 than that of $G_A^Z(Q^2)$, even possibly an increase with Q^2, as well as quite different behavior for the isoscalar and isovector pieces. These could substantially enhance the effects of radiative corrections at momentum transfers in the range of the G0 experiment. It would thus be very useful to have some experimental information on $G_A^{e\,(T=1)}$ at higher momentum transfers. A program of backward angle measurements with a deuterium target is part of the planned running for the G0 experiment, and aerogel Čerenkov detectors have been added to the detector array

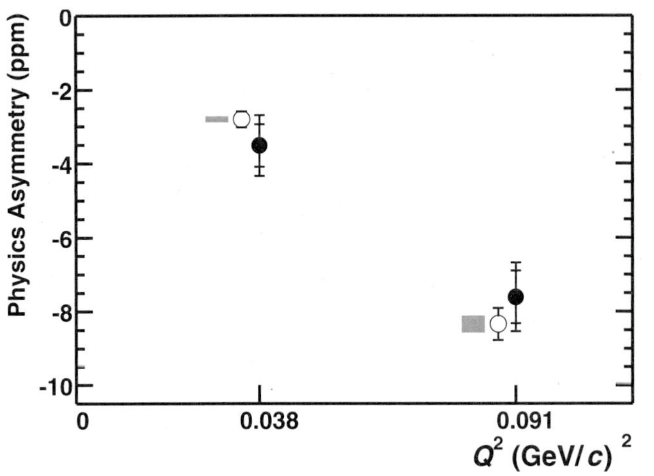

Fig. 1. Asymmetry results from the two SAMPLE deuterium experiments (*solid circles*, see [16]), compared to expectation from theory using the axial radiative corrections of [1] (*open circles*). The theory also assumes a value of G_M^s of 0.15 nuclear magnetons, and the *grey band* represents a change in G_M^s of ± 0.6 n.m.

in order to identify and separate charged pions produced in the deuterium target from the desired quasielastically scattered electrons. These data will not only reduce the model uncertainties in the determination of G_E^s and G_M^s from the hydrogen data, but will also allow the first experimental information on G_A^e away from the static limit. Shown in Fig. 2 are the projected uncertainties from the G0 experiment [20] in the difference between $G_A^{e\,(T=1)}$ and the tree-level G_A^Z that is seen in neutrino scattering, along with the calculation of [1] and the two SAMPLE measurements.

As an aside, it should be noted that, assuming that a determination of the nucleon axial form factor can straightforwardly be related to electron-quark interactions, the two SAMPLE measurements can be recast in terms of the two electron-quark couplings C_{2u} and C_{2d}. Prior to the SAMPLE measurements, experimental limits on these were from the original SLAC DIS parity-violation experiment [21], and from the parity-violating quasielastic elecron scattering experiment on ^{9}Be carried out at Mainz [22]. The two SAMPLE measurements are sensitive to the combination $C_{2u} - C_{2d}$. These are modified by 1-quark radiative corrections, and in the case of elastic e-N scattering the multiquark corrections as well. In order to compare directly to the SLAC DIS data, the multi-quark radiative corrections must be removed, which although small, dominate the uncertainty. The resulting values from the 200 MeV and 125 MeV data sets, respectively, are

$$C_{2u} - C_{2d} = -0.042 \pm 0.040 \pm 0.035 \pm 0.02$$
$$C_{2u} - C_{2d} = -0.12 \pm 0.05 \pm 0.05 \pm 0.02 \pm 0.01\,,$$

where the first two uncertainties are statistical and experimental systematic, the third is that due the radiative corrections, and, for the 125 MeV case, the last corresponds to variations in G_M^s by ± 0.6 because it is unde-

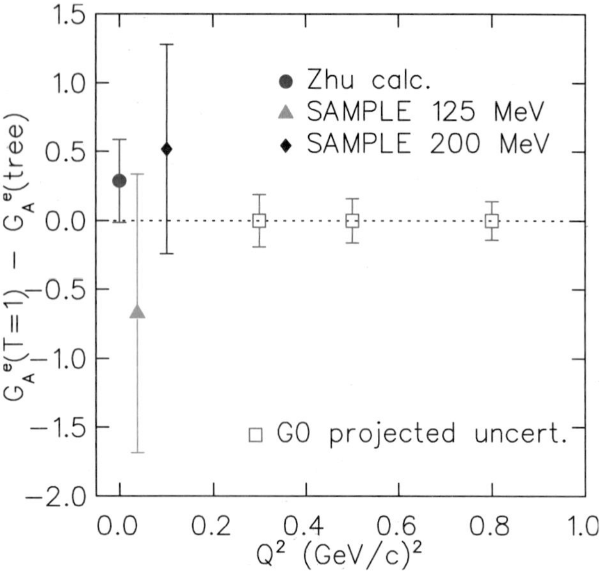

Fig. 2. The difference between the axial form factor seen by an electron in parity-violating electron scattering and $G_A^Z(Q^2)$, the tree-level form factor. Shown are the calculation of [1], SAMPLE data, and projected uncertainties in the G0 experiment

termined at this momentum transfer. These values are in good agreement with the Standard Model prediction [4], and represent a significant improvement over the earlier data.

4 Conclusion

In summary, while much attention has been focused on determination of the neutral weak vector form factors, in order to extract strange quark effects in the nucleon, there are variety of experimental avenues to pursue in the near future to improve our knowledge of the nucleon's axial form factor. The tree-level form factor is now known reasonably well at low momentum transfers from neutron beta decay, from quasielastic neutrino scattering and from pion electroproduction. Its knowledge at higher momentum transfer, including potential deviations from a generic dipole behavior, can be improved with new neutrino scattering experiments, and these experiments will help provide the required precision cross section information needed for the next generation of neutrino oscillation measurements. In PV electron scattering, the axial form factor is substantially modified and very little is known about the Q^2 behavior of the higher order terms. The G0 experiment will uniquely be able to provide the higher Q^2 data through quasielastic scattering from a deuterium target.

References

1. S.-L. Zhu, S.J. Puglia, B.R. Holstein, M.J. Ramsey-Musolf: Phys. Rev. D **62**, 033008 (2000)
2. D.O. Riska: Nucl. Phys. A **678**, 79 (2000)
3. C.M. Maekawa, U. van Kolck: Phys. Lett. B **478**, 73 (2000); C.M. Maekawa, J.S. Viega, U. van Kolck: Phys. Lett. B **488**, 167 (2000)
4. K. Hagiwara et al.: *Review of Particle Properties*, Phys. Rev. D **66**, 010001 (2002)
5. H. Budd, A. Bodek, J. Arrington: arXiv:hep-ex/0308005
6. Y. Nambu, D. Lurié: Phys. Rev. **125**, 1429 (1962); Y. Nambu, E.Shrauner: Phys. Rev. **128**, 862 (1962)
7. A. Liesenfeld et al.: Phys. Lett. B **468**, 20 (1999)
8. V. Bernard, L. Elouadrhiri, U.-G. Meissner: J. Phys. G **28**, R1 (2002)
9. Minerνa proposal: K. McFarland, spokesperson
10. Jefferson Laboratory PAC25 Letter of Intent LOI–04–006: A. Deur, contact
11. M.J. Musolf, T.W. Donnelly, J. Dubach, S.J. Pollock, S. Kowalski, E.J. Beise: Physics Reports **239**, 1 (1994)
12. L.A. Ahrens et al.: Phys. Rev. D **35**, 785 (1987)
13. G.T. Garvey, W.C. Louis, D.H. White: Phys. Rev. C **48**, 761 (1993)
14. S. Pate: Phys. Rev. Lett. **92**, 082002 (2004)
15. FiNeSSE proposal, R. Tayloe, B. Fleming: contacts
16. T. Ito et al.: Phys. Rev. Lett. **92**, 102003 (2004)
17. L. Diaconescu, R. Schiavilla, U. van Kolck: Phys. Rev. C **63**, 044007 (2001)
18. R. Schiavilla, J. Carlson, M. Paris: Phys. Rev. C **67**, 032501 (2003)
19. C.P. Liu, G. Prezeau, M.J. Ramsey-Musolf: Phys. Rev. C **67**, 035501 (2003)
20. G0 Backward angle proposal: D. Beck, contact. Uncertainties were projected based on 80 μA of (normal time structure) beam on a 20 cm deuterium target, along with uncertainties achieved in the forward angle measurement that was carried out in 2004
21. C.Y. Prescott et al.: Phys. Lett. B **84**, 524 (1979)
22. W. Heil et al.: Nucl. Phys. B **327**, 1 (1989)

Eur Phys J A (2005) **24**, s2, 47–50
DOI: 10.1140/epjad/s2005-04-010-6

EPJ A direct
electronic only

Parity violating electron scattering at the MAMI facility in Mainz

The strangeness contribution to the form factors of the nucleon

F.E. Maas, for the A4-Collaboration

Johannes Gutenberg Universität Mainz,
Institut für Kernphysik, J.J.Becherweg 45, 55299 Mainz, Germany
e-mail: maas@kph.uni-mainz.de

Received: 15 December 2004 / Published Online: 8 February 2005

Abstract. We report here on a new measurement of the parity violating (PV) asymmetry in the scattering of polarized electrons on unpolarized protons performed with the setup of the A4-collaboration at the MAMI accelerator facility in Mainz. This experiment is the first to use counting techniques in a parity violation experiment. The kinematics of the experiment is complementary to the earlier measurements of the SAMPLE collaboration at the MIT Bates accelerator and the HAPPEX collaboration at Jefferson Lab. After discussing the experimental context of the experiments, the setup at MAMI and preliminary results are presented.

PACS. 12.15 Electroweak interactions – 11.30.Er Charge conjugation, parity, time reversal, and other discrete – 13.40.Gp Electromagnetic form factors – 25.30.Bf Elastic electron scattering

1 Strangeness in the nucleon

The understanding of quantum chromodynamics (QCD) in the nonperturbative regime of low energy scales is crucial for understanding the structure of hadronic matter like protons and neutrons (nucleons). The successful description of a wide variety of observables by the concept of effective, heavy ($\approx$ 350 MeV) constituent quarks, which are not the current quarks of QCD, is still a puzzle. There are other equivalent descriptions of hadronic matter at low energy scales in terms of effective fields like chiral perturbation theory (χPT) or Skyrme-type soliton models. The effective fields in these models arise dynamically from a sea of virtual gluons and quark-antiquark pairs. In this context the contribution of strange quarks plays a special role since the nucleon has no net strangeness, and any contribution of strange quarks to the nucleon structure observables is a pure sea-quark effect. Due to the heavier current quark mass of strangeness (m_s) as compared to up (m_u) and down (m_d) with $m_s \approx 140$ MeV $\gg m_u, m_d \approx$ 5-10 MeV, one expects a suppression of strangeness effects in the creation of quark-antiquark pairs. On the other hand the strange quark mass is in the range of the mass scale of QCD ($m_s \approx \Lambda_{QCD}$) so that the dynamic creation of strange sea quark pairs could still be substantial. The topic of the strangeness content of the nucleon has been widely discussed in other contributions to this conference from J. Ellis, A. Silva, B. Kubis and others. Parity violating (PV) electron scattering off nucleons provides experimental access to the strange quark vector current in the nucleon $\langle N|\bar{s}\gamma_\mu s|N\rangle$ which is parameterized in the electromagnetic form factors of proton and neutron, G_E^s and G_M^s [1]. Recently the SAMPLE-, HAPPEX- and A4-collaboration have published experimental results. A direct separation of electric (G_E^s) and magnetic (G_M^s) contribution at forward angle has been impossible so far, since the measurements have been taken at different Q^2-values. The experimental details of the SAMPLE, HAPPEX, A4 and G0 collaboration is discussed in greater detail in many long and short contributions of this conference proceedings.

A determination of the weak vector form factors of the proton ($\tilde{G}_E^p$ and $\tilde{G}_M^p$) by measuring the PV asymmetry in the scattering of longitudinally polarized electron off unpolarized protons allows the determination of the strangeness contribution to the electromagnetic form factors G_E^s and G_M^s. The weak vector form factors $\tilde{G}_{E,M}^p$ of the proton can be expressed in terms of the known electromagnetic nucleon form factors $G_{E,M}^{p,n}$ and the unknown strange form factors $G_{E,M}^s$ using isospin symmetry and the universality of the quarks in weak and electromagnetic interaction $\tilde{G}_{E,M}^p = (G_{E,M}^p - G_{E,M}^n) - 4\sin^2\theta_W\, G_{E,M}^p - G_{E,M}^s$. The interference between weak (Z) and electromagnetic (γ) amplitudes leads to a PV asymmetry $A_{LR}(\mathbf{e}p) = (\sigma_R - \sigma_L)/(\sigma_R + \sigma_L)$ in the elastic scattering cross section of right- and left-handed electrons (σ_R and σ_L respectively), which is given in the framework of the Standard Model [2]. $A_{LR}(\mathbf{e}p)$ is of order parts per million (ppm). The asymmetry can be expressed as a sum of three terms, $A_{LR}(\mathbf{e}p) = A_V + A_s + A_A$. A_V represents the vector coupling at the proton vertex where the possible strangeness contribution has been taken out and has been put into A_s, a term arising only from a contribution of strangeness to the electromagnetic vector form factors. The term A_A represents the contribution from the axial coupling at the proton vertex due to the neutral current weak axial form factor $\tilde{G}_A^p$. We average $A_0 = A_V + A_A$ over the acceptance of the detector and the target length in order to calculate the expected asymmetry. The largest contribution to the uncertainty of A_0 comes from the uncertainty in the axial form factor G_A, the electric form factor of the proton G_E^p, and the magnetic form factor of the neutron G_M^n.

2 The A4 experimental setup and analysis

The A4 experiment at MAMI is complementary to other experiments since for the first time counting techniques have been used in a scattering experiment measuring a PV asymmetry. Therefore possible systematic contributions to the experimental asymmetries and the associated uncertainties are of a different nature as compared to previous experiments, which use analogue integrating techniques. More details on the A4 apparatus can be found in the contributions of J. Diefenbach, B. Gläser, Y. Imai, J. H. Lee and C. Weinrich.

The PV asymmetry was measured at the MAMI accelerator facility in Mainz [3] using the setup of the A4 experiment [4]. The polarized 570.4 and 854.3 MeV electrons were produced using a strained layer GaAs crystal that is illuminated with circularly polarized laser light [5]. Average beam polarization was about 80 %. The helicity of the electron beam was selected every 20.08 ms by setting the high voltage of a fast Pockels cell according to a randomly selected pattern of four helicity states, either $(+P - P - P + P)$ or $(-P + P + P - P)$. A 20 ms time window enabled the histogramming in all detector channels and an integration circuit in the beam monitoring and luminosity monitoring systems. The exact window length was locked to the power frequency of 50 Hz in the laboratory by a phase locked loop. For normalization, the gate length was measured for each helicity. Between each 20 ms measurement gate, there was an 80 μs time window for the high voltage at the Pockels cell to be changed. The intensity $I = 20$ μA of the electron current was stabilized to better than $\delta I/I \approx 10^{-3}$. An additional $\lambda/2$-plate in the optical system was used to rotate small remaining linear polarization components and to control the helicity correlated asymmetry in the electron beam current to the level of < 10 ppm in each five minute run.

From the source to the target, the electron beam develops fluctuations in beam parameters such as position, energy and intensity which are partly correlated to the reversal of the helicity from +P to -P. We have used a system of microwave resonators in order to monitor beam current, energy, and position in two sets of monitors separated by a drift space of about 7.21 m in front of the hydrogen target. In addition, we have used a system of 10 feed-back loops in order to stabilize current, energy [6], position, and angle of the beam. The polarization of the electron beam was measured with an accuracy of 2 % using a Møller polarimeter

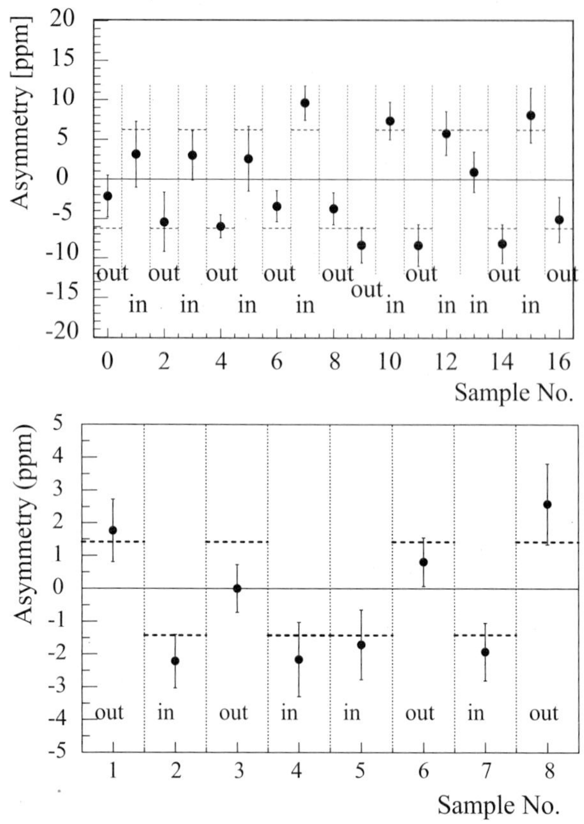

Fig. 1. The *top plot* shows the data samples of 854.3 MeV data with the $\lambda/2$-plate in and out. The *lower plots* represents the data sample for the 570.4 MeV data with the $\lambda/2$-plate in and out as described in the text

which is located on a beam line in another experimental hall [7]. Due to the fact that we had to interpolate between the weekly Møller measurements, the uncertainty in the knowledge of the beam polarization increased to 4 %. The 10 cm high power, high flow liquid hydrogen target was optimized to guarantee a high degree of turbulence with a Reynolds-number of $R > 2 \times 10^5$ in the target cell in order to increase the effective heat transfer. For the first time, a fast modulation of the beam position of the intense CW 20 μA beam could be avoided. It allowed us to stabilize the beam position on the target cell without target density fluctuations arising from boiling. The total thickness of the entrance and exit aluminum windows was 250 μm. The luminosity L was monitored for each helicity state (R, L) during the experiment using eight water-Cherenkov detectors (LuMo) that detect scattered particles symmetrically around the electron beam for small scattering angles in the range of $\theta_e = 4.4° - 10°$, where the PV asymmetry is negligible. The photomultiplier tube currents of these luminosity detectors were integrated during the 20 ms measurement period by gated integrators and then digitized by customized 16-bit analogue-to-digital converters (ADC). The same method was used for all the beam parameter signals. A correction was applied for the nonlinearity of the luminosity monitor photomultiplier tubes. From the beam current helicity pair data $I^{R,L}$ and lumi-

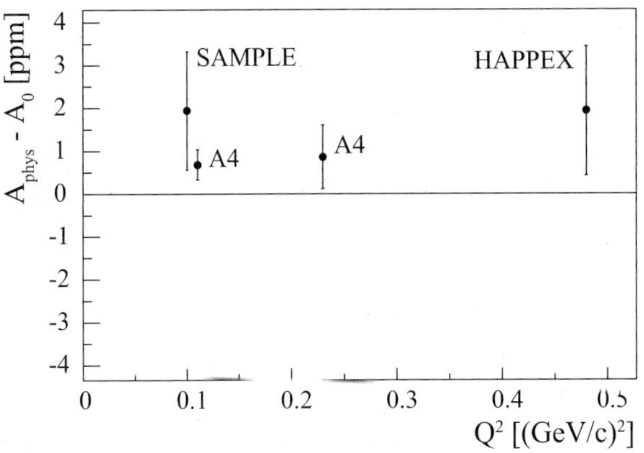

Fig. 2. Difference between the measured parity violating asymmetry in electron proton scattering $A_{LR}(\mathbf{e}p)$ and the asymmetry A_0 without vector strangeness contribution from the Standard Model for the SAMPLE experiment at *backward angles*, and the HAPPEX and the two A4 results at *forward angles*. The new experimental result at $Q^2{=}0.108$ (GeV/c)^2 presented here is the most accurate measurement

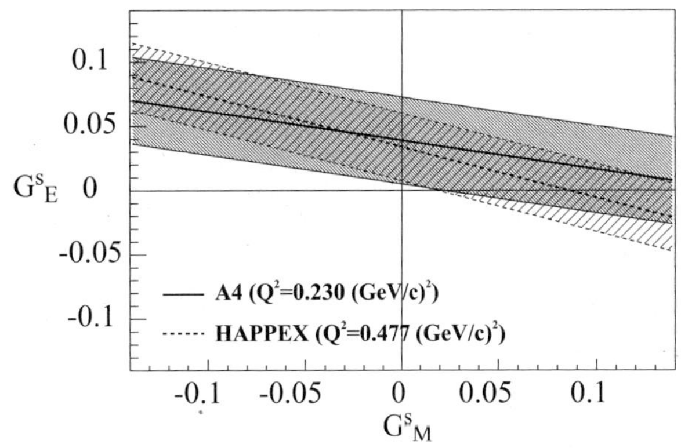

Fig. 3. The *solid line* represents all possible combinations of $G^s_E + 0.225 G^s_M$ as extracted from the work presented here at a Q^2 of 0.230 (GeV/c)^2. The densely hatched region represents the 1-σ uncertainty. The recalculated result from the HAPPEX published asymmetry at Q^2 of 0.477 (GeV/c)^2 is indicated by the *dashed line*, the less densely hatched area represents the associated error of the HAPPEX result

nosity monitor helicity pair $L^{R,L}$ data we calculated the target density $\rho^{R,L} = L^{R,L}/I^{R,L}$ for the two helicity states independently.

To detect the scattered electrons we developed a new type of a very fast, homogeneous, total absorption calorimeter consisting of individual lead fluoride (PbF_2) crystals [8,9]. The material is a pure Cherenkov radiator and has been chosen for its fast timing characteristics and its radiation hardness. This is the first time this material has been used in a large scale calorimeter for a physics experiment. The crystals are dimensioned so that an electron deposits 96 % of its total energy in an electromagnetic shower extending over a matrix of 3×3 crystals. Together with the readout electronics this allows us a measurement of the particle energy with a resolution of $3.9\,\%/\sqrt{E}$ and a total dead time of 20 ns. For the data taken at 854.3 MeV only 511 out of 1022 channels of the detector and the readout electronics were operational, for the 570.4 MeV data all the 1022 channels were installed. The particle rate within the acceptance of this solid angle was $\approx 50 \times 10^6\ \mathrm{s}^{-1}$. Due to the short dead time, the losses due to double hits in the calorimeter were 1 % at 20 μA. This low dead time is only possible because of the special readout electronics employed. The signals from each cluster of 9 crystals were summed and integrated for 20 ns in an analogue summing and triggering circuit and digitized by a transient 8-bit ADC. There was one summation, triggering, and digitization circuit per crystal. The energy, helicity, and impact information were stored together in a three dimensional histogram.

The number of elastic scattered electrons is determined for each detector channel by integrating the number of events in an interval from 1.6 σ_E above pion production threshold to 2.0 σ_E above the elastic peak in each helicity histogram, where σ_E is the energy resolution for nine crystals. These cuts ensure a clean separation between elastic scattering and pion production or Δ-excitation which has an unknown PV cross section asymmetry. The linearity of the PbF_2 detector system with respect to particle counting rates and possible effects due to dead time were investigated by varying the beam current. We calculate the raw normalized detector asymmetry as $A_{\mathrm{raw}} = (N_e^R/\rho^R - N_e^L/\rho^L)/(N_e^R/\rho^R + N_e^L/\rho^L)$. The possible dilution of the measured asymmetry by background originating from the production of π^0's that subsequently decays into two photons where one of the photons carries almost the full energy of an elastic scattered electron was estimated using Monte Carlo simulations to be much less than 1 % and is neglected here. The largest background comes from quasi-elastic scattering at the thin aluminum entrance and exit windows of the target cell. We have measured the aluminum quasi-elastic event rate and calculated in a static approximation a correction factor for the aluminum of 1.030 ± 0.003 giving a smaller value for the corrected asymmetry.

Corrections due to false asymmetries arising from helicity correlated changes of beam parameters were applied on a run by run basis. The analysis was based on the five minute runs for which the counted elastic events in the PbF_2 detector were combined with the correlated beam parameter and luminosity measurements. In the analysis we applied reasonable cuts in order to exclude runs where the accelerator or parts of the PbF_2 detector system were malfunctioning. The analysis is based on a total of 7.3×10^6 histograms corresponding to 4.8×10^{12} elastic scattering events for the 854.3 MeV data and $4.8 \cdot 10^6$ histograms corresponding to $2 \cdot 10^{13}$ elastic events for the 570.4 MeV data.

For the correction of helicity correlated beam parameter fluctuations we used multi-dimensional linear regression analysis using the data. The regression parameters have been calculated in addition from the geometry of the precisely surveyed detector geometry. The two different methods agree very well within statistics. The experimental asymmetry is normalized to the electron beam po-

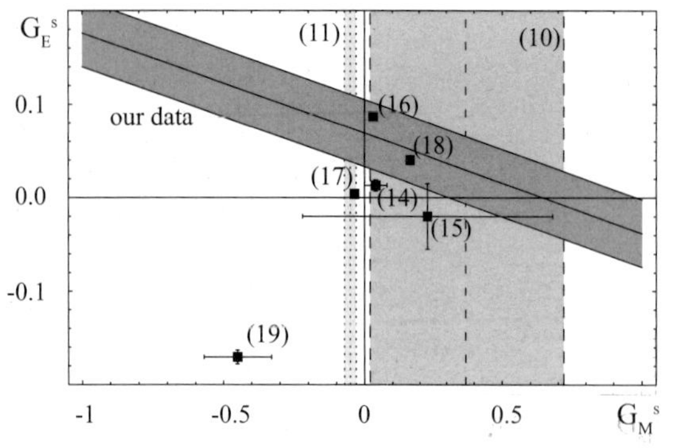

Fig. 4. The *solid lines* represent the result on $G_E^s + 0.106 G_M^s$ as extracted from our new data at $Q^2 = 0.108$ (GeV/c)^2 presented here. The hatched region represents in all cases the one-σ-uncertainty with statistical and systematic and theory error added in quadrature. The *dashed lines* represent the result on G_M^s from the SAMPLE experiment [10]. The *dotted lines* represent the result of a recent lattice gauge theory calculation for μ_s [11]. The boxes represent different model calculations and the numbers denote the references

larization P_e to extract the physics asymmetry, $A_{\mathrm{phys}} = A_{\mathrm{exp}}/P_e$. We have taken half of our data with a second $\lambda/2$-plate inserted between the laser system and the GaAs crystal. This reverses the polarization of the electron beam and allows a stringent test of the understanding of systematic effects. The effect of the plate can be seen in Fig. 1: the observed asymmetry extracted from the different data samples changes sign, which is a clear sign of parity violation if, as in our case, the target is unpolarized. Our measured result for the PV physics asymmetry in the scattering cross section of polarized electrons on unpolarized protons at an average Q^2 value of 0.230 (GeV/c)^2 is $A_{LR}(\mathbf{e}p) = (-5.44 \pm 0.54_{stat} \pm 0.26_{syst})$ ppm for the 854.3 MeV data [12] and $A_{LR}(\mathbf{e}p) = (-1.36 \pm 0.29_{stat} \pm 0.13_{syst})$ ppm for the 570.4 MeV data [13]. The first error represents the statistical accuracy, and the second represents the systematical uncertainties including beam polarization. The absolute accuracy of the experiment represents the most accurate measurement of a PV asymmetry in the elastic scattering of longitudinally polarized electrons on unpolarized protons. Figure 2 shows the present situation of all the published asymmetry measurements in elastic electron proton scattering. From the difference between the measured $A_{LR}(\mathbf{e}p)$ and the theoretical prediction in the framework of the Standard Model, A_0, we extract a linear combination of the strange electric and magnetic form factors for the 570.4 MeV data at a Q^2 of 0.108 (GeV/c)^2 of $G_E^s \pm 0.106\ G_M^s = 0.071 \pm 0.036$. For the data at 854.3 MeV corresponding to a Q^2 value of 0.230 (GeV/c)^2 we extract $G_E^s + 0.225\ G_M^s = 0.039 \pm 0.034$. Statistical and systematic error of the measured asymmetry and the error in the theoretical prediction of A_0 been added in quadrature. In Fig. 3 the results for the 570.4 MeV data are displayed. A recent very accurate determination of the strangeness contribution to the magnetic moment of the proton $\mu_s = G_M^s(Q^2 = 0\ (\mathrm{GeV/c})^2)$ from lattice gauge theory [11] would yield a larger value of $G_E^s = 0.076 \pm 0.036$ if the Q^2 dependence from 0 to 0.108 (GeV/c)^2 is neglected. The theoretical expectations for another quenched lattice gauge theory calculation [14], for SU(3) chiral perturbation theory [15], from a chiral soliton model [16], from a quark model [17], from a Skyrme-type soliton model [18] and from an updated vector meson dominance model fit to the form factors [19] are included into Fig. 4.

Our results concerning the measurement of the transverse beam spin asymmetry have been presented in the contribution of S. Baunack. We are also analyzing our data in the region of the $\Delta(1232)$ resonance, which has been presented in the contribution of L. Capozza. We are preparing a series of measurements of the parity violating asymmetry in the scattering of longitudinally polarized electrons off unpolarized protons and deuterons under backward scattering angles of $140° < \theta_e < 150°$ with the A4 apparatus in order to separate the electric (G_E^s) and magnetic (G_M^s) strangeness contribution to the electromagnetic form factors of the nucleon.

Acknowledgements. This work has been supported by the DFG in the framework of the SFB 201 and SPP 1034. We are indebted to K.H. Kaiser and the whole MAMI crew for their tireless effort to provide us with good electron beam. We are also indebted to the A1-Collaboration for the use of the Moeller polarimeter.

References

1. D.B. Kaplan et al.: Nucl. Phys. B **310**, 527 (1988)
2. M. Musolf et al.: Phys. Rep. **239**, 1 (1994)
3. H. Euteneuer et al.: Proc of the EPAC 1994 **1**, 506 (1994)
4. F.E. Maas et al.: Eur. Phys. J. A **17**, 339 (2003)
5. K. Aulenbacher et al.: Nucl. Ins. Meth. A **391**, 498 (1997)
6. M. Seidl et al.: Proc. of the EPAC 2000 , 1930 (2000)
7. P. Bartsch: Dissertation Mainz (2001)
8. F.E. Maas et al.: *Proc. of the ICATPP-7*, (World Scientific, 2002), p. 758
9. P. Achenbach et al.: Nucl. Ins. Meth. A **465**, 318 (2001)
10. D.T. Spayde et al.: Phys. Lett. B **583**, 79 (2004)
11. D.B. Leinweber et al.: hep-lat/0406002 (2004)
12. F.E. Maas et al.: Phys. Rev. Lett. **93**, 022002 (2004)
13. F.E. Maas et al.: nucl-ex/0412030 (2004)
14. R. Lewis et al.: Phys. Rev. D **67**, 013003 (2003)
15. T.R. Hemmert et al.: Phys. Rev. C **60**, 045501 (1999)
16. A. Silva et al.: Eur. Phys. J. A **22**, 481 (2004)
17. V. Lyubovitskij et al.: Phys. Rev. C **66**, 055204 (2002)
18. H. Weigel et al.: Phys. Lett. B **353**, 20 (1995)
19. H.-W. Hammer et al.: Phys. Rev. C **60**, 045204 (1999)

Eur Phys J A (2005) **24**, s2, 51–54
DOI: 10.1140/epjad/s2005-04-011-5

EPJ A direct
electronic only

Updated results from the SAMPLE experiment

Damon T. Spayde[a], for the SAMPLE Collaboration

Loomis Laboratory of Physics, University of Illlinois, 1110 West Green Street, Urbana, IL 61801, USA

Received: 15 November 2004 / Published Online: 8 February 2005

Abstract. The SAMPLE collaboration has recently completed its measurements of the parity violating elastic electron scattering asymmetry from hydrogen and the quasielastic asymmetry from deuterium [1, 2,3,4,5]. Neutral weak form factors of the nucleon, both vector and axial, can be extracted from these data and used to determine the strange quark contribution to electromagnetic form factors. The results of the original measurements at a Q^2 of 0.1 (GeV/c)^2 [2,3] have recently been reanalyzed [4], incorporating improvements in simulation and new background data, and combined with a new measurement of the quasielastic deuteron asymmetry at a Q^2 of 0.038 (GeV/c)^2 [5]. Final results from this new analysis and data set will be presented.

PACS. 13.60.-r Photon and charged-lepton interactions with hadrons – 11.30.Er Charge conjugation, parity, time reversal, and other discrete symmetries – 13.40.Gp Electromagnetic form factors – 14.20.Dh Protons and neutrons

1 Introduction

In the simplest, nonrelativistic quark model protons and neutrons are composed of three heavy ($\approx$ 300 MeV/c^2), valence up (u) and down (d) quarks. In this model the various static properties of the nucleons such as mass and intrinsic spin are explained quite well in terms of the properties of the valence quarks. However, the full theory that describes the structure of nucleons and folds in the Standard Model and quantum chromodynamics (QCD) has the valence quarks as very light ($\approx$ 10 MeV/c^2) particles. Furthermore, the valence quarks are surrounded by a sea of gluons and quark-antiquark pairs. This quark sea can contain the heavier quarks such as the strange (s) as well as the lighter u and d. A full description of the properties of the nucleon must account for the presence of these quark-antiquark pairs.

A variety of experiments have detected nonzero strange quark contributions to static nucleon properties such as spin, mass, and momentum (see [6] for a recent review). Given that other low energy nucleon properties contain non-zero contributions from strange quarks it is natural to wonder how these quarks contribute to the electric and magnetic properties of the proton: the charge radius and the magnetic moment. These properties arise from the proton's vector strange matrix element $\langle p|\bar{s}\gamma^\mu s|p\rangle$, a quantity that can be determined by measuring the proton's neutral weak magnetic form factor G_M^Z [7]. A method for measuring G_M^Z using parity violating elastic electron scattering was proposed by Beck and McKeown [8,9]. This experimental technique has formed the basis for several past and present experiments trying to probe the strange quark contribution to the electric and magnetic properties of the nucleon. This paper discusses results from one of these experiments, SAMPLE, that was performed at the MIT/Bates Linear Accelerator Center in the late 1990s/early 2000s. For a recent review of this experiment see [6].

In the scattering of electrons from nucleons via the electromagnetic interaction (exchange of a virtual photon γ) the finite extant of the nucleon is parameterized by the electric and magnetic form factors G_E^γ and G_M^γ. These form factors are functions of the momentum transferred by the exchanged photon, or four-momentum transfer squared Q^2. An analogous pair of form factors, G_E^Z and G_M^Z, arise from the same scattering process when a neutral weak Z^0 boson is exchanged instead of a γ. The electromagnetic and neutral weak form factors for the proton (p) and the neutron (n) can be expressed in terms of contributions from the three lightest quarks u, d, and s [10]. If charge symmetry is assumed (see [11] for a discussion of the validity of this assumption), that is to say that the distribution of u quarks in the proton is the same as that of d quarks in the neutron, then the neutral weak form factors can be written in terms of the electromagnetic form factors and a strange quark contribution (to lowest order)

$$G_{E,M}^Z = \left(1 - 4\sin^2\theta_W\right) G_{E,M}^{\gamma,p} - G_{E,M}^{\gamma,n} - G_{E,M}^s. \quad (1)$$

The neutral weak interaction brings in a third form factor known as the axial form factor G_A^e; to lowest order it is written as follows

$$G_A^e = -\tau_3 G_A + \Delta s \quad (2)$$

[a] *Present address:* Department of Physics, Grinnell College, 1116 8th Avenue, Grinnell, IA 50112 USA

where $\tau_3 = 1(-1)$ for $p(n)$, $G_A(Q^2 = 0) = -(g_A/g_V) = 1.2670 \pm 0.0035$ as measured in β-decay experiments [12], and Δs is the strange quark contribution to the nucleon spin. Equations 1 and 2 both have higher order electroweak radiative corrections that are not considered here (see [10] for details).

It would be extremely difficult to measure $G_{E,M}^Z$ with a straight cross-section measurement due to the difference in coupling strengths between EM and neutral weak interactions. However, the parity violating nature of the weak interaction can be exploited to make the measurement feasible. If a beam of longitudinally polarized electrons is scattered from a target of unpolarized electrons and the helicity (product of the electron spin and momentum directions) of the electrons is changed, then the cross section asymmetry A_{PV} is sensitive to the neutral weak interaction at leading order

$$\begin{aligned} A_{PV} &= \frac{d\sigma_R - d\sigma_L}{d\sigma_R + d\sigma_L} \\ &= -\frac{G_F Q^2}{4\pi\alpha\sqrt{2}} \frac{A_E + A_M + A_A}{\left[\epsilon\left(G_E^\gamma\right)^2 + \tau\left(G_M^\gamma\right)^2\right]} \end{aligned} \tag{3}$$

where G_F is the Fermi coupling constant and α is the fine structure constant. The other terms in 3 are defined as follows

$$A_E = \epsilon G_E^Z(Q^2) G_E^\gamma(Q^2) \tag{4}$$

$$A_M = \tau G_M^Z(Q^2) G_M^\gamma(Q^2) \tag{5}$$

$$\begin{aligned} A_A = &-\left(1 - 4\sin^2\theta_W\right)\sqrt{\tau(1+\tau)(1-\epsilon^2)} \\ &\times G_A^e(Q^2) G_M^\gamma(Q^2) \end{aligned} \tag{6}$$

$$\tau = \frac{Q^2}{4M_N^2} \tag{7}$$

$$\epsilon = \frac{1}{1 + 2(1+\tau)\tan^2\frac{\theta}{2}} \tag{8}$$

where M_N is the mass of the nucleon.

At backward scattering angles the A_E term in 3 is negligible relative to the other two terms so a measurement of A_{PV} on the proton A_p is sensitive to G_M^s and G_A^e. The isovector piece of G_A^e is not well-constrained by theory or experiment, necessitating an additional measurement to extract it. This additional measurement comes from measuring A_{PV} for the deuteron A_d. If one assumes the static approximation is true for the deuteron, i.e. it is essentially a free proton and neutron, then one can write A_d in terms of A_p and A_n

$$A_d = \frac{\sigma_p A_p + \sigma_n A_n}{\sigma_p + \sigma_n} \tag{9}$$

where $\sigma_{p(n)}$ is the cross-section for elastic $e - p(n)$ scattering. The deuteron asymmetry is sensitive to G_M^s and $G_A^e(T = 1)$ (the isovector piece of G_A^e) as well, therefore, a measurement of A_p and A_d at backward scattering angles and a fixed Q^2 should allow G_M^s to be uniquely determined.

2 Experiment

The SAMPLE experiment was performed at the MIT/Bates Linear Accelerator Center in Middleton, MA; the target and detector apparatus were installed in the North Hall. The proton asymmetry was measured at a beam energy of 200 MeV during the summer of 1998 [2]. The deuteron asymmetry was measured at two different beam energies: 200 MeV during the summer of 1999 [3] and 125 MeV during the fall of 2001 and spring of 2002 [5]. The original proton and deuteron results at 200 MeV were updated after the completion of the 125 MeV data-taking [4,5].

Longitudinally polarized electrons were generated in the polarized source by photoemission from a bulk gallium arsenide (GaAs) crystal; a Ti:Sapphire laser provided the circularly polarized photons used to emit the electrons. The helicity Pockels cell (HPC) was used to reverse the laser and electron polarizations 600 times a second. A correction Pockels cell (CPC), set to function as an intensity adjuster, was also placed in the laser line to reduce the helicity correlated charge asymmetry measured in the experimental hall. A position feedback system consisting of a piece of optical glass on a piezoelectric mount was used to reduce the helicity correlated position differences measured in the hall by adjusting the position of the laser spot on the cathode in a helicity-correlated way [13]. The electron spins were precessed away from the longitudinal in a Wien filter prior to injection into the main accelerator to compensate for the 36.5° bend the electrons went through as they entered the experimental hall.

In the accelerator the electrons were accelerated to the final beam energy of 125 or 200 MeV. At the end of the accelerator was an energy feedback system that reduced the fluctuations in the electron beam energy [14]. Dipole magnets bent the electrons through 36.5° onto the North Hall beam line. Prior to entering the experimental hall the electrons passed through a region containing a Møller polarimeter that could be used to measure the beam polarization in dedicated, low-current runs. Upon entering the hall the electron beam passed through a series of toroids and RF cavities used to measure the beam current and position of each beam pulse. A set of Čerenkov detectors (lucite coupled to photomultiplier tubes) were placed symmetrically about the beamline upstream of the target to monitor the halo about the electron beam and another set were placed downstream to measure the luminosity.

The experimental apparatus, shown in Fig. 1 consisted of a high-power cryogenic target encased in a lead-lined scattering chamber [15]. The scattering chamber was enclosed in a large, light-tight box that also contained the detector package. The detector package consisted of an array of ten, large ellipsoidal mirrors symmetrically placed about the scattering chamber at backward angles. The mirrors focused Čerenkov light from back-scattered particles onto a corresponding array of 8 inch photomultiplier tubes (PMT) encased in lead shielding. Each PMT was read out by a channel of integrating electronics; the current from each PMT anode was sent to a current-to-voltage converter before being integrated and digitized.

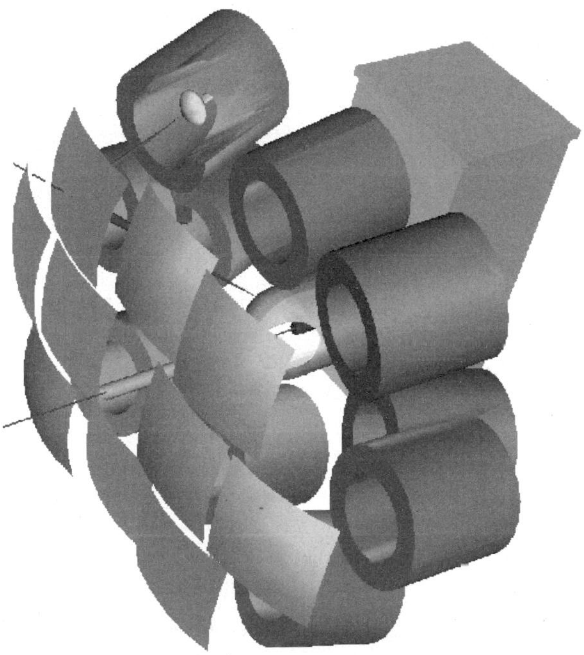

Fig. 1. This is a schematic of the SAMPLE target and main detector apparatus. The electron beam enters from the *left* and strikes the target which is shown inside the scattering chamber in a cutaway view

In this mode of operation there was no way to separate the background events from the signal. Dedicated runs of various types were used to measure the background contribution.

3 Analysis

The first step in the data analysis was to calculate the raw detector asymmetry for beam pulses of positive and negative helicity. Simultaneously, the sensitivity of the detectors to changes in the various beam parameters were calculated. In the next step of the analysis this information on detector sensitivity was used to remove any beam parameter dependence from the measured asymmetry.

After this correction had been performed, corrections for various background processes were applied as given by the following expression

$$A_{exp} = \frac{R_C}{P_B f_l f_e (1 - f_\pi)} [A_O - (1 - f_l) A_C] \qquad (10)$$

where R_C is an electromagnetic radiative correction, P_B is the measured beam polarization, f_l is the fraction of the yield due to light on the PMTs, f_e is the fraction of the light signal due to Čerenkov events, and f_π is the fraction of the Čerenkov light due to pion production. The quantity A_O is the beam parameter corrected asymmetry measured under normal running conditions; A_C is the same asymmetry measured with shutters drawn over the PMTs to block out the light signal. The final quantity A_{exp}, where exp is p or d depending on whether the target is full of hydrogen or deuterium, was combined with the theoretical asymmetry in order to extract G_M^s and $G_A^e(T = 1)$.

A variety of improvements were made to the data analysis and theoretical calculations between the publication of the original results [2,3] and the final results [4, 5]. In both the hydrogen and deuterium 200 MeV data sets an improved Monte Carlo code was used, based on GEANT [16], that incorporated the geometry of the target and detectors. This code was used to calculate the electromagnetic radiative corrections R_C due to internal and external bremsstrahlung processes as well as the theoretical asymmetry used for the form factor extraction [17]. The pion dilution factor was a new addition to the analysis; it was not present in the original publications. New calculations by Schiavilla were incorporated into the theoretical deuteron asymmetry to account for both quasielastic scattering and threshold breakup. Finally, the treatment of a ϕ dependent background process in the hydrogen measurement was altered (see [6] for details) The effect was to increase the experimental asymmetry in each data set by about 10-15% and reduce the theoretical asymmetry by 3-5%.

The third set of SAMPLE data, on deuterium at a beam energy of 125 MeV, were taken to serve as a cross-check on the original 200 MeV deuteron result. The original deuteron measurement [3] resulted in a value for $G_A^e(T = 1)$ that deviated significantly from the theoretical value [18].

4 Results

Updated results for the measured asymmetries on the proton and deuteron are listed below. The results are taken from [5] and [4] (all Q^2 are in units of (GeV/c)^2)

$$A_p(Q^2 = 0.1) = (-5.61 \pm 0.67 \pm 0.88) \text{ ppm} \qquad (11)$$
$$A_d(Q^2 = 0.091) = (-7.77 \pm 0.73 \pm 0.72) \text{ ppm} \qquad (12)$$
$$A_d(Q^2 = 0.038) = (-3.51 \pm 0.57 \pm 0.58) \text{ ppm} \qquad (13)$$

where the first error is statistical and the second is systematic. The updated theoretical asymmetries, taking into account the detector geometry, are

$$A_p(Q^2 = 0.1) = -5.56 + 3.37 G_M^s + 1.54 G_A^{e(T=1)} \qquad (14)$$
$$A_d(Q^2 = 0.091) = -7.06 + 0.72 G_M^s + 1.66 G_A^{e(T=1)} \qquad (15)$$
$$A_d(Q^2 = 0.038) = -2.14 + 0.27 G_M^s + 0.76 G_A^{e(T=1)} \qquad (16)$$

from [5] and [4].

Figure 2 shows the deuteron asymmetry measured at 125 and 200 MeV as a function of Q^2. The result of theoretical calculations for A_d based on [18] are shown as well. The good agreement between theory and experiment shown in this plot from [5] motivated the use of the theoretical calculation for $G_A^e(T = 1)$ in the final result for G_M^s quoted in [4]. Figure 3 is taken from [4] and shows the region allowed by the measurement of A_p on hydrogen at

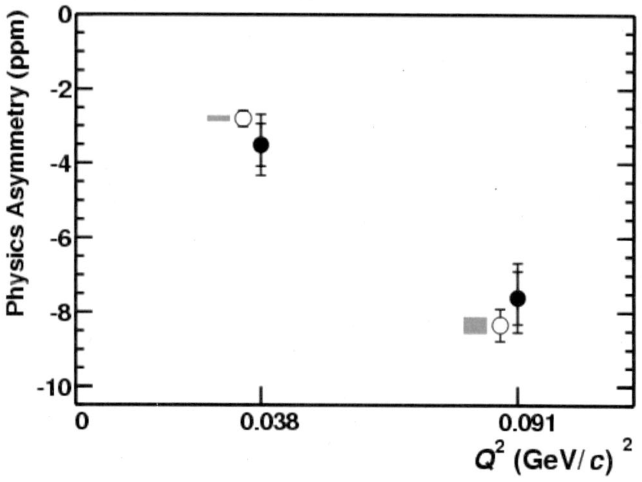

Fig. 2. This is a plot of the deuteron asymmetry A_d as a function of Q^2. The *filled circles* are the data from the 125 and 200 MeV SAMPLE runs (statistical and systematic error bars are shown). The *empty circles* are the theoretical prediction for A_d using [18] for the multi-quark radiative corrections and a value of 0.15 for G^s_M. The *grey bands* show the change in A_d as G^s_M is changed by ± 0.3 n.m. The figure is from [5]

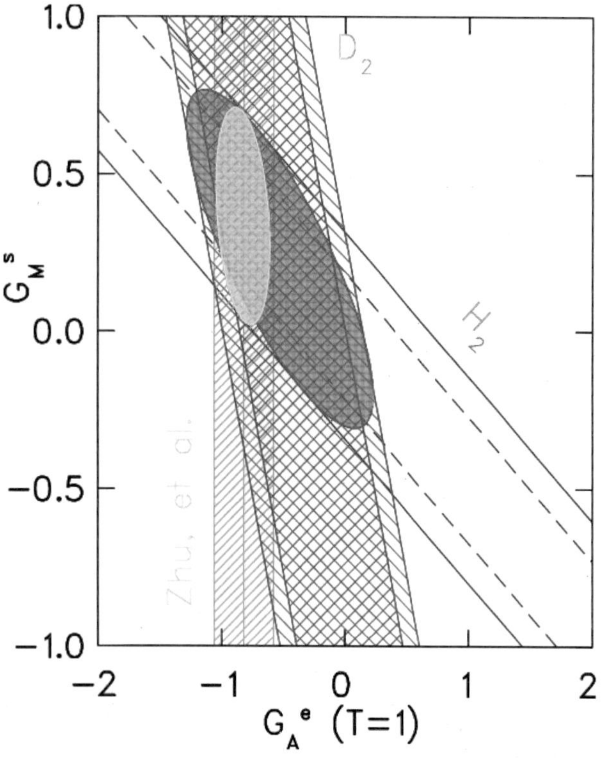

Fig. 3. This is a plot of G^s_M versus $G^e_A(T=1)$ and shows the region allowed by the various SAMPLE measurements. The *open* and *cross-hatched bands* are the regions allowed at the one sigma level by the 200 MeV hydrogen and deuterium measurements, respectively. The *vertical, hatched band* is the theoretical calculation of $G^e_A(T=1)$ based on [18]. The *small oval* is the final result reported in [4]

200 MeV and the theoretical calculation of $G^e_A(T=1)$. Using the theoretical value of -0.83 ± 0.26 for $G^e_A(T=1)$ results in

$$G^s_M(Q^2 = 0.1) = 0.37 \pm 0.20 \pm 0.26 \pm 0.07 \tag{17}$$

where the errors are due to statistics, systematics, and the uncertainty on the electroweak radiative corrections [4]. This is lower than, but consistent with, the earlier result quoted in [3].

5 Conclusion

The SAMPLE collaboration has performed an important series of measurements of the parity violating electron scattering asymmetries from the proton and the deuteron. These measurements have made it possible to extract, for the first time, the strange quark contribution to the magnetic form factor of the nucleon; a quantity that is directly related to the magnetic moment of the proton and the neutron. This strange quark contribution was found to be slightly positive. The measurements of the deuteron asymmetry can also be used to place constraints on the quantities C_{2u} and C_{2d} which represent electron-quark axial couplings in the Standard Model [6]. The SAMPLE collaboration has also performed the first ever measurement of a single spin asymmetry for scattering of transversely polarized electrons from an unpolarized proton target [19]. These measurements are sensitive to two-photon exchange diagrams in $e - N$ scattering.

References

1. B. Mueller et al.: Phys. Rev. Lett. **78**, 3824 (1997)
2. D.T. Spayde et al.: Phys. Rev. Lett. **84**, 1106 (2000)
3. R. Hasty et al.: Science **290**, 2021 (2000)
4. D.T. Spayde et al.: Phys. Lett. B **583**, 79 (2004)
5. T. Ito et al.: Phys. Rev. Lett. **92**, 102003 (2004)
6. E.J. Beise, M.L. Pitt, D.T. Spayde: Prog. Part. Nucl. Phys. **54**, 289 (2005)
7. D.B. Kaplan, A. Manohar: Nucl. Phys. B **310**, 527 (1988)
8. R.D. McKeown: Phys. Lett. B **219**, 140 (1989)
9. D.H. Beck: Phys. Rev. D **39**, 3248 (1989)
10. M.J. Musolf et al.: Phys. Rep. **239**, 1 (1994)
11. G.A. Miller: Phys. Rev. C **57**, 1492 (1998)
12. K. Hagiwara et al.: Phys. Rev. D **66**, 010001 (2002)
13. T. Averett et al.: Nucl. Inst. Meth. A **438**, 246 (1999)
14. D.H. Barkhuff et al.: Nucl. Inst. Meth. A **450**, 187 (2000)
15. E. Beise et al.: Nucl. Inst. Meth. A **378**, 383 (1996)
16. GEANT: CERN Program Library
17. D.T. Spayde: Ph.D. thesis, University of Maryland, (2001)
18. S.-L. Zhu, S.J. Puglia, B.R. Holstein, M.J. Ramsey-Musolf: Phys. Rev. D **62**, 033008 (2000)
19. S.P. Wells et al.: Phys. Rev. C **63**, 064001 (2001)

Eur Phys J A (2005) **24**, s2, 55–58
DOI: 10.1140/epjad/s2005-04-012-4

EPJ A direct
electronic only

The next generation HAPPEX experiments

Richard S. Holmes, for the HAPPEX collaboration

Syracuse University, Syracuse, NY 13244, USA

Received: 15 October 2004 / Published Online: 8 February 2005

Abstract. A new generation of parity violating electron scattering experiments began running in Summer 2004 at the Thomas Jefferson National Accelerator Facility. HAPPEX-H will measure the parity violating asymmetry in polarized electron scattering from protons at $Q^2 = 0.11$ (GeV/c)^2. This asymmetry is sensitive to a linear combination of the strange electric and magnetic form factors of the proton. HAPPEX-He will measure the parity violating asymmetry in polarized electron scattering from a ^{4}He target at the same value of Q^2, using the same spectrometers and similar detectors, accessing the strange electric form factor cleanly. PREX will use parity violating electron scattering to determine the neutron radius of the ^{206}Pb nucleus.

PACS. 13.60.Fz Elastic and Compton scattering – 11.30.Er Charge conjugation, parity, time reversal, and other discrete symmetries – 13.40.Gp Electromagnetic form factors – 14.20.Dh Protons and neutrons – 24.80.+y Nuclear tests of fundamental interactions and symmetries – 21.10.Gv Mass and neutron distributions – 25.30.Bf Elastic electron scattering

1 Introduction

The question of how sea quarks contribute to nucleon observables has been of great interest since the European Muon Collaboration's observation [1] that valence quarks contribute less than half of the proton spin. Kaplan and Manohar [2] suggested that strange quarks (necessarily sea quarks) could contribute to the vector matrix elements of the nucleon, giving rise to nonzero strange electric and magnetic form factors.

The size of these effects can be deduced from measurements of the parity-violating asymmetry in elastic scattering of longitudinally polarized electrons from protons or nuclei [3,4,5]. These asymmetries are functions of the conventional electromagnetic form factors of the nucleons, an axial form factor, and the strange form factors. Asymmetries measured in three such experiments [6,7,8,9] have been reported.

Polarized electron scattering from a high-Z nucleus can be used to address very different physics. The parity violating asymmetry measured in such an experiment is sensitive to the neutron radius of the target nucleus [10], an observable which has to date not been measured in a clean, model independent way.

I report here on a set of second-generation parity violating electron scattering experiments which will be performed at the Thomas Jefferson National Accelerator Facility (Jlab). They will provide improved sensitivity to the strange form factors at low Q^2, and a clean measurement of the neutron radius of the ^{206}Pb nucleus.

2 Experimental goals

The parity violating asymmetry in scattering of polarized electrons from protons is given by

$$A^{\mathrm{PV}} = -\frac{G_F|Q|^2}{4\pi\alpha\sqrt{2}} \times \rho' \left[(1 - 4\kappa' \sin^2\theta_W) - \frac{\epsilon G_E^{\gamma p}(G_E^{\gamma n} + G_E^s) + \tau G_M^{\gamma p}(G_M^{\gamma n} + G_M^s)}{\epsilon (G_E^{\gamma p})^2 + \tau (G_M^{\gamma p})^2} \right] + A_A \quad (1)$$

where $G_{E(M)}^{\gamma p}$ is the electric (magnetic) Sachs form factor for the proton, $G_{E(M)}^{\gamma n}$ is the same for the neutron, G_E^s and G_M^s are the electric and magnetic strange form factors, θ_W is the weak mixing angle, G_F is the Fermi constant, and α is the fine-structure constant. A_A is a term proportional to G_A^{Zp}, the neutral weak axial form factor. The kinematic factors are $Q^2 = -q_\mu^2 > 0$, the square of the four-vector momentum transfer, $\tau = Q^2/4M^2$ where M is the proton mass, and $\epsilon = [1 + 2(1+\tau)\tan^2(\theta/2)]^{-1}$ where θ is the scattering angle. Finally, $\rho' = 0.9879$ and $\kappa' = 1.0029$ are parameters arising from electroweak radiative corrections [11]. At forward angles A_A is small, and a linear combination of the strange form factors can be calculated from the asymmetry and measured values of the nucleon electromagnetic form factors. One can express a strangeness radius ρ_s and a strange magnetic moment μ_s in terms of the strange form factors as $\rho_s = [dG_E^s/d\tau]_{\tau=0}$, $\mu_s = G_M^s(0)$.

The original Hall A Proton Parity Experiment (HAPPEX) measured parity violating asymmetry in the elastic scattering of polarized electrons from a liquid hydrogen target at $\theta = 12.3$ degrees, $Q^2 = 0.5$ (GeV/c)^2. The result [7] was $A^{\mathrm{PV}} = -14.92 \pm 0.98$ (stat) ± 0.56 (syst) parts per million (ppm); from this we obtain $G_E^s + 0.392 G_M^s = 0.014 \pm 0.020 \pm 0.010$ where the first error is experimental and the second is due to errors in the values of the electromagnetic form factors. The strange form factors may be small at this Q^2, or there may be an accidental cancellation. It is of interest to measure G_E^s and G_M^s separately, and particularly at a smaller value of Q^2 to better allow extrapolation to zero. These considerations have motivated two new experiments, HAPPEX-H (Experiment E99-115) [12] and HAPPEX-He (Experiment E00-114) [13].

HAPPEX-H, like HAPPEX, scatters polarized electrons elastically from a hydrogen target. However, with the use of the new septum magnet upgrades to the HRS spectrometers the average laboratory scattering angle is reduced to approximately 6 degrees. The Q^2 for this experiment is about 0.1 GeV^2 and the theoretical value for the parity violating asymmetry in the absence of strange quark effects is 1.6 ppm. The goal is to measure this asymmetry with statistical and systematic errors of 5% and 2.5%, respectively.

Obtaining values for the two strange form factors separately requires a measurement at different kinematics or on a different target. A particularly attractive choice is to use a spinless isoscalar target in which case the parity violating asymmetry for elastic scattering of polarized electrons is simply

$$A^{\mathrm{PV}} = -\frac{A_0}{2}\left(2\sin^2\theta_W + \frac{G_E^s}{G_E^{\gamma p} + G_E^{\gamma n}}\right) \qquad (2)$$

and A^{PV} is sensitive only to G_E^s. HAPPEX-He scatters polarized electrons from a high pressure ^{4}He target with the same kinematics as HAPPEX-H: $\theta \sim 6$ degrees, $Q^2 \sim 0.1$ GeV^2. Without strange quark effects, the predicted asymmetry is 8.4 ppm, and the goal is to measure this asymmetry to 2.2% statistical and 2.1% systematic. Both HAPPEX-H and HAPPEX-He began data taking in Summer 2004.

The parity violating asymmetry in elastic scattering of polarized electrons from a heavy nucleus is sensitive to the ratio R_n/R_p of the neutron and proton radii for that nucleus. The lead radius experiment, PREX (Experiment E00-114) [14], will measure this asymmetry with a ^{208}Pb target. Kinematics will be $\theta \sim 6$ degrees, $Q^2 \sim 0.01$ GeV^2, with an expected asymmetry $A^{\mathrm{PV}} \sim 0.5$ ppm. Our goal is to measure this asymmetry with statistical and systematic errors of 3% and 1.1%, respectively, yielding a 1% measurement of R_n/R_p. PREX is conditionally approved.

A schematic of the HAPPEX-H experiment is shown in Fig. 1. Polarized electrons are produced at the source using circularly polarized light incident on a GaAs crystal. Polarization is set at 30 Hz, with a randomly chosen polarization followed by its complement to make a pair of opposite helicity "windows" each about 30 ms in length.

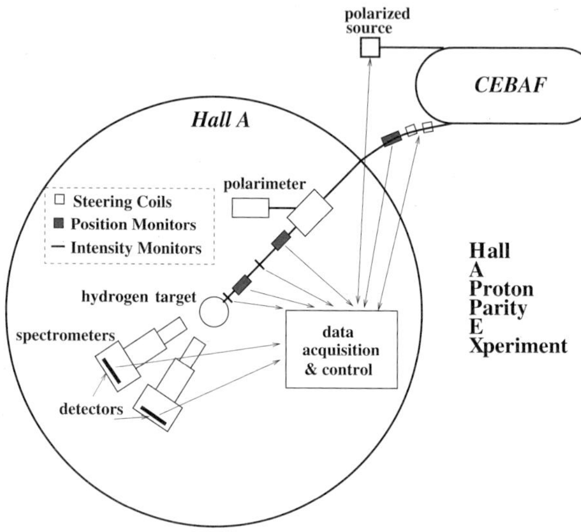

Fig. 1. Schematic of HAPPEX-H experiment

In Hall A, elastically scattered electrons are focused onto two detectors by two identical magnetic spectrometers. Inelastic events are cleanly rejected by being focused out of the detector acceptance.

The high rate in the detectors would result in significant dead time issues for a counting experiment. Instead, we integrate the detector signals over each helicity window using custom built integrating ADCs and a data acquisition system triggered at 30 Hz. The beam position, energy, and intensity are measured using various detectors in the Hall A beam line and these signals are also integrated over each helicity window.

The physics asymmetry is given by

$$A_{phys} \sim A_{det} - A_Q + \alpha A_E + \sum i\beta_i \Delta x_i . \qquad (3)$$

Here $A_{det} = (S^R - S^L)/(S^R + S^L)$ is the helicity correlated asymmetry in the integrated detector signals $S^{R(L)}$ for right (left) polarized beam. The remaining terms are corrections for systematic effects. A_Q and A_E are the helicity correlated relative differences in the beam intensity and energy, respectively. Δx_i is the helicity correlated difference in beam parameter x_i, where the four parameters are horizontal and vertical positions and angles at the target. α and β_i are the responses of the detector to changes in beam energy and position or angle x_i, respectively. This equation is correct under the assumptions that the detectors and beam monitors are linear and that there are no additional sensitivities to higher order parameters, such as spot size.

3 Beam and instrumentation upgrades

In order to meet the challenging goals of these experiments, we must control and understand our systematics at a more precise level than in the past. A number of upgrades and improvements to the polarized source, beam transport, and experimental instrumentation are required,

Table 1. HAPPEX beam requirements. "Jitter" is RMS width for signals integrated over 30 ms; "Difference" is size of helicity correlated difference averaged over the data set

Property	Nominal	Jitter	Difference
Energy	3.2 GeV	< 80 ppm	< 13 ppb
Current	100 μA	< 1000 ppm	< 600 ppb
Position	0	< 12 μm	< 2 nm
Angle	0	< 12 μrad	< 2 nrad
Halo @ 3 mm	<100 Hz/μA		

including: new target cells; various improvements at the polarized source; better polarimetry; new Čerenkov detectors; new luminosity monitors; new profile scanners; new beam intensity and position monitors; and an improved data acquisition system.

The correction terms in 3 must be kept small and precisely measured. Potentially the largest correction is A_Q, which can easily reach levels of several hundred ppm or larger. A standard technique for reducing this term is to introduce a feedback system, by which the helicity correlated beam intensity difference measured in the experimental hall is used to govern modulation of the intensity of the laser light at the source cathode, driving A_Q to a value consistent with zero.

While it is possible to feed back on beam energy, position, and angle differences, correlations between these parameters and fluctuations in their measurement introduce complications best avoided if possible. Careful control of position and angle of the laser spot on the source cathode can keep these beam parameter differences small; then, by measuring the differences in the hall and calibrating the detector responses to parameter differences by introducing deliberate modulations of the beam during a portion of the data taking, the required small corrections can be made with good precision.

The revamped laser table setup at the polarized source features two complementary systems for modulating the laser intensity. A PZT mirror is available for modulating position of the laser beam, if required. As in HAPPEX there is an insertable half wave plate to provide slow helicity reversal, which suppresses certain systematics; there also is a rotatable half wave plate for control of position and intensity systematics. A lens is used to image the beam onto the photocathode. The cathode provides better than 80% polarization with intensities up to 100 μA.

Some of the requirements for HAPPEX-H and HAPPEX-He quality beam are shown in Table 1. Thanks to hard work by the source group in collaboration with HAPPEX personnel, these requirements seem to be achievable now. PREX requirements will be even more stringent. For further details on control of source systematics see Cates's talk at this conference.

A significant portion of the final systematic error comes from polarimetry. Two polarimeters are available in Hall A, a Møller polarimeter [15,16] and a Compton polarimeter [17,15]. For the Møller polarimeter the main uncertainty is the foil polarization; at the HAPPEX-H and HAPPEX-He energy of ~3 GeV, a total relative error of 3 to 3.4% is expected.

The Compton polarimeter, which was commissioned during the first HAPPEX experiment, has since been upgraded with the addition of an electron recoil detector. In this new configuration total relative errors of 1.4% per 1 hour measurement have been seen with a 4.5 GeV beam, and 2.0% is probably achievable at 3 GeV.

Additional improvements to polarimetry will be required for PREX, where the combination of high beam intensity and low energy (850 MeV) means neither of the existing polarimeters can obtain precise measurements under running conditions. To address this the Compton polarimeter will be upgraded with a green laser that should enable measurements at the 1% level in under a day.

The HAPPEX-H detectors are brass/quartz sandwich Čerenkov total absorption detectors. Elastically scattered electrons are contained in a ~100 cm × 600 cm region of the focal plane in each spectrometer arm. To cover this region with minimal transmission losses, each detector consists of two segments at right angles, oriented to put the Čerenkov peak in line with the quartz layers and aimed at the single photomultiplier tube coupled to each segment. For HAPPEX-He, where the elastic peak region is smaller in size, we disassemble one of these detectors into its two segments and use one segment in each spectrometer arm. Detector design for PREX, where the elastic peak region will be very small, has not been finalized.

Interpretation of our results requires precise knowledge of our effective Q^2. For HAPPEX this was done using the standard Hall A drift chamber package at low beam intensity. For the new experiments we supplement this method with a pair of quartz Čerenkov profile detectors which can be scanned across the elastic peak at high beam intensity. This enables us to verify the drift chamber results and to monitor the Q^2 stability during the experiment.

The stripline position monitors used for HAPPEX have a resolution of $\sim$ 1.8 μm. This is adequate for HAPPEX-H and HAPPEX-He, but the stringent limits required for PREX demand better instrumentation. We have installed new cavity based position and intensity monitors which will be commissioned during the HAPPEX-H/HAPPEX-He running.

Target density fluctuations give rise to increased widths and correlations in the detector signals. New luminosity monitors mounted on the downstream beamline at very small scattering angles enable us to monitor such target density fluctuations. They also provide high sensitivity to helicity correlated beam parameters and enable better measurements of electronic noise. The monitors have been tested to ~200 ppm resolution per 30 ms helicity window at low beam intensity, and performance of ~100 ppm at high intensity is anticipated.

A new target cell design is being commissioned for HAPPEX-H and HAPPEX-He. Tests in early 2004 demonstrated that the existing design suffered from density fluctuations in a high power beam which would have

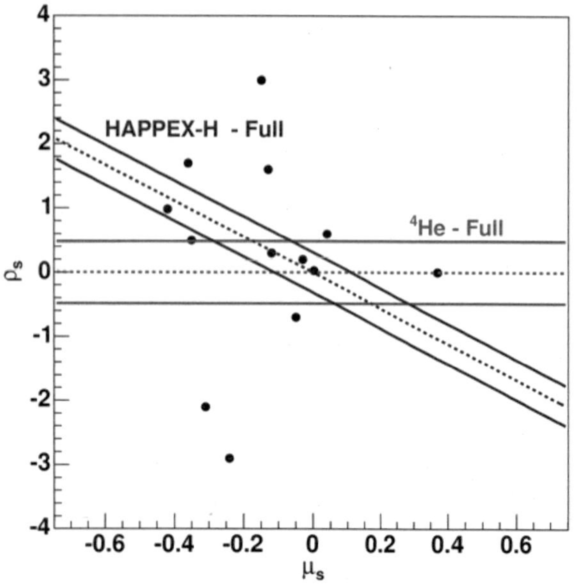

Fig. 2. Expected impact of HAPPEX-H (*diagonal band*) and HAPPEX-He (*horizontal band*). Points are predictions of various theoretical models. Expected limits on ρ_s and μ_s are computed assuming the full data set and an extrapolation to $Q^2 = 0$, and are centered on $\rho_s = 0$, $\mu_s = 0$

been unacceptable for parity experiments. The new "racetrack" design features fluid flow transverse to the beam direction to reduce such boiling behavior.

For PREX, a thin lead target able to withstand beam intensity of tens of μA without melting is obviously a challenge. A design in which the foil is sandwiched between thin diamond films in a liquid helium cooled copper frame has been successfully tested. Diamond has high thermal conductivity and scattering from ^{12}C is well understood, making this an ideal solution for this experiment.

4 Summary and prospects

HAPPEX demonstrated the need for an experimental program to provide information on the electric and magnetic strange form factors separately, with high precision, at low Q^2. HAPPEX and HAPPEX-He are designed to significantly limit the space of allowed values for these strange quark effects. These are challenging experiments, but feasible with the improvements in equipment and techniques developed in the past several years. The expected impact is shown in Fig. 2. PREX is even more challenging, but we are confident our experience in HAPPEX-H and HAPPEX-He will establish our readiness to undertake it. HAPPEX-H and HAPPEX-He began data taking in Summer 2004, and preliminary results for both have been presented [18]. We expect to complete data taking for these experiments in 2005, with PREX to be scheduled subsequently.

References

1. J. Ashman et al.: Phys. Lett. B **206**, 364 (1988); Nucl. Phys. B **328**, 527 (1988); Nucl. Phys. B **328**, 1 (1989)
2. D.B. Kaplan, A. Manohar: Nucl. Phys. B **310**, 527 (1988)
3. R.D. McKeown: Phys. Lett. B **219**, 140 (1989)
4. E.J. Beise, R.D. McKeown: Comments Nucl. Part. Phys. **20**, 105 (1991)
5. D.H. Beck: Phys. Rev. D **39**, 3248 (1989)
6. K. Aniol et al. [HAPPEX Collaboration]: Phys. Rev. Lett. **82**, 1096 (1999); K. Aniol et al., [HAPPEX Collaboration]: Phys. Lett. B **509**, 211 (2001)
7. K. Aniol et al. [HAPPEX Collaboration]: Phys. Rev. C **69**, 065501 (2004)
8. T.M. Ito et al. [SAMPLE Collaboration]: Phys. Rev. Lett. **92**, 102003 (2004), and references therein
9. F.E. Maas et al. [A4 Collaboration]: Phys. Rev. Lett. **93**, 022002 (2004)
10. C.J. Horowitz, S.J. Pollock, P.A. Souder, R. Michaels: Phys. Rev. C **63**, 025501 (2001)
11. R.M. Barnet et al. [Particle Data Group]: Phys. Rev. D **54**, 1 (1996)
12. R. De Leo et al. [HAPPEX Collaboration]. JLab Experiment E99-115 Proposal (1999)
13. P.A. Souder et al. [HAPPEX Collaboration]: JLab Experiment E00-114 Proposal (2000)
14. K.A. Aniol et al. [HAPPEX Collaboration]: JLab Experiment E00-003 Proposal (1999); K.A. Aniol et al., [HAPPEX Collaboration]: JLab Experiment E03-011 Proposal Update (2002)
15. J. Alcorn et al.: Nucl. Instrum. Meth. A **522**, 294 (2004)
16. A.V. Glamazdin et al.: Fizika B **8**, 91 (1999)
17. M. Baylac, et al.: Phys. Lett. B **539**, 8 (2002)
18. K. Paschke: Baryons 2004 conference (2004)

Eur Phys J A (2005) **24**, s2, 59–63
DOI: 10.1140/epjad/s2005-04-013-3

EPJ A direct
electronic only

The G0 experiment: Parity violation in e-N elastic scattering

Philip G. Roos, for the G0 Collaboration[a]

University of Maryland, College Park, MD 20742, USA

Received: 1 December 2004 / Published Online: 8 February 2005

Abstract. The G^0 experiment will measure the parity-violating asymmetries in elastic electron-nucleon scattering. The experiment is being performed in Hall C at the Jefferson Laboratory using a polarized electron beam and a dedicated experimental setup. Measurements of the electron-proton asymmetries will be made at both forward and backward angles, and electron-deuteron asymmetries at the backward angles. These measurements will cover a momentum transfer range of 0.1 – 1.0 GeV^2/c^2. ¿From these data the vector neutral weak form factors, G_E^Z and G_M^Z, and the effective axial current of the nucleon, G_A^e, may be extracted. When combined with the known electromagnetic form factors, one will be able to extract the contributions of u, d, and s quarks to the proton's charge and magnetization distributions. The first measurements at forward angles for the full momentum transfer range have very recently been successfully completed and preliminary results are presented here.

PACS. 1 3.60.Fz Elastic and Compton Scattering – 1 3.40.Gp Electromagnetic Form Factors – 1 4.20.Dh Protons and Neutrons

1 Introduction

A crucial aspect to our knowledge of the structure of the nucleon is the understanding of the role of the quark-anti-quark sea in contributing to the various properties of the nucleon - charge, magnetism, spin, etc. To this end, over the past decade several parity-violating electron scattering experiments have been carried out in an effort to identify and quantify the strange quark contribution to the nucleon sea. Experiments for which data have already been published are the SAMPLE experiment at MIT-Bates [1, 2,3,4], the HAPPEX experiment in Hall A at Jefferson Lab [5], and the PVA4 experiment at Mainz [6,7]. Each of these experiments has been discussed at this meeting, and the contributions are included in these proceedings. In each of these experiments a measurement is made for a single Q^2 value.

A new parity-violating electron scattering experiment currently being carried out in Hall C at Jefferson Lab is the G0 experiment [8]. The G0 collaboration consists of approximately 100 Ph.D. scientists from 20 institutions in the United States, Canada, France, and Armenia. The G0 experiment will measure parity violation (PV) in the scattering of polarized electrons from nucleons at both forward and backward angles using hydrogen and deuterium targets. The ultimate goal of G0 is to use the measurements of the parity-violating asymmetries in elastic electron scattering from the nucleon to provide a comprehensive and precise map of the neutral weak form factors of the proton over the range of momentum transfers 0.1-1.0 (GeV/c)^2,

The interaction between electrons and nucleons involves interference between the dominant electromagnetic interaction (γ exchange) and the parity-violating weak interaction (Z^0 exchange). This interference leads to a helicity dependence in the cross section for elastic scattering of polarized electrons from the nucleon, the observable measured by the G0 experiment, as well as the other experiments mentioned above. The asymmetry has three contributing terms that reflect this interference. Specifically, the parity-violating asymmetry for elastic electron scattering from the proton can be written as a sum of three terms:

$$A = -\frac{G_F Q^2}{4\pi\alpha\sqrt{2}} \times \frac{\left[\varepsilon G_E^p G_E^Z + \tau G_M^p G_M^Z - (1-4\sin^2\theta_W)\,\varepsilon' G_M^p G_A^e\right]}{\varepsilon\,(G_E^p)^2 + \tau\,(G_M^p)^2} \quad (1)$$

where $\tau = Q^2/4M_p^2$, $\varepsilon = \left(1+2(1+\tau)\tan^2\frac{\theta}{2}\right)^{-1}$ and $\varepsilon' = \sqrt{(1-\varepsilon^2)\,\tau\,(1+\tau)}$. When combined with known information on the nucleon electromagnetic form factors and assuming isospin conservation, it is possible to extract explicit contributions of $s\bar{s}$ pairs to the nucleon form factors. Away from the static limit, this information is characterized by the strange quark form factors, G_M^s and G_E^s, which contribute to the nucleon electric and magnetic form factors, defined by

$$G_{E,M}^Z = (1-4\sin^2\theta_W)(1+R_V^p)G_{E,M}^p - (1+R_V^n)G_{E,M}^n - G_{E,M}^s \quad (2)$$

[a] Jefferson Lab Experiment E-00-006. A list of collaborators and information about the experiment can be found at http://www.npl.uiuc.edu/exp/G0/

The strange form factors are of particular interest because they are a direct probe of the quark sea contribution to ground state properties of the nucleon. A complication arises from the third term in 1, the axial form factor G_A^e which has contributions both from the first-order exchange of a Z-boson, but also from higher order processes that produce a parity violating electromagnetic, or "anapole" interaction. For example, there can be a Z-exchange between two quarks in the proton and a photon exchange between the electron and proton. Theoretical estimates of these axial radiative corrections have been made for the SAMPLE experiment [9,10], but not for higher Q^2 values. In order to measure G_A^e and thus address the uncertainty in these corrections, a third measurement is needed. To obtain a third measurement the G0 experiment will adopt the method used by the SAMPLE collaboration; specifically, we will carry out measurements of parity-violating quasielastic scattering from deuterium at back angles. The different sensitivities to the axial form factor between the proton and the deuteron permit one to extract all three terms G_M^s, G_E^s, and G_A^e. The sensitivities of the three measurements of the G0 experiment to the various form factors are indicated for one value of Q^2 in the following.

G0 elastic scattering program

A_F: one measurement for all Q^2 → detect recoil protons
A_B: three measurements for three Q^2 values
→ detect electrons at 108°
A_d: Quasielastic scattering from deuterium
→ detect electrons at 108°

$$\begin{pmatrix} A_F \\ A_B \\ A_d \end{pmatrix} = \begin{pmatrix} \xi_F & \chi_F & \psi_F \\ \xi_B & \chi_B & \psi_B \\ \xi_d & \chi_d & \psi_d \end{pmatrix} \begin{pmatrix} G_E^s \\ G_M^s \\ G_A^e \end{pmatrix} + \begin{pmatrix} \eta_F \\ \eta_B \\ \eta_d \end{pmatrix}$$

at $Q^2 = 0.44\ (\mathrm{GeV/c})^2$.

	η(ppm)	ξ(ppm)	χ(ppm)	ψ(ppm)
A_F	-13.77	51.80	18.63	1.01
A_B	-25.01	16.10	31.41	6.96
A_d	-34.00	13.13	7.07	8.41

2 G0: Forward angle scattering

To carry out the G0 measurement program on a reasonable time scale, it was necessary to meet certain specifications. Because the asymmetries were expected to be as small as a few ppm, it is necessary to achieve counting statistics of the order of 10^{13} to 10^{14} counts. This necessitated the design of a large acceptance device with high count rate detectors and electronics, and the capability of isolating elastic scattering from inelastic processes. In addition an electron beam with high intensity, high polarization and small helicity-correlated beam properties is highly desirable, if not absolutely essential. These requirements led to the design and construction of specialized equipment by the G0 collaboration, as detailed next.

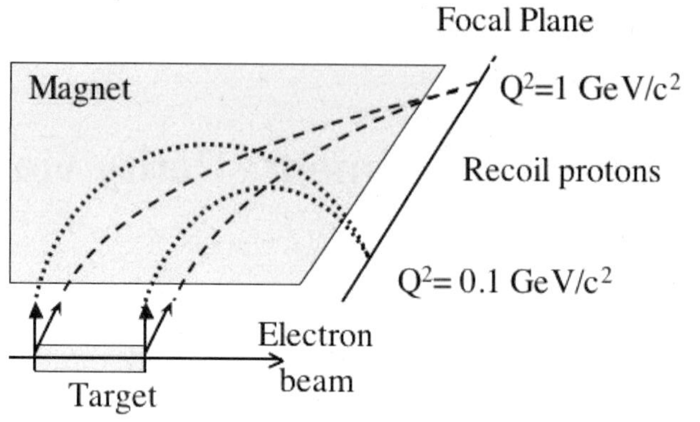

Fig. 1. A schematic diagram of the G0 magnet and detection scheme for the "forward" angle mode

The heart of the G0 apparatus is a superconducting toroidal magnet that focuses scattered particles through collimators onto a focal plane array of scintillators. The magnet contains eight superconducting coils splitting the magnet and detector system into octants. Each octant, as defined by collimators near the cryogenic target, accepted 20° in azimuthal angle. Combined with the polar angle acceptance, defined by the momentum defining collimators, the device had a total solid angle of about 0.9 sr. Figure 1 shows a schematic representation of the principle of operation for the forward angle scattering mode ("forward" mode). In its "forward" mode, the apparatus detects and counts recoiling protons from forward angle electron scattering (the protons are ejected at angles between approximately 60° and 75°). In this mode a simultaneous measurement for all Q^2 values (0.1-1.0 $(\mathrm{GeV/c})^2$) is made. Note that for a given Q^2 the magnet focuses protons from any point along the target length, at least at the center of each octant. After delays in delivery of the magnet and the necessity for significant modifications, in the end the magnet operated very reliably at the design current of 5000 A.

The Focal Plane Detectors (FPD) for each octant consisted of 16 pairs of scintillation detectors (Bicron BC400) which were shaped and segmented to detect protons corresponding to a limited band of Q^2 and to limit the e-p elastic count rate to less than approximately 750 kHz. These were placed at, or near, the focal plane of the magnet. FPD1-14 accepted relatively narrow bands of Q^2, whereas FPD15 accepted a broad band of Q^2 at the end of the range of acceptance. No elastic scattering events were recorded by FPD16, and it was used as a monitor of background and magnet current. Pairs of scintillators were used to suppress background from neutral particles. Light from both ends of each scintillator was transmitted through light guides to photomultiplier tubes which were mounted in a region of very low magnetic field. A picture of one of the detector octants is shown in Fig. 2.

The signals from the FPD photomultiplier tubes were sent to constant fraction discriminators, mean timing was done between the two ends of the scintillators, and a coincidence was done between the front and back scintillator.

Fig. 2. Forward proton detector octant

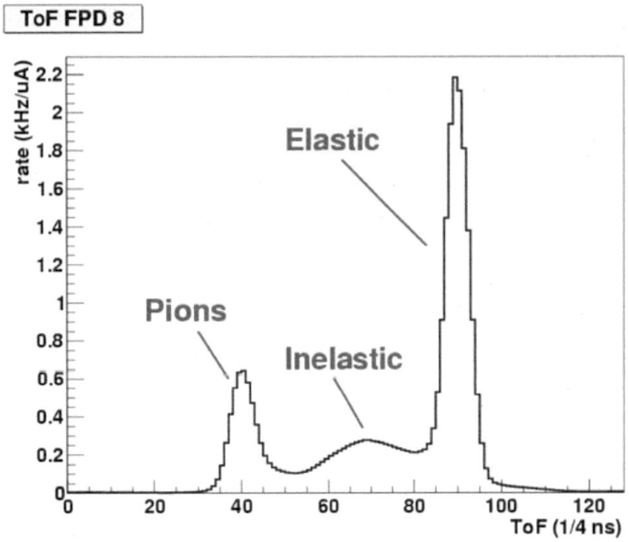

Fig. 3. Typical time spectra for the G0 experiment

Fig. 4. G0 experiment installed in Hall C at Jefferson Lab

The resultant signal was sent to custom time-encoding electronics which measured the flight time of the particle using a signal associated with the beam arrival on target. This time measurement is the primary method used to separate elastic protons from inelastic protons and pions. Two different sets of time-encoding electronics were used. Four of the octants used latching time digitizers to extract the flight time with a resolution of 1 ns. This information was recorded by fast scalers, so that rates of several MHz could be handled. The other four octants utilized electronics designed and built in France that were based on flash TDCs. This system had a time resolution of 0.25 ns, and could operate up to about 4 MHz. The basic data transfered to the computer consisted of 128 bit timing histograms. The time histograms were read out every 33 msec, corresponding to the helicity reversal period. A typical time spectrum is shown in Fig. 3.

The cryogenic target is 20 cm long and was designed to have high flow rate to minimize target density fluctuations. At the 40 μA current and 3 GeV electron energy for the forward angle mode, the heat load on the target was 250 W. In tests during G0 running the observed target density fluctuations were negligible at 40 μA, and we believe that we can run at currents up to 80 μA for the back angle mode.

The final requirement for the forward angle G0 run is the time structure of the beam. Since we require a measurement of the time-of-flight of the particles, we requested a time structure of 32 MHz. Producing a 40 μA beam with this time structure, and hence very high charge density, required extensive development work on the part of the JLab accelerator group, and I would like to acknowledge their crucial contributions to the G0 experiment. After lengthy development time, the accelerator was ultimately able to provide parity quality beam with the required time structure.

After two commissioning/engineering runs in 2002 and 2003, followed by a production run, the forward angle measurement of the G0 experiment has now been completed, and analysis is proceeding. The desired statistical uncertainties were achieved, and most subsystems of the apparatus performed at or above their design goals. Figure 4 shows the G0 experiment as installed for the forward angle mode. The success of this run was due in large part to the heroic efforts of the postdocs and graduate students associated with G0.

Two problems that could potentially impact the error in the final results were encountered during the production running. Both were due to small background yields which had an unexpectedly large asymmetry. The first was due to leakage current from the other two halls at JLab.

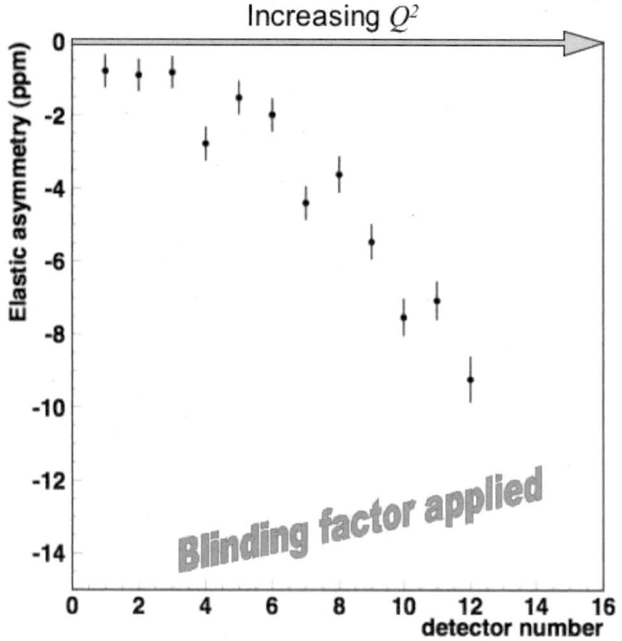

Fig. 5. Preliminary G0 forward angle data. The data are blinded by an overall multiplicative factor

These leakage beams, although small, had the standard 499 MHz time structure and a helicity dependent charge asymmetry not under our control. With some dedicated running under various conditions, we are able to correct for this contribution to the measured asymmetry to a precision of approximately 0.1 ppm, insignificant compared to our statistical error.

The second problem encountered is due to an unexpected background lying under the elastic peak in the time spectra. This background has a significant asymmetry which changes over the time spectrum. A great deal of effort has gone into attempting to understand this background, and to devise methods of subtracting it from the elastic scattering. In addition to expected background from inelastic protons and target windows, we believe that the large asymmetry in the background arises from hyperon decay, and a large simulation effort is underway.

Overall we believe that we have achieved the goals of the G0 proposal for the forward angle measurement with the caveat that our understanding, or lack thereof, of the background may increase the overall error bar somewhat, particularly for the higher Q^2 values. Shown in Fig. 5 are preliminary results for the first 12 FPDs for which the background is less severe. These results are blinded by an overall multiplicative factor to reduce possible analysis biases, and should not be interpreted as anything other than an indication of the overall quality and rough magnitude of the asymmetry.

3 G0: Backward angle scattering

The second phase of the G0 experiment is a measurement of the parity-violating asymmetry for backward angle elastic scattering. For these measurements the magnet-detector system is rotated by 180° – a process which has already been completed. In the "backward" mode (first run tentatively scheduled to begin in late 2005), there are a number of differences compared to the forward angle mode. Firstly, elastically scattered electrons are detected rather than recoil protons. The optics of the magnet are such that the scattered electrons are at an average angle of $108°(\pm 10°)$, corresponding to a single value of Q^2 for each beam energy. Therefore, to cover the full range of Q^2, measurements will be required at a variety of incident energies. The G0 collaboration has proposed measurements at three values of Q^2: 0.3, 0.5 and 0.8 $(\mathrm{GeV/c})^2$, corresponding to incident electron energies of 0.434, 0.585 and 0.799 GeV, respectively.

Another major difference is that the elastically scattered electrons are not focused on specific focal plane detectors (FPD), but rather on each focal plane detector there are both elastic and inelastic events. To separate these, a second scintillator array (labeled the Cryostat Exit Detectors or CEDs) will be mounted near the magnet cryostat exit window for each octant. The CED array consists of 9 arch shaped scintillators, similar in shape to the FPDs but smaller, since they lie closer to the target. By recording all possible combinations of coincidences between the 9 CEDs and 14 FPDs, we will be able to separate the elastic and inelastic events. In the worst case the contamination of the elastic events due to inelastic electrons will be less than a few percent. While elastic and inelastic electrons can be separated this way, the FPD/CED combination does not provide any discrimination between electrons and negative pions, both of which have velocities close to that of the speed of light. This is not a serious issue for the hydrogen run, since at the G0 energies the 2-pion production rate is very low, resulting in a contamination of less than a percent even at the highest proposed energy. However, there will be a small π^- rate from the target windows. More importantly the π^- rate from the neutron in the deuterium target will be similar to or greater than that of the electrons. For this reason 8 aerogel Čerenkov detectors have been added to the design to separate electrons and pions. The CED scintillators and light guides have been fabricated at TRIUMF and are currently being assembled on the support structure at JLab. The Čerenkov detectors, provided by the Canadian and Grenoble collaborators, are complete and at JLab awaiting mounting. A cartoon of the backangle setup and its operation are shown in Fig. 6. The diagram on the bottom shows the combinations of CEDs and FPDs and the relative contributions from elastic and inelastic electrons.

Rather than use time-of-flight to identify the particles of interest, we will now simply count the particles detected in every possible FPD/CED combination that did not also fire the Čerenkov. Custom electronics modules which record and sort all possible coincidences between the 9 CEDs and 14 FPDs were designed in Grenoble and Louisiana Tech. Prototypes have been fabricated and tested, and production of the final modules is nearing completion with delivery expected by Spring 2005.

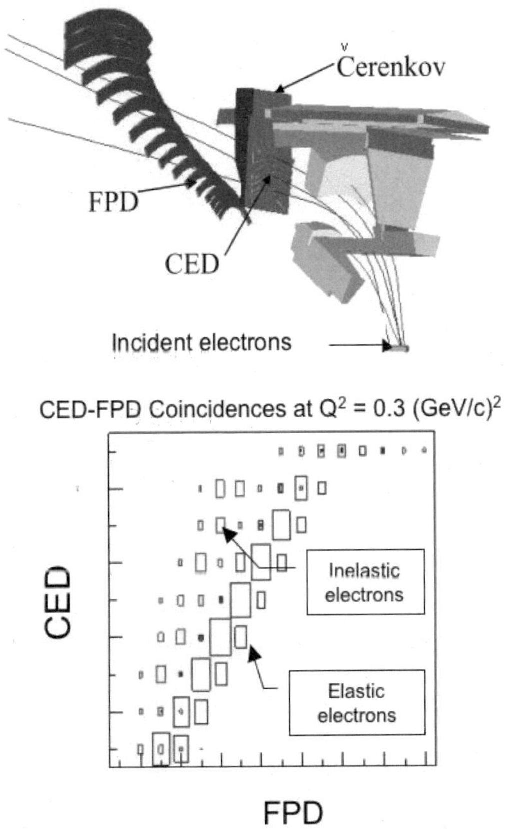

Fig. 6. G0 setup for backangle e-N scattering

With the above differences some distinct benefits have come. Since we no longer will be taking time-of-flight information, we do not require the special 32 MHz time structure needed for the forward angle measurements for which the beam current limit was 40 μA. We therefore plan to make use of the excellent performance of the cryotarget to run with 80 μA of beam current with the standard JLab 2 ns beam structure. This will improve our expected statistics by a factor of two, thereby reducing the statistical error, the largest contributor to the uncertainties at the backward angle.

The inelastically scattered electrons in the G0 backward mode are primarily due to excitation of the Δ resonance. As stated above these events are separated from the elastic events in the space mapped out by different combinations of FPD/CED pairs. As a result, it will be possible to simultaneously measure the parity violation asymmetry in the N-Δ transition [11]. The dominant contribution to the inelastic asymmetry is expected to be from the one-body, axial transition form factor, $G^A_{N\Delta}$. This transition form factor has been measured in charged current reactions, but the G0 measurement would be the first determination of this form factor in the neutral current sector. Since we can tag the pions, the collaboration is also considering a simultaneous measurement of the parity violating asymmetry in pion production.

4 Conclusion

After many years and a great deal of effort on the part of many G0 collaborators, we are seeing the payoff from this work. We have successfully completed the forward angle measurements, as described above, and are embarking on measurements at back angle. With the completion of G0 and the other measurements by the HAPPEX and PVA4 collaborations, we will have a new view into the underlying quark makeup of the nucleon. It promises to be an exciting next few years.

References

1. T. Ito et al.: Phys. Rev. Lett. **92**, 102003 (2004)
2. D.T. Spayde et al.: Phys. Lett. B **583**, 79 (2004)
3. D.T. Spayde et al.: Phys. Rev. Lett. **84**, 1106 (2000)
4. R. Hasty et al.: Science, **290**, 2021 (2000)
5. K. Aniol et al.: Phys. Rev. C **69**, 065501 (2004), arXiV:nucl-ex/0402004. See also ibid, Phys. Lett B **509**, 211 (2001); Phys. Rev. Lett. **82**, 1096 (1999)
6. F. Maas et al.: Phys. Rev. Lett. **93**, 022002 (2004)
7. F. Maas: private communication. Results not yet published but shown at, for example, the PAVI04 workshop in Grenoble, France, June 2004
8. For many details of the G0 experiment see the paper of C. Furget: Proceedings of the International Workshop on Parity Violation and Hadronic Structure, Mainz, Germany, June 2002. eds. F.E. Maas, S. Kox, D. Lhuillier, J. Van de Wiele: (World Scientific 2004) in print
9. M.J. Musolf, B.R. Holstein: Phys. Lett. B **242**, 461 (1990)
10. S.-L. Zhu, S.J. Puglia, B.R. Holstein, M.J. Ramsey-Musolf: Phys. Rev. D **62**, 033008 (2000)
11. Jefferson Lab Experiment E-04-101: T. Forest, K. Johnston, N. Simicevic, S. Wells, Spokespersons, 2004

Eur Phys J A (2005) **24**, s2, 65–65
DOI: 10.1140/epjad/s2005-04-014-2

EPJ A direct
electronic only

Study of the parity violation in the $\Delta(1232)$ region

Response of the A4 calorimeter in the region of the $\Delta(1232)$ resonance using MC simulations

Luigi Capozza[a], for the A4 Collaboration

Institut für Kernphysik, Universität Mainz, J. J. Becherweg 45, D-55099 Mainz, Germany

Received: 1 November 2004 / Published Online: 8 February 2005

Abstract. A measurement of the parity violation (PV) asymmetry in electron-proton scattering using a polarised electron beam is performed at MAMI. The experimental apparatus is able to detect electrons, that are scattered exciting the $\Delta(1232)$ resonance. In order to study the PV in such process, it is important to understand the energy spectrum in the corresponding region. The aim of this work is a simulation including all relevant processes in order to extract an estimate of the PV asymmetry in the $\Delta(1232)$ region.

PACS. 12.15.-y Electroweak interactions – 11.30.Er Charge conjugation, parity, time reversal, and other discrete – 14.20.Gk Baryon resonances with S=0 – 13.40.-f Electromagnetic processes and properties

1 Introduction

Up to now, the measurements of the A4 experiments have been used for studying spin observables in the elastic scattering of polarised electrons off unpolarised protons [1]. Nevertheless, the A4 experiment offers also the possibility to study the inelastic electron scattering off the proton. In fact, both elastically and inelastically scattered electrons are detected simultaneously by the lead fluoride calorimeter. In particular, the A4 detector allows the detection of the scattered electrons exciting the $\Delta(1232)$ resonance. Therefore, the measurement of spin observables in the electroproduction of the $\Delta(1232)$ is possible within the A4 experiment. Among these observables the parity violation (PV) asymmetry can be measured by scattering longitudinally polarised electrons on protons. It has been shown [2, 3,4], that the PV asymmetry in the cross section for the electroexcitation of the $\Delta(1232)$ resonance can yield an important insight into the proton structure.

The scattering cross section asymmetry is obtained, in the A4 experiment, counting the number of electrons scattered with each state of helicity. The energy of the scattered electrons is measured by the lead fluoride calorimeter and histogrammed, giving an energy spectrum. By analysing this energy spectrum, it is possible to distinguish between electrons scattered elastically or inelastically, e.g. exciting the $\Delta(1232)$. For the analysis of the data in the energy region of the $\Delta(1232)$ resonance, a detailed study of the energy spectrum is necessary for having a better knowledge of the background. This is in particular necessary in this energy range, where the ratio of the signal to the background is smaller than in the case of the elastic scattering range. This results from the smaller absolute value of the signal cross section and the larger background.

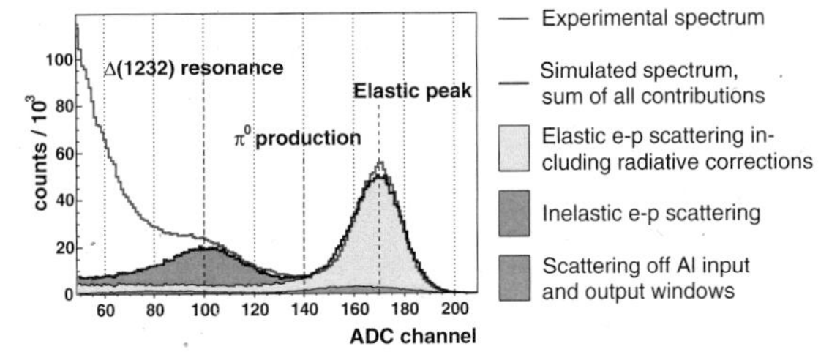

Fig. 1. Comparison of simulated and experimental spectra

Within this work a detailed study of the lead fluoride detector response to the scattered electrons has been carried out by means of Monte Carlo simulations. In addition, an event generator has been implemented, which includes the elastic *e-p* scattering considering radiative corrections, the inelastic *e-p* scattering and the electron scattering off the aluminium input and output windows. These simulations can reproduce quite well the experimental energy spectrum in the range between the pion production threshold and the maximum of the $\Delta(1232)$ peak, as shown in Fig. 1. For this energy region, in principle, the PV asymmetry could be already extracted and compared with the theoretical predictions. More effort has to be provided in order to extend our understanding of the spectrum to include the whole $\Delta(1232)$ peak.

[a] Comprises part of PhD thesis

References

1. F.E. Maas et al.: Phys. Rev. Lett. **93**, 022002 (2004)
2. L.M. Nath et al.: Phys. Rev. D **25**, 9 (1982) 2300–2305
3. H.W. Hammer, D. Drechsel: Z. Phys. A **353**, 321–331 (1995)
4. N.C. Mukhopadhyay et al.: Nucl. Phys. A **633**, 481–518 (1998)

Eur Phys J A (2005) **24**, s2, 67–70
DOI: 10.1140/epjad/s2005-04-015-1

EPJ A direct
electronic only

Don't forget to measure Δs

Stephen F. Pate

Physics Department, New Mexico State University, Las Cruces NM 88003, USA

Received: 15 October 2004 / Published Online: 8 February 2005

Abstract. This talk explores our lack of knowledge of the strange quark contribution to the nucleon spin, Δs. Data on Δs from inclusive and semi-inclusive polarized deep-inelastic scattering will be reviewed, followed by a discussion of how the ongoing program of parity-violating elastic electron-nucleon scattering experiments, that seek out the strange electromagnetic form factors of the nucleon, need to have an estimate for the strange axial form factor to carry out that program, and how the value of Δs extracted from the DIS experiments has filled that role. It is shown that elastic νp, $\bar{\nu} p$, and parity-violating $\vec{e}p$ data can be combined to extract the strange electric, magnetic and axial form factors simultaneously. A proposed experiment that could address this important issue if briefly previewed.

PACS. 14.20.Dh Protons and neutrons – 13.40.Gp Electromagnetic form factors of elementary particles

1 Introduction

In the current experimental program of nucleon structure studies, we find two broad areas of experimentation. On the one hand, elastic scattering of electrons from nucleons is used to measure the electromagnetic and axial form factors of the nucleon, over a range of momentum transfer of $0.1 < Q^2 < 10$ (GeV/c)^2. These experiments have taken place at a variety of laboratories over the years, with the current program focused at MIT-Bates, JLab, and Mainz. One of the highlights of the current program is the emphasis on nailing down the strange quark contributions to the electromagnetic form factors, through the exploitation of the interference between photon and Z-boson exchange processes. On the other hand, deep-inelastic scattering of muons and electrons from nucleon and nuclear targets, historically responsible for the discovery of the partonic structure of matter, continues to play a role in the exploration of the distribution of quarks and gluons in nucleons. One of the highlights here is the focus, over the last 15 years, on the spin structure of the nucleon. The deep-inelastic exploration of nucleon spin takes place now at both leptonic and hadronic facilities, the spin program at RHIC being the most notable example of an hadronic facility taking on this physics topic.

QCD provides a simple framework in which these two experimental programs are joined together. The asymmetries observed in the polarized deep-inelastic scattering experiments arise from the antisymmetric part of the virtual Compton amplitude, which contains at its heart the nucleon axial current, $\bar{q}\gamma_\mu\gamma_5 q$. In the quark-parton model, inclusive scattering of leptons from nucleon targets measures the nucleon structure function F_1,

$$F_1(x) = \frac{1}{2}\sum_q e_q^2 q(x)$$

where e_q and $q(x)$ are respectively the charge and parton distribution function for quarks of flavor q. Inclusive scattering of *polarized* leptons from *polarized* nucleon targets measures the *spin-dependent* nucleon structure function g_1,

$$g_1(x) = \frac{1}{2}\sum_q e_q^2 \Delta q(x)$$

where now $\Delta q(x)$ is a polarized parton distribution function. In QCD, these distribution functions take on a scale dependence: $\Delta q(x, Q^2)$. At the same time, the axial form factors $G_A^q(Q^2)$ measured in elastic scattering are themselves matrix elements of the axial current,

$${}_N\langle p'|\bar{q}\gamma_\mu\gamma_5 q|p\rangle_N = \bar{u}(p')\gamma_\mu\gamma_5 G_A^q(Q^2)u(p)$$

where the matrix element has been taken between two nucleon states of momenta p and p', and $Q^2 = -(p'-p)^2$. The diagonal matrix elements of the axial current are called the axial charges,

$${}_N\langle p|\bar{q}\gamma_\mu\gamma_5 q|p\rangle_N = 2Ms_\mu\Delta q$$

where M and s_μ are respectively the mass and spin vector of the nucleon. The quantities Δq are called "axial charges" because they are the value of the axial form factors at $Q^2 = 0$; that is to say, for example, $G_A^s(Q^2 = 0) = \Delta s$. The connection between the two sets of observables lies in a well-known QCD sum rule for the axial current, namely that the value of the axial form factor at $Q^2 = 0$ is equal to the integral over the polarized parton distribution function measured at $Q^2 = \infty$. For example,

$$\Delta s = G_A^s(Q^2 = 0) = \int_0^1 \Delta s(x, Q^2 = \infty)dx.$$

In this way, the axial charges Δq provide the link between the low-energy elastic scattering measurements of axial form factors and the high-energy deep-inelastic measurements of polarized parton distribution functions.

Of course, there are practical difficulties in the full exploration of this sum rule. No scattering experiment can reach $Q^2 = 0$ or $Q^2 = \infty$, and no deep-inelastic experiment can ever reach $x = 0$. However, the consequences of these difficulties are more severe in some cases than in others. Our inability to reach $Q^2 = \infty$ in the deep-inelastic program means that QCD corrections enter into the sum rule written above, and there is much theoretical experience in calculating these corrections. While the low-energy elastic experiments cannot reach $Q^2 = 0$, there is not expected any divergent behavior of the form factors near $Q^2 = 0$ and so the idea of extrapolating to $Q^2 = 0$ from measurements at low, non-zero Q^2 is not met with alarm. On the other hand, the limitations imposed by our inability to reach $x = 0$ in the deep-inelastic experiments are more problematic. The unpolarized parton distribution functions $q(x)$ are all known to increase rapidly as $x \to 0$ and there is no calculation of the expected behavior near $x = 0$ to rely upon for an extrapolation from measurements made at $x \neq 0$. Similar comments apply to the polarized parton distributions $\Delta q(x)$. Unpolarized measurements of the parton distributions at HERA have reached very low values of x, nearing $x = 3 \times 10^{-5}$, but the corresponding measurements of the polarized distribution functions, from data at SLAC, CERN, and DESY, only reach $x = 3 \times 10^{-3}$. Therefore, measurements of the axial charges place important constraints on the behavior of the distributions $\Delta q(x)$ in the unmeasureable low-x region.

2 Δs from inclusive leptonic deep-inelastic scattering

As mentioned earlier, the double-spin asymmetries in polarized inclusive leptonic deep-inelastic scattering measure the spin-dependent nucleon structure function g_1:

$$g_1(x) = \frac{1}{2} \sum_q e_q^2 \Delta q(x).$$

In leading order QCD, these functions take on a scale dependence:

$$g_1(x, Q^2) = \frac{1}{2} \sum_q e_q^2 \Delta q(x, Q^2).$$

In next-to-leading order (NLO) QCD, there are significant radiative corrections and the relation between g_1 and the Δq becomes more complex. In the discussion here, we will limit our attention to the leading-order QCD analysis because the NLO version of the analysis does not change the result (nor the uncertainty) for Δs very much, and the problems to be pointed out exist at all orders, because they are problems coming from the data itself.

We will use the analysis from the SMC Collaboration [2] as a model. They measured $g_1(x, Q^2)$ over a wide kinematic range, $0.003 < x < 0.70$ and $1.3 < Q^2 < 58.0$. This coverage is not a rectangle, *i.e.* there is a correlation between x and Q^2 in the acceptance of the experiment, and so for a reasonable analysis it is necessary to use QCD to evolve all the data to a single value of Q^2, in this case $Q^2 = 10$ GeV^2. In the process of performing this evolution, a fit function for g_1 is produced. Then, to integrate the distribution g_1 over $0 < x < 1$, it is necessary to extrapolate to $x = 1$ and $x = 0$. The extrapolation to $x = 1$ makes use of the fact that g_1, being a difference of two quark distributions, must go to 0 as $x \to 0$. This requirement is satisfied in this analysis by assuming the measured experimental asymmetry to be constant for $x > 0.7$. The extrapolation to $x = 0$, on the other hand, is not straightforward, as the expected behavior of $g_1(x)$ for $x \to 0$ is unknown. In this analysis, two methods were used. In one, the QCD evolution fit was simply extrapolated to $x = 0$. In another, called the "Regge extrapolation," the value of g_1 was assumed to be constant for $x < 0.003$. The two values of the integral of g_1 from these two extrapolations are

$$\begin{aligned} \Gamma_1 = \int_0^1 g_1(x)dx &= 0.142 \pm 0.017 \quad \text{"Regge"} \\ &= 0.130 \pm 0.017 \quad \text{QCD fit.} \end{aligned}$$

This integral is related to the axial charges:

$$\begin{aligned} \Gamma_1 = \int_0^1 g_1(x)dx &= \frac{1}{2} \sum_q e_q^2 \int_0^1 \Delta q(x)dx \\ &= \frac{1}{2} \left[\frac{4}{9}\Delta u + \frac{1}{9}\Delta d + \frac{1}{9}\Delta s \right]. \end{aligned}$$

Now, assuming that $SU(3)_f$ is a valid symmetry of the baryon octet, and using hyperon β decay data, then two other relations between the three axial charges are determined:

$$\Delta u - \Delta d = g_A = F + D \quad \text{and} \quad \Delta u + \Delta d - 2\Delta s = 3F - D$$

where $g_A = 1.2601 \pm 0.0025$ and $F/D = 0.575 \pm 0.016$ (in 1997). Now one may solve for the axial charges, and the results are shown in Table 1. It is well to note that, of course, the error bars quoted here do not include any estimate of the theoretical uncertainty underlying the assumption of $SU(3)_f$ symmetry. They do include an estimate of the uncertainty due to the extrapolations, but of course that is only an estimate because the actual behavior of g_1 is

Table 1. Results for the axial charges from the SMC analysis [2] of their inclusive DIS data

	"Regge"	QCD fit
Δu	0.84 ± 0.06	0.80 ± 0.06
Δd	-0.42 ± 0.06	-0.46 ± 0.06
Δs	-0.08 ± 0.06	-0.12 ± 0.06

unknown in the $x \to 0$ region. The only conclusion to be drawn for Δs from this analysis is that it may be negative, with a value anywhere in the range 0 to -0.2.

3 $\Delta s(x)$ from semi-inclusive leptonic deep-inelastic scattering

In semi-inclusive deep-inelastic scattering experiments, a leading hadron is observed in coincidence with the scattered lepton, allowing a statistical identification of the struck quark, and thus a measurement of the x-dependence of the individual $\Delta q(x)$ distributions. (Inclusive scattering only measures the total structure function $g_1(x)$.) The HERMES Experiment [3] at DESY was especially designed to make this measurement. HERMES measured double-spin asymmetries in the production of charged hadrons in polarized deep-inelastic scattering of positrons from polarized targets; specifically, the asymmetry in the production of charged pions on targets of hydrogen and deuterium, and of charged kaons in scattering from deuterium. There is no assumption of $SU(3)_f$ symmetry in their analysis. They extract the following quark polarization distributions, over the range $0.023 < x < 0.30$ [4]:

$$\frac{\Delta u}{u}(x) \quad \frac{\Delta d}{d}(x) \quad \frac{\Delta \bar{u}}{\bar{u}}(x) \quad \frac{\Delta \bar{d}}{\bar{d}}(x) \quad \frac{\Delta s}{s}(x)$$

where $\frac{\Delta s}{s}(x)$ is defined to be the sum of $\frac{\Delta s}{s}(x)$ and $\frac{\Delta \bar{s}}{\bar{s}}(x)$.

Within the measured uncertainties, and within the measured x-region, the valence quarks (u and d) are polarized and the sea quarks ($\bar{u}$, $\bar{d}$, and s) are unpolarized. The integral value of the measured polarized strange quark distribution is

$$\text{“}\Delta s\text{”} = \int_{x=0.023}^{0.30} \Delta s(x)dx = +0.03 \pm 0.03(\text{stat}) \pm 0.01(\text{syst}).$$

[Note this would only be the true Δs if the integral was over the range $0 < x < 1$.]

Given the fact that the inclusive analysis described in the previous section produced a negative value of Δs, it is natural to ask "where did the negative Δs go?" If the analyses shown of the inclusive and semi-inclusive data are both correct, then all the negative contribution to the value of Δs must come from the unmeasured x-region, that is from $x < 0.023$. That would imply an average value of $\Delta s(x)$ of approximately -5 in the range $x < 0.023$, which is not impossible, as $s(x)$ is of order 20-300 in the range $0.01 < x < 0.001$ [5]. Some very interesting physics indeed would be revealed, if the "turn on" of the strange quark polarization in the low-x region was this dramatic.

Of course, there are other explanations. The invocation of $SU(3)_f$ symmetry in the analysis of the inclusive data is known to be problematic. And the extrapolations to $x = 0$ in those analyses do not have firm theoretical support. It is clear that a *direct* measurement of Δs would serve to clarify these issues.

4 Parity-violating eN elastic scattering

One of the highlights of the current low- and medium-energy electron scattering program is the measurement of the strange vector form factors of the nucleon via parity-violating eN scattering. These measurements are sensitive as well to the non-strange part of the axial form factor, but rather insensitive to the strange axial form factor due to the relative sizes of kinematic factors multiplying the various form factors that contribute to the asymmetry. To be specific, the parity-violating asymmetry observed in these experiments, when the target is a proton, can be expressed as [6]

$$A_p = \left[\frac{-G_F Q^2}{4\pi\alpha\sqrt{2}}\right] \times \frac{\epsilon G_E^\gamma G_E^Z + \tau G_M^\gamma G_M^Z - (1 - 4\sin^2\theta_W)\epsilon' G_M^\gamma G_A^e}{\epsilon \left(G_E^\gamma\right)^2 + \tau \left(G_M^\gamma\right)^2}$$

where $G^\gamma_{E(M)}$ are the traditional electric (magnetic) form factors of the proton and $G^Z_{E(M)}$ are their weak analogs, $\tau = Q^2/4M_p^2$, M_p is the mass of the proton, $\epsilon = [1 + 2(1+\tau)\tan^2(\theta/2)]^{-1}$, θ is the electron scattering angle, and $\epsilon' = \sqrt{\tau(1+\tau)(1-\epsilon^2)}$. Lastly, G_A^e is the effective axial form factor seen in electron scattering:

$$G_A^e = -G_A^{CC}(1 + R_A^{T=1}) + G_A^s + R_A^{T=0}.$$

Here, G_A^{CC} is the non-strange (CC) axial form factor, G_A^s is the strange axial form factor, and the terms $R_A^{T=0,1}$ represent electroweak radiative corrections [6,7,8,9]. The presence of these radiative corrections clouds the interpretation of the axial term extracted from these experiments. To solve this problem, the SAMPLE [10] experiment measured also the same asymmetry on a deuterium target, in which case the relative kinematic factors of the non-strange ($T = 1$) and strange ($T = 0$) parts of the axial form factor are changed, allowing a separation of the two. However, one may show that this does not help in identifying the value of G_A^s, because the relative size of the kinematic factors for G_M^s and G_A^s remain the same for either target:

$$\frac{\partial G_M^s}{\partial G_A^s} = -(1 - 4\sin^2\theta_W)\frac{\epsilon'}{\tau} \approx -\frac{1}{2} \text{ for SAMPLE.}$$

Therefore, parity-violating eN scattering experiments can only establish a relationship between the strange magnetic and axial form factors, they cannot measure them separately.

5 Combining νN and parity-violating eN elastic data

Recently it has been demonstrated [11] that the best (perhaps only) way to obtain all three strange quark form factors (electric, magnetic and axial) is through a combination of low energy νN and parity violating eN elastic

Table 2. Two solutions for the strange form factors at $Q^2 = 0.5$ GeV^2 produced from the E734 and HAPPEX data. (Table from [11])

	Solution 1	Solution 2
G_E^s	0.02 ± 0.09	0.37 ± 0.04
G_M^s	0.00 ± 0.21	-0.87 ± 0.11
G_A^s	-0.09 ± 0.05	0.28 ± 0.10

scattering. This prescription is briefly summarized here. Using the known values for the electric, magnetic and non-strange (CC) axial form factors of the proton and neutron, one may take the difference of the νp and $\bar{\nu} p$ elastic cross sections and show it to be a function only of the strange magnetic and axial form factors, G_M^s and G_A^s. At the same time, the sum of the νp and $\bar{\nu} p$ elastic cross sections can be shown to be a function only of the strange electric and magnetic form factors, G_E^s and G_M^s. Measurements of forward-angle parity-violating $\mathbf{e}p$ elastic scattering are largely functions only of G_E^s and G_M^s as well. Therefore, combining these three kinds of data can determine all three strange form factors. At the present time, there is only sufficient data at $Q^2 = 0.5$ GeV^2 to make such a determination. In [11] it is shown that by combining the E734 [12] results with the HAPPEX [13] data from JLab, there are two possible solutions at $Q^2 = 0.5$ GeV^2, as summarized in Table 2. In the long run, additional experimentation (already on the schedule at JLab) will select one of these solution sets, but there are already several good reasons to favor Solution 1 over Solution 2. The value of the strange electric form factor G_E^s must approach zero as $Q^2 \to 0$, and Solution 1 is consistent with that requirement. The value of G_A^s in Solution 1 is consistent with the estimated value of $G_A^s(Q^2 = 0) = \Delta s \approx -0.1$ from deep-inelastic data, whereas that found in Solution 2 is much larger and of a different sign. The value of G_M^s in Solution 1 is consistent with that measured by SAMPLE [10] at $Q^2 = 0.1$ (GeV/c)^2, whereas the value in Solution 2 is much larger in magnitude. Finally, the value of G_M^s in Solution 1 is consistent with the value of $G_M^s(Q^2 = 0) = -0.051 \pm 0.021 \mu_N$ predicted recently from lattice QCD [14]. It seems nearly certain that Solution 1 is the true physical solution. Future experimentation will in any event select the correct solution set.

Additional data from the G^0 Experiment [15], recently collected and in the process of analysis, will allow the extraction of the three strange form factors when combined with the E734 data. However, it is unlikely that knowledge of the strange axial form factor over the range $0.5 < Q^2 < 1.0$ GeV^2 will prove sufficient for the extrapolation to $Q^2 = 0$ needed for a determination of Δs. New neutrino data are needed at lower Q^2 to permit a good determination of Δs.

6 A future experiment to measure the strange axial charge

Even if the program I have described determines the strange axial form factor down to $Q^2 = 0.45$ GeV^2 successfully, it almost certainly will not determine the Q^2-dependence sufficiently for an extrapolation down to $Q^2 = 0$. Also, questions remain about the normalization of the E734 data. Most of their target protons were inside of carbon nuclei, and there was not much known about nuclear transparency in the mid-1980's. The E734 collaboration did make a correction for transparency effects, but this issue needs to be revisited if we continue to use the E734 data.

A new experiment [16] has been proposed to measure elastic and quasi-elastic neutrino-nucleon scattering to sufficiently low Q^2 to determine the strange axial charge, Δs. The FINeSSE Collaboration proposes to measure the NC to CC neutrino scattering ratio

$$R_{NC/CC} = \frac{\sigma(\nu p \to \nu p)}{\sigma(\nu n \to \mu^- p)}$$

and from it extract the strange axial form factor down to $Q^2 = 0.2$ GeV^2. The numerator in this ratio is sensitive to the full axial form factor, $-G_A^{CC} + G_A^s$, while the denominator is sensitive to only G_A^{CC}. The processes in the numerator and denominator have unique charged particle final state signatures. This ratio is largely insensitive to uncertainties in neutrino flux, detector efficiency and nuclear target effects. A 6% measurement of $R_{NC/CC}$ down to $Q^2 = 0.2$ GeV^2 provides a ± 0.04 measurement of Δs.

Acknowledgements. This work was supported by the US Department of Energy.

References

1. L.A. Ahrens et al.: Phys. Rev. D **34**, 75 (1986)
2. B. Adeva et al.: Phys. Lett. B **412**, 414 (1997)
3. K. Ackerstaff et al.: Nucl. Instrum. Meth. A **417**, 230 (1998)
4. A. Airapetian et al.: Phys. Rev. Lett. **92**, 012005 (2004)
5. J. Pumplin et al.: JHEP **0207**, 012 (2002)
6. M.J. Musolf et al.: Phys. Rep. **239**, 1 (1994)
7. M.J. Musolf, B.R. Holstein: Phys. Lett. B **242**, 461 (1990)
8. M.J. Musolf, T.W. Donnelly: Nucl. Phys. A **546**, 509 (1992)
9. S.-L. Zhu et al.: Phys. Rev. D **66**, 034021 (2002)
10. D.T. Spayde et al.: Phys. Lett. B **583**, 79 (2004); T.M. Ito et al.: Phys. Rev. Lett. **92**, 102003 (2004)
11. S.F. Pate: Phys. Rev. Lett. **92**, 082002 (2004)
12. L.A. Ahrens et al.: Phys. Rev. D **35**, 785 (1987)
13. K.A. Aniol et al.: Phys. Lett. B **509**, 211 (2001)
14. D.B. Leinweber et al.: submitted to Phys. Rev. Lett. [arXiv:hep-lat/0406002]
15. G^0 Experiment, `http://www.npl.uiuc.edu/exp/G0/`, Doug Beck, spokesperson
16. FINeSSE Proposal, B. Fleming, R. Tayloe et al.: [arXiv:hep-ex/0402007]

III Weak form factors of the nucleon

III-2 Theory

Eur Phys J A (2005) **24**, s2, 73–78
DOI: 10.1140/epjad/s2005-04-016-0

EPJ A direct
electronic only

Getting to grips with hadrons

Damir Becirevic

Laboratoire de Physique Theorique (Unité mixte de Recherche du CNRS - UMR8627),
Université Paris Sud, F-91405 Orsay Cedex, France

Received: 7 December 2004 / Published Online: 8 February 2005

Abstract. A short discussion concerning the lattice QCD approach to physics of hadrons is made to non-specialists. A special attention is given to topics that are of particular interest to the nuclear physics community.

PACS. 11.15.Ha Lattice Gauge Theory – 12.39.Fe Chiral Lagrangians

1 Why lattice QCD?

One of the most important questions that still remains to be answered is to explain how hadrons arise from the QCD lagrangian

$$\mathcal{L}_{\rm QCD} = \frac{1}{4} F^a_{\mu\nu} F^a_{\mu\nu} + \sum_{q=u,d,s,\dots} \bar{q}\left\{\gamma_\mu \left(\partial_\mu + g A^a_\mu t^a\right) + m_q\right\} q\,. \quad (1)$$

The dynamics that governs the confinement of quarks and gluons into hadrons is of notoriously nonperturbative nature for which an analytic treatment is still missing. Although various quark models help understanding quite a number of phenomena of hadronic interactions, it should be stressed that a covariant quark model that solves simultaneously confinement and spontaneous chiral symmetry breaking has never been constructed. Other than quark models, much effort has been put in building the effective theories of QCD, valid for specific ranges of low energy scale. Those theories are built upon some symmetry property of the QCD lagrangian in some specific limit. The most prominent example is the chiral (left↔right) symmetry, $SU(N_f)_L \otimes SU(N_f)_R$, that is manifest when the quarks are massless. That symmetry is spontaneously broken down to $SU(N_f)_V$, resulting in the appearance of $N_f^2 - 1$ Goldstone bosons ('pions'). Chiral perturbation theory (ChPT) provides us with an effective description of QCD that incorporates these features and, in addition, allows one to account for the explicit chiral symmetry breaking corrections, namely those generated by the non-zero quark mass terms in the QCD lagrangian. The computation of such corrections, unfortunately, generates a bunch of low energy constants that are supposed to be obtained from the matching procedure of appropriately chosen amplitudes computed both in ChPT and in QCD, at some energy scale at which ChPT can be trusted and at which direct QCD computations can be made. This is where lattice QCD is expected to provide information to the QCD piece in this matching. In the above discussion N_f stands for the number of light quark flavors. Today we are confident that ChPT provides an adequate framework to describe the dynamics of strange-less hadrons ($N_f = 2$), whereas the situation with the strange quark (m_s) is still unclear [1]. This is not only because m_s is about 25 times larger than $m_q = (m_u + m_d)/2m_u$ [2], but also because it is not much smaller than the hadronic QCD scale, $\Lambda_{\rm QCD}^{\overline{\rm MS}\,(N_f=3)} = 336^{+42}_{-38}$ MeV [3]. What do we know about m_s? This is one of the highlights of the lattice QCD achievements over the past decade which is why I decided to briefly discuss it here. That discussion will also allow me to introduce the methodology but also the challenges of lattice QCD.

1.1 Lattice QCD and the strange quark mass

The numerical solution to the problem in hands, namely to compute the hadronic spectra numerically from the QCD lagrangian only (1), does exist. The crucial first step in that direction is to make the analytic continuation to the euclidean metric ($x_0^M \to i x_0^E$), in which the QCD generating functional reads

$$Z[A,q,\bar{q}] = \int \mathcal{D}A\mathcal{D}q\mathcal{D}\bar{q} \exp\{-S[A,q,\bar{q}]\}\,. \quad (2)$$

In euclidean space the QCD action is real and bounded from below. Discretization of the euclidean space and time, $L = N_S a$ and $T = N_t a$, allows for an equivalence between the generating functional and the partition function, so that the Monte Carlo methods can be employed. $SU(3)$ gauge fields are placed on the links of the lattice whereas the quark degrees of freedom are sitting on the sites. A particularly important feature, while discretizing the QCD action, is that the gauge invariance is preserved at every stage of calculation. The price to pay is that the lattice spacing $a \neq 0$ breaks the Lorentz invariance, which is however recovered once we take the continuum limit, $a \to 0$ (i.e. after we send the UV cut-off to infinity). Finally, after the continuum limit has been taken appropriately, we should worry about the finite volume effects and work out the limit $L, T \to \infty$ (i.e. send the IR regulator to zero). This is a very challenging task for numerical simulations, and it requires a lot of clever ideas and a huge computing power. What is important to retain is that –in principle– the QCD simulations on the lattice offer a first principle approach to the physics of hadrons. In other words the solution to nonperturbative QCD is provided without introducing any additional parameter except for those that appear in the QCD lagrangian, namely the quark masses and the $SU(3)_c$ gauge coupling. In practice, however, various approximations are often necessary in order to make the calculation feasible on the present day computing resources. Importantly though, all those approximations are controllable and, for the most part, we can get rid of them by increasing the computing power. The most infamous (least controllable) is the so-called quenched approximation. It consists in neglecting the dynamical quark loops while producing the gauge field configuration. This is certainly a serious drawback of the most of currently available lattice results, but it nevertheless make a good case for developing the methodology for the computation of various physical quantities on the lattice. One way to compute the quark mass on the lattice is via the axial Ward identity [1], $\partial_\mu A_\mu(x) = 2m_q P_5(x)$. One computes the following two correlation functions:

$$\langle \sum_{\mathbf{x}} \partial_\mu \underbrace{\bar{q}(x)\gamma_\mu\gamma_5 q(x)}_{A_\mu(x)}\ \mathcal{O}(0)\rangle \quad \text{and} \quad \langle \sum_{\mathbf{x}} \underbrace{\bar{q}(x)\gamma_5 q(x)}_{P_5(x)}\ \mathcal{O}(0)\rangle \ ,$$

where $\mathcal{O}$ is a bilinear quark operator with quantum numbers $J^P = 0^-$, and after having properly renormalized the axial current and the pseudoscalar density, the ratio of these two correlation functions gives the quark mass. Various ways to nonperturbatively renormalize the composite quark operators on the lattice have been developed (see [4]) and they are implemented in most of the present day quark mass calculations. Besides the ratio of the above correlation functions, from the exponential dependence of the second correlator one can read off the corresponding pseudoscalar meson mass. At this point it should be stressed that the lattice results are consistent with the Gell-Mann–Oakes–Renner (GMOR) formula, $m_{PS}^2 = 2B_0 m_q$. Surprisingly, however, although the GMOR formula is expected to be valid for very small quark masses (it receives the m_q^2-corrections and higher), the lattice QCD results (with Wilson fermions) display a rather impressive consistency with the leading GMOR formula while working with heavy pions ($m_{PS}^2 \geq 500$ MeV). The strategy to reach the physical quark mass is to tune the quark mass in the QCD action in such a way that the corresponding pseudoscalar meson mass coincides with the physical kaon mass. The resulting strange quark mass

[1] For alternative strategies and lattice actions to compute the strange quark mass, please see [5].

Table 1. Strange quark mass obtained from the quenched QCD simulations on the lattice. Results by various collaborations [6] refer to the continuum limit ($a \to 0$)

collaboration	$m_s^{\overline{\rm MS}}(2\ \text{GeV})$
JLQCD	106(7) MeV
Alpha & UKQCD	97(4) MeV
QCDSF	105(4) MeV
CP-PACS	111^{+3}_{-4} MeV
SPQcdR	106(2)(8) MeV

by various lattice collaborations have been obtained by means of high statistics simulations, by implementing the non-perturbative renormalization on fine grained lattices, so that the continuum limit could be taken. Finite volume effects have also been examined and shown to be tiny, i.e., at the level much smaller than the errors they quote. The results, listed in Table 1, are obtained in the quenched approximation. Important qualitative outcome from the lattice studies is that the quark masses are indeed small and that the light hadron masses are mostly due to QCD interaction rather than to their valence quark content. Finally notice that the first lattice studies in which the effects of dynamical quarks are included show that the strange quark mass gets even smaller [7], but we are not yet at the stage of providing the precision unquenched computation of m_s.

2 Hadron spectrum

Lattice QCD is particularly well suited to study the spectra of hadrons. In the previous section we already mentioned that the pseudoscalar meson masses were necessary to identify the strange quark mass. One can also study the correlation functions with the interpolating bilinear quark operators carrying other quantum numbers and thus extract the vector, axial-vector, tensor and even scalar mesons (for the last the valence quarks should be non-degenerate in order to have the correlator with a discernible signal).

2.1 Glueballs

In spite of the quenched approximation, some long standing problems can still be tackled. One such a problem is the existence of the mass gap in the pure Yang–Mills theory. This problem is stated as one of Seven Millenium Math Prize Problems [9], to which an analytical solution is missing. On the other hand, many lattice analyses performed so far show that the glueball states indeed exist. Nowadays even the spectrum of such states has been established. This is a very important prediction of lattice QCD. The spectrum shown in Fig. 1, is given in multiples of r_0, a quantity that is defined from the condition $r \cdot dV(r)/dr|_{r=r_0} = 1.65$ [10], where $V(r)$ is the potential

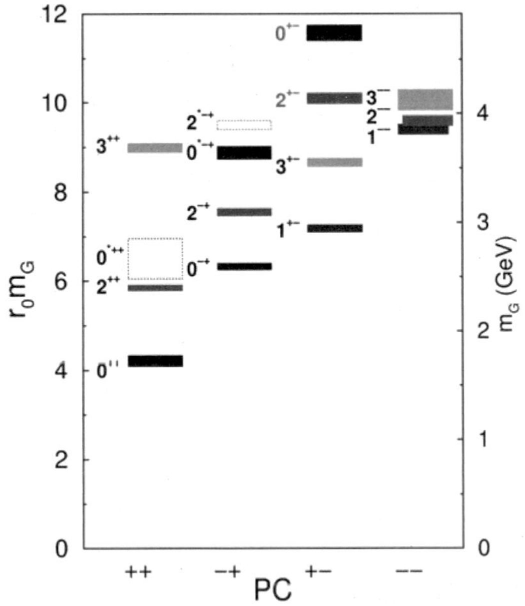

Fig. 1. The spectrum of glueball states as established from the extensive quenched lattice QCD simulations in [8]. The widths of the lines reflect the *error bars* of lattice results

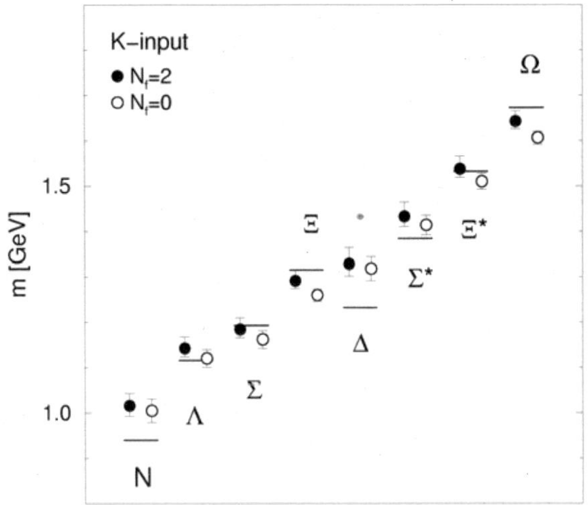

Fig. 2. The spectrum of baryons produced by JLQCD both in quenched ($N_f = 0$) and in unquenched ($N_f = 2$) QCD [11]. Physical (experimentally established) masses [12] are marked by the *horizontal lines*

between two infinitely heavy quarks, which can be (and has been) accurately studied on the lattice. To convert to physical units, a commonly assumed value is $r_0 = 0.5$ fm (or $r_0 = 2.5\ \mathrm{GeV}^{-1}$).

2.2 Baryons

As we already mentioned, from the exponential fall-off of various correlation functions (made with various interpolating operators consisting of quark and gluon fields), one can extract the hadron masses with quantum numbers carried by the considered composite operator. The operators used to extract the proton mass (and its coupling to these interpolating operator) are

$$J(x) = \varepsilon^{abc}\left[u_a^T(x) C\gamma_5 d_b(x)\right] u_c(x),$$
$$\tilde{J}(x) = \varepsilon^{abc}\left[u_a^T(x) C d_b(x)\right] \gamma_5 u_c(x), \qquad (3)$$

where C stands for the charge conjugation operator. Neutron mass is simply obtained by replacing one u quark by d, whereas the Ξ state arise after replacing u and d by two s quarks, and so on. The spectrum of lowest baryon states computed on the lattice is shown in Fig. 2. Strange quark mass is fixed from the physical kaon mass, as explained in the previous section. The most striking feature of that plot is that the baryon spectrum, as deduced from the quenched simulations is essentially unchanged after unquenching the QCD vacuum fluctuation by $N_f = 2$ dynamical quarks. This probably indicates that the most significant effect of quenching has been absorbed in the conversion of results from the lattice to physical units. Second important feature is the nucleon mass that in both cases is larger than that of the physical nucleon. To discuss the reasons for that effect we should stress that the nucleon mass is not obtained directly on the lattice but rather after a long extrapolation. This is so because the lattice simulations are performed with the light quarks $m_q \geq m_s^{\mathrm{phys}}/2$, with $m_q \equiv m_u = m_d$, while the physical limit is $m_q/m_s = 0.04$. Since the sector of light quark masses over which one has to extrapolate is expected to be highly sensitive to the effects of spontaneous chiral symmetry breaking, it is not enough to extrapolate the linear (or quadratic) quark mass dependence observed with the directly accessible quarks (i.e. in the 'heavy pion world'). Therefore, the task, that the lattice community approached very seriously, is to reduce the value of simulated quark ('pion') masses. The trouble is that reduction is very costly in computing power. Even if we manage to create clever algorithms to work closer to the chiral limit the artifacts due to finiteness of the lattice size (L) become more pronounced and the chiral extrapolations should be made by using the formulae derived by using the chiral perturbation theory in the finite volume. That issue attracted quite a bit of attention in the lattice community over the past couple of years [13]. In the case of nucleon, the leading order chiral lagrangian

$$\mathcal{L}_N^{(1)} = \bar{\Psi}\left(i\gamma_\mu D^\mu - m_0\right)\Psi + \frac{1}{2} g_A \bar{\Psi}\gamma_\mu\gamma_5\xi^\mu\Psi, \qquad (4)$$

is consisted of the nucleon field Ψ, the covariant derivative $D_\mu = \partial_\mu + \frac{1}{2}[\xi^\dagger, \partial_\mu\xi]$, in which the Goldstone boson fields enter via $\Sigma = \xi^2$, and $\xi_\mu = i\xi^\dagger\partial_\mu\Sigma\xi^\dagger$. The standard axial coupling is used, $g_A = 1.2$. One-loop chiral corrections to the self energy of the nucleon propagator produce the shift to the bare nucleon mass, m_0, as

$$m_N = m_0 - 4c_1 m_\pi^2 - \frac{3\pi g_A^2}{(4\pi f_\pi)^2} m_\pi^3$$
$$+ \left[C(\Lambda) - \frac{3g_A^2}{2m_0}\frac{1}{(4\pi f_\pi)^2}\left(1 + \log\frac{m_\pi^2}{\Lambda^2}\right)\right] m_\pi^4 + \ldots \qquad (5)$$

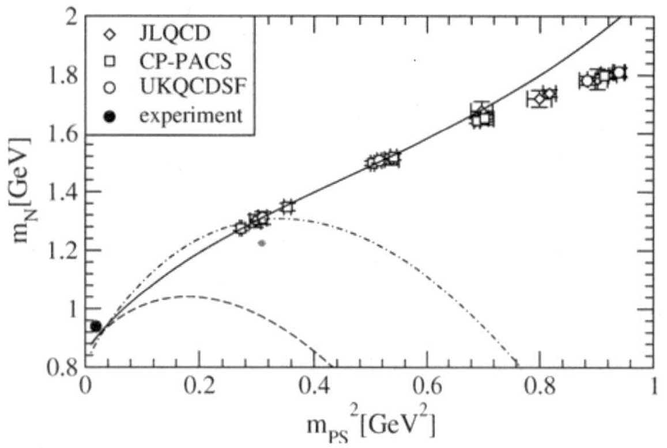

Fig. 3. The chiral extrapolation of the lattice QCD results with $N_f = 2$. *Solid* and *dashed curves* correspond to the so-called infrared regularization and to the non-relativistic treatment of the nucleon. For more information, please see [15]

where $C(\Lambda)$ is the counter-term which cancels the Λ-dependence that otherwise arises from the renormalization of the ultraviolet divergences in the chiral loops. It turns out, however, that the two conventional descriptions lead to quite different results when applied to the lattice data with $N_f = 2$, and that they coincide only for very light pions, namely $m_\pi \ll m_N$ [14,15]. This is shown in Fig. 3 where the solid curve indicates the fit to lattice data and by using the expressions derived by means of the so-called infrared regularization [16]. The non-relativistic treatment [17] is depicted by the two dashed lines corresponding to two specific choices of parameters c_1 and m_0. Besides providing the guidelines to extrapolate to the physically interesting limit for nucleons, ChPT is also useful in estimating the impact of the effects of the finiteness of the lattice box. It verifies the general Lüscher formula [18] and provides the corrections to it. That highly useful aspect of ChPT has been extended to other quantities involving baryons [19]. Important to retain is that the chiral logarithmic behavior gets enhanced by the finiteness of the lattice box. This makes the chiral extrapolations ever more delicate: physical chiral logarithms are mostly drowned in the finite size effects, and therefore one does not only seek the range of quark masses in which the ChPT are valid description of the lattice data, but one also has to disentangle the finite size effects from physical chiral logarithms (former being often overwhelming compared to the size of the latter).

Finally, it should be noted that the above discussion refers to the so-called p-regime (i.e., with $m_\pi L \gg 1$). A probably viable alternative has been recently proposed in [20] where it is claimed that the ϵ-regime ($m_\pi L \ll 1$) might be helpful in discerning the finite volume behavior and thus extract the terms involving the low energy constants, such as $C(\Lambda)$ in (5). Clearly, whatever the regime is considered (p or ϵ), one should make sure that the ChPT formulas provide the fiducial description of the lattice data, and only then claim the extraction of the low energy constants reliable. That is very difficult and very CPU-consuming, so that the question of reliable chiral extrapolation will remain a very hot research topic in the lattice community for quite some time.

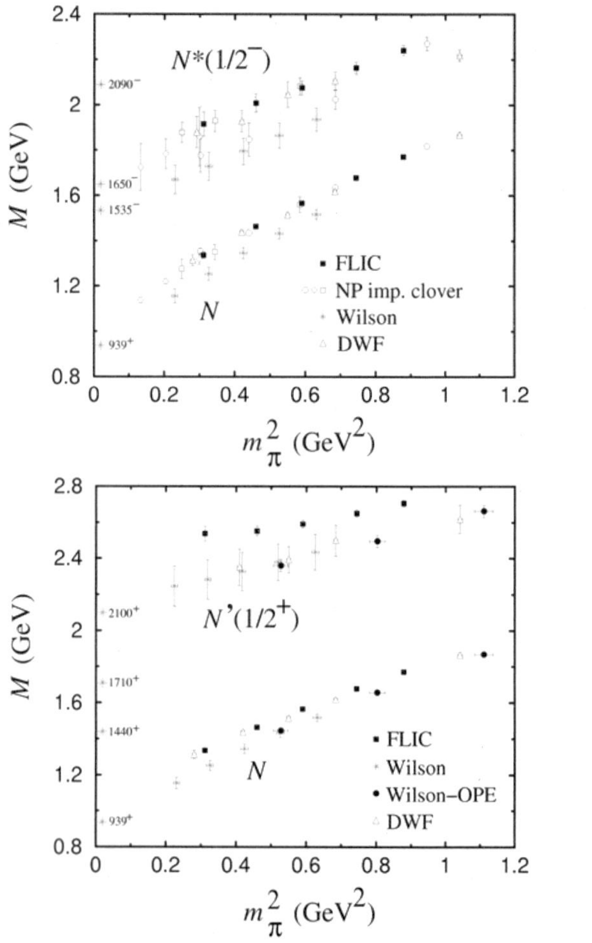

Fig. 4. Results for the orbital (*upper plot*) and radial (*lower plot*) excitations as a function of the pion mass directly accessible from the quenched QCD simulations on the lattice. For more details please see [21]

2.3 Fuss about the Roper resonance

The experimental phenomenon for which the hadron physics phenomenologists do not have a viable explanation is the lightness of the radially excited state with quantum numbers of the nucleon ($J^P = \frac{1}{2}^+$), also known as Roper resonance. The puzzle is that its mass, $m_{N'} \approx 1.44$ GeV, is smaller than that of the first orbital excitation, $m_{\tilde{N}} \approx 1.535$ GeV, which contradicts most of the mass formulae derived by various forms of constituent quark model which suggest $m_N(\frac{1}{2}^+) < m_{\tilde{N}}(\frac{1}{2}^-) < m_{N'}(\frac{1}{2}^+)$. That issue attracted quite a bit of attention recently since some lattice QCD simulations claimed to have found the solution to that puzzle. A special care should be devoted to choosing a good interpolating field, i.e. the one that allows a good overlap with the orbital excitation $J^P = \frac{1}{2}^-$ (also known as S_{11}), as discussed in [21]. The results of that paper, along with those provided by other lattice groups [22], are shown in Fig. 4. In the region of quark masses directly accessed from the lattice studies, the mass pattern is consistent with quark models, i.e., $m_N(\frac{1}{2}^+)$

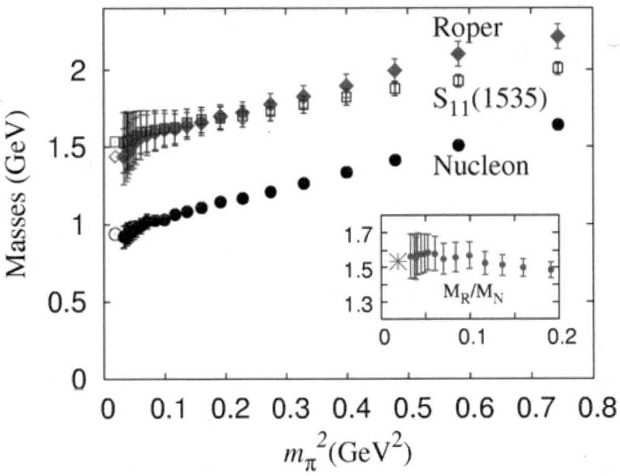

Fig. 5. Results of [23]. Plot similar to those shown in Fig. 4

$< m_{\tilde{N}}(\frac{1}{2}^-) < m_{N'}(\frac{1}{2}^+)$, although the mass difference, $m_{N'}(\frac{1}{2}^+) - m_{\tilde{N}}(\frac{1}{2}^-)$, appears to be smaller as the quark mass is lowered. If naively extrapolated to the chiral limit, the level crossing can be envisaged too, i.e. after extrapolation $m_{N'}(\frac{1}{2}^+) - m_{\tilde{N}}(\frac{1}{2}^-)$ may become negative. That is what [23] claimed to see from their lattice study (see Fig. 5). However, as we saw with the nucleon mass, one should make sure that the finite volume effects and the chiral behavior is well under control. While for the nucleon mass the help in that respect comes from ChPT, there is no such effective theory that provides a similar help for the nucleon excitations. For those reasons I personally think that one should make a better control over various sources of systematic uncertainties in the lattice study of excited nucleons before claiming to observe the level crossing between the nucleon's orbital and radial excitations. This is the point at which one should mention that the clean extraction of the mass of radially excited state has been for a long time a subject to controversies in the lattice community. A recent proposal of [24] provides a possible remedy. The idea is simple and it consists to consider the standard correlation function

$$G(t) = \langle \sum_{\mathbf{x}} J(x) J^\dagger(0) \rangle = \sum_{i \geq 0}^{\infty} Z_i^2 e^{-m_i t} . \tag{6}$$

We need not only to disentangle the excitations from the leading/dominant contribution (i.e. the one to the ground state), but – of all excitations– we want to isolate the piece corresponding to the first radial excitation only. The proposal of [24] is to consider

$$\begin{aligned}\hat{G}(t) &= G(t+1)G(t-1) - G(t)G(t) \\ &= 2 \sum_{j>i=0}^{\infty} Z_i Z_j e^{-(m_i+m_j)t} ,\end{aligned} \tag{7}$$

so that for large time separation the ground state contributes less, and the first radial excitation is then more accessible. The first numerical studies also seem to be encouraging in that respect.

3 Generalized parton distributions (GPD)

Lattice QCD also offers the possibility to study the matrix elements of the local operators sandwiched by the hadron states. This is particularly important for the studies of the CP-violation in the Standard Model. For the recent review on that topic, see [25].

A particularly interesting case to the nuclear physics community is the possibility to get some information about the GPD's of the nucleon. In particular, the matrix elements needed to study the 1st moment have been studied by two lattice groups [26,27].

$$\begin{aligned}\langle P'|O^q_{\{\mu\nu\}}|P\rangle &\equiv \frac{i}{2}\langle P'|\bar{q}\gamma_{\{\mu}\overleftrightarrow{D}_{\nu\}}q|P\rangle \\ &= A_2^q(\Delta^2)\bar{u}(p')\gamma_{\{\mu}\bar{p}_{\nu\}}u(p) \\ &\quad - B_2^q(\Delta^2)\frac{i}{2m_N}\bar{u}(p')\Delta^\alpha\sigma_{\alpha\{\mu}\bar{p}_{\nu\}}u(p) \\ &\quad + C_2^q(\Delta^2)\frac{1}{m_N}\bar{u}(p')u(p)\Delta_{\{\mu}\Delta_{\nu\}} ,\end{aligned} \tag{8}$$

where $\Delta = p - p'$, and the operator is traceless and symmetrised over the indices in the curly brackets. The form factors A_2, B_2 and C_2 can be extracted for several kinematic configurations which then allows one to study their Δ^2-dependence. Both groups fit their data ($X = A_2, B_2, C_2$) to a dipole ansatz

$$X(\Delta^2) = \frac{X(0)}{(1 - M_X^2/\Delta^2)^2} , \tag{9}$$

that unfortunately does not provide us with more insight in physical mechanism that governs the Δ^2-dependence. [2] It should be stressed, however, that the calculation of the above matrix elements on the lattice is demanding for many reasons. One of the most involving problems is renormalization of the operators on the lattice containing the covariant derivative. The reason is that at non-zero lattice spacing the Lorentz group in Euclidean space $SO(4)$ is replaced by the group of discrete hypercubic rotations $H(4)$, which additionally complicates the renormalization mixing pattern among various combinations of operators with covariant derivatives.

Particularly interesting physics information from the first lattice QCD studies of GPD's is the fraction of the total angular momentum of the nucleon carried by the valence quarks. The total angular momentum of a quark q in the nucleon can be expressed via [28]

$$J_q = \frac{1}{2}\left[A_2^q(0) + B_2^q(0)\right] . \tag{10}$$

In [26], the following values have been reported: $A_2^u(0) = 0.40(2)$, $A_2^d(0) = 0.15(1)$, and $B_2^u(0) = 0.33(11)$, $B_2^d(0) = -0.23(8)$, and therefore $J_u = 0.37(6)$ and $J_d = -0.04(4)$.

[2] Resonances in the crossed t-channel are poles in the dispersion relations for these form factors. Apart from the convenient fit formula, no reasonable physical significance could be given to the resulting M_X^{dipole}.

In other words, about 30% of the (quenched) proton's angular momentum is carried by the gluons. The situation in the world with $N_f = 2$ dynamical quarks seems to show that the fraction of the total angular momentum carried by the valence quarks is even smaller. A more definite conclusion on this issue necessitates a more control over the chiral extrapolations that are involved in these calculations.

4 Conclusion

In conclusion I would like to point out the main benefits of the lattice QCD approach:

- Lattice QCD offers a unique possibility to study the physics of hadrons on the basis of the QCD lagrangian only.
- The high statistics numerical simulations of QCD on the lattice have so far been done in the so called quenched approximation. Nowadays more and more studies are made by including the effects of dynamical quarks.
- The methodology to extract the physically interesting information from the data produced on the lattice is developed in the world with heavy pions. It is highly important to extend the range of directly accessible quark masses on the lattice to lighter ones in order to confront the quark mass dependence observed on the lattice with the expressions obtained by means of ChPT.
- Many phenomenologically relevant question in particle physics have been studied by using lattice QCD. If we are to make the precision calculation of hadronic quantities on the lattice, we first need to solve the above mentioned problems. More computing power, better algorithms, more clever physical ideas and the combination of all three aspects are essential in reaching that goal.

References

1. S. Descotes-Genon, N.H. Fuchs, L. Girlanda, J. Stern: Eur. Phys. J. C **34**, 201 (2004)
2. H. Leutwyler: Phys. Lett. B **378**, 313 (1996)
3. S. Bethke: Nucl. Phys. Proc. Suppl. **121**, 74 (2003)
4. M. Luscher et al.: Nucl. Phys. B **384**, 168 (1992) [hep-lat/9207009]; K. Jansen et al.: Phys. Lett. B **372**, 275 (1996) [hep-lat/9512009]; G. Martinelli et al.: Nucl. Phys. B **445**, 81 (1995) [hep-lat/9411010]; M. Gockeler et al.: Nucl. Phys. B **544**, 699 (1999) [hep-lat/9807044]
5. R. Gupta: eConf C0304052 (2003) WG503 [hep-ph/0311033]
6. S. Aoki et al., [JLQCD Collaboration]: Phys. Rev. Lett. **82**, 4392 (1999) [hep-lat/9901019]; J. Garden et al., [ALPHA Collaboration]: Nucl. Phys.B **571**, 237 (2000) [hep-lat/9906013]; M. Gockeler et al.: Phys. Rev. D **62**, 054504 (2000) [hep-lat/9908005]; D. Becirevic, V. Lubicz, C. Tarantino, [SPQ(CD)R Collaboration]: Phys. Lett. B **558**, 69 (2003) [hep-lat/0208003]; S. Aoki et al., [CP-PACS Collaboration]: Phys. Rev. D **67**, 034503 (2003) [hep-lat/0206009]
7. P.E.L. Rakow: hep-lat/0411036
8. C.J. Morningstar, M.J. Peardon: Phys. Rev. D **60**, 034509 (1999)
9. Please see the following URL: `http://www.claymath.org/millennium/Yang-Mills_Theory/Official_Problem_Description.pdf`
10. R. Sommer: Nucl. Phys.B **411**, 839 (1994) [hep-lat/9310022]
11. S. Aoki et al., [JLQCD Collaboration]: Phys. Rev. D **68**, 054502 (2003) [hep-lat/0212039]
12. S. Eidelman et al., [Particle Data Group Collaboration]: Phys. Lett. B **592**, 1 (2004)
13. G. Colangelo, S. Durr: Eur. Phys. J. C **33**, 543 (2004) [hep-lat/0311023]; D. Becirevic, G. Villadoro: Phys. Rev. D **69**, 054010 (2004) [hep-lat/0311028]; D. Arndt, C.J.D. Lin: Phys. Rev. D **70**, 014503 (2004) [hep-lat/0403012]; G. Colangelo, C. Haefeli: Phys. Lett. B **590**, 258 (2004) [hep-lat/0403025]; C.J.D. Lin et al.: Phys. Lett. B **581**, 207 (2004) [hep-lat/0308014]
14. D.B. Leinweber, A.W. Thomas, R.D. Young: Phys. Rev. Lett. **92**, 242002 (2004) [hep-lat/0302020]; T.R. Hemmert, M. Procura, W. Weise: Nucl. Phys. A **721**, 938 (2003) [hep-lat/0301021]; M. Frink, U.G. Meissner: JHEP **0407**, 028 (2004) [hep-lat/0404018]
15. A. Ali Khan et al., [QCDSF-UKQCD Collaboration]: Nucl. Phys. B **689**, 175 (2004) [hep-lat/0312030]
16. T. Becher, H. Leutwyler: Eur. Phys. J. C **9**, 643 (1999) [hep-ph/9901384]
17. N. Fettes, U.G. Meissner, M. Mojzis, S. Steininger: Annals Phys. **283**, 273 (2000) [Erratum-ibid. **288**, 249 (2001)] [hep-ph/0001308]
18. M. Luscher: Commun. Math. Phys. **104**, 177 (1986)
19. S.R. Beane, M.J. Savage: Phys. Rev. D **70**, 074029 (2004) [hep-ph/0404131]; S.R. Beane: Phys. Rev.D **70**, 034507 (2004) [hep-lat/0403015]; R.D. Young, D.B. Leinweber, A.W. Thomas: hep-lat/0406001
20. P.F. Bedaque, H.W. Griesshammer, G. Rupak: hep-lat/0407009; see also W. Detmold, M.J. Savage: Phys. Lett. B **599**, 32 (2004) [hep-lat/0407008]
21. W. Melnitchouk et al.: Phys. Rev. D **67**, 114506 (2003) [hep-lat/0202022]
22. T. Burch et al., [BGR Collaboration]: Phys. Rev.D **70**, 054502 (2004) [hep-lat/0405006]; S. Sasaki, T. Blum, S. Ohta: Phys. Rev. D **65**, 074503 (2002) [hep-lat/0102010]; Y. Nemoto et al.: Nucl. Phys. A **721**, 879 (2003) [hep-lat/0312033]; K. Sasaki et al.: Nucl. Phys. Proc. Suppl. **129**, 212 (2004) [hep-lat/0309177]; M. Gockeler et al., [QCDSF and UKQCD]: Phys. Lett. B **532**, 63 (2002) [hep-lat/0106022]
23. S.J. Dong et al.: hep-ph/0306199
24. D. Guadagnoli, M. Papinutto, S. Simula: Phys. Lett. B **604**, 74 (2004) [hep-lat/0409011]
25. S. Hashimoto: hep-ph/0411126
26. M. Gockeler et al., [QCDSF Collaboration]: Phys. Rev. Lett. **92**, 042002 (2004) [hep-ph/0304249]; see also hep-lat/0410023
27. P. Hagler et al., [LHPC and SESAM]: Phys. Rev. Lett. **93**, 112001 (2004) [hep-lat/0312014]; see also hep-ph/0410017
28. X.D. Ji: Phys. Rev. Lett. **78**, 610 (1997) [hep-ph/9603249]

Eur Phys J A (2005) **24**, s2, 79–84
DOI: 10.1140/epjad/s2005-04-017-y

EPJ A direct
electronic only

Systematic uncertainties in the precise determination of the strangeness magnetic moment of the nucleon

D.B. Leinweber[1,2], S. Boinepalli[1], A.W. Thomas[1,2], A.G. Williams[1], R.D. Young[1,2], J.B. Zhang[1], and J.M. Zanotti[3]

[1] Special Research Center for the Subatomic Structure of Matter, and Department of Physics, University of Adelaide, Adelaide SA 5005, Australia
[2] Jefferson Lab, 12000 Jefferson Ave., Newport News, VA 23606, USA
[3] John von Neumann-Institut für Computing NIC, Deutsches Elektronen-Synchrotron DESY, D-15738 Zeuthen, Germany

Received: 1 December 2004 / Published Online: 8 February 2005

Abstract. Systematic uncertainties in the recent precise determination of the strangeness magnetic moment of the nucleon are identified and quantified. In summary, $G_M^s = -0.046 \pm 0.019\ \mu_N$.

PACS. 13.40.Em Electric and magnetic moments – 12.38.Gc Lattice QCD calculations – 12.39.Fe Chiral Lagrangians

1 Introduction

Recent low-mass lattice QCD simulation results combined with new chiral extrapolation techniques and the principle of charge symmetry have enabled a precise determination of the strangeness magnetic moment of the nucleon G_M^s [1]. In this paper, the systematic errors of the approach are explored and quantified. In particular, we examine the sensitivity of G_M^s and the magnetic moments of the baryon octet to the regulator-mass scale of finite-range regularized chiral effective field theory, the lattice scale determination, the finite-volume of the lattice, and the quenched approximation.

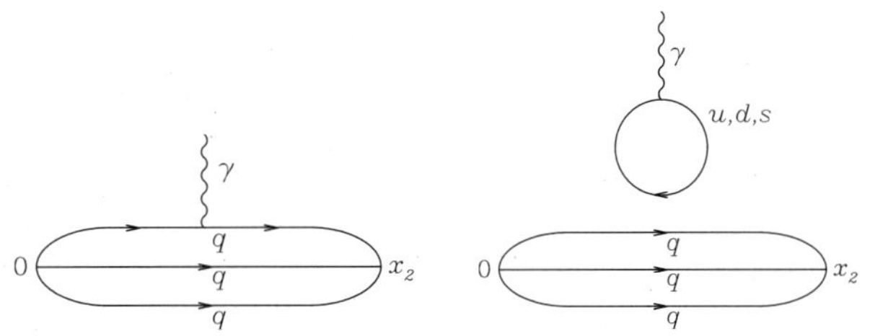

Fig. 1. Diagrams illustrating the two topologically different insertions of the current within the framework of lattice QCD. These skeleton diagrams for the connected (*left*) and disconnected (*right*) current insertions are dressed by an arbitrary number of gluons and quark loops

2 Charge symmetry

The approach [1] centres around two equations for the strangeness magnetic moment of the nucleon, G_M^s, obtained from charge symmetry

$$G_M^s = \left(\frac{{}^{\ell}R_d^s}{1 - {}^{\ell}R_d^s}\right)\left[2p + n - \frac{u^p}{u^{\Sigma}}\left(\Sigma^+ - \Sigma^-\right)\right], \qquad (1)$$

$$G_M^s = \left(\frac{{}^{\ell}R_d^s}{1 - {}^{\ell}R_d^s}\right)\left[p + 2n - \frac{u^n}{u^{\Xi}}\left(\Xi^0 - \Xi^-\right)\right]. \qquad (2)$$

Here the baryon labels represent the experimentally measured baryon magnetic moments and ${}^{\ell}R_d^s \equiv G_M^s / {}^{\ell}G_M^d$ is the ratio of s- and d-quark sea-quark loop contributions, depicted in the right-hand diagram of Fig. 1. ${}^{\ell}R_d^s$ lies in the range (0,1). The ratios u^p/u^{Σ} and u^n/u^{Ξ} are ratios of valence-quark contributions to baryon magnetic moments in full QCD as depicted in the left-hand diagram of Fig. 1. The latter are determined by lattice QCD calculations [1, 3,4,5,6].

Equating (1) and (2) provides a linear relationship between u^p/u^{Σ} and u^n/u^{Ξ} which must be obeyed within QCD under the assumption of charge symmetry – itself typically satisfied at the level of 1% or better [2]. There are no other systematic uncertainties associated with this constraint. Figure 2 displays this relationship by the dashed and solid line. Since this line does not pass through the point (1.0, 1.0), corresponding to the simple quark model assumption of universality, there must be an environment effect exceeding 12% in both ratios or approaching 20% or more in at least one of the ratios. To determine the sign of G_M^s, it is sufficient to determine where on this constraint curve, QCD resides.

3 u^p/u^{Σ} and u^n/u^{Ξ} determination

Our present precise analysis has been made possible by a significant breakthrough in the regularization of the chiral-loop contributions to hadron observables [7,8,9]. Through

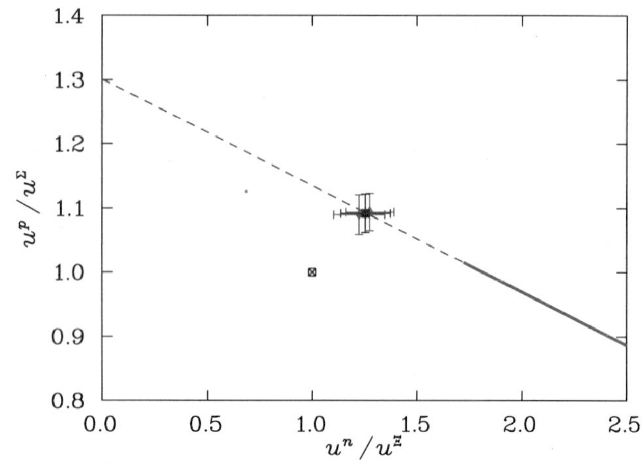

Fig. 2. The straight line (*dashed* $G_M^s(0) < 0$, *solid* $G_M^s(0) > 0$) indicates the constraint on the ratios u^p/u^Σ and u^n/u^Ξ implied by charge symmetry and the experimentally measured magnetic moments. The assumption of environment independent quark moments is indicated by the square. The dependence of the extrapolated ratios from lattice QCD simulations on the parameter $\Lambda = 0.7$, 0.8 and 0.9 GeV, (governing the size of pion-cloud corrections) is illustrated by the cluster of points with Λ increasing from left to right

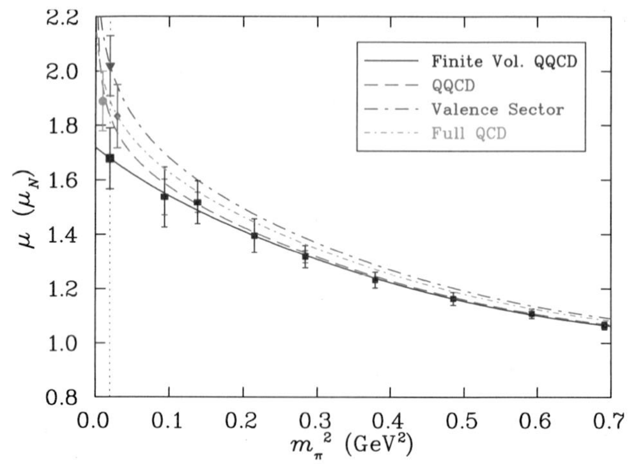

Fig. 3. The contribution of a single u quark (*with unit charge*) to the magnetic moment of the proton. Lattice simulation results (*square symbols for* $m_\pi^2 > 0.05$ GeV^2) are extrapolated to the physical point (*vertical dashed line*) in finite-volume QQCD as well as infinite volume QQCD. Estimates of the valence u quark contribution in full QCD and the full u-quark sector contribution in full QCD are also illustrated. Extrapolated values at the physical pion mass (*vertical dashed line*), are offset for clarity

the process of regulating loop integrals via a finite-range regulator (FRR) [8,10], the chiral expansion is effectively re-summed to produce an expansion with vastly improved convergence properties.

The chiral expansion for the u-quark contribution to the proton magnetic moment in quenched QCD (QQCD), has the form

$$u^p = a_0^\Lambda + u^p\,\chi_{\eta'}\,I_{\eta'}(m_\pi,\Lambda) + \chi_{\pi B}\,I_B(m_\pi,\Lambda) + \chi_{KB}\,I_B(m_K,\Lambda) + a_2^\Lambda\,m_\pi^2 + a_4^\Lambda\,m_\pi^4, \quad (3)$$

where the repeated index, B, sums over allowed baryon octet and decuplet intermediate states. The dependence of the unrenormalized coefficients, a_i^Λ, and the associated dipole-vertex regulated loop integrals, $I(m_\pi,\Lambda)$, on the regulator parameter, Λ, is emphasized by the explicit appearance of Λ. The loop integrals are defined as

$$I_B(m,\Lambda) = \quad (4)$$
$$-\frac{2}{3\pi}\int dk\,\frac{\left(2\sqrt{k^2+m^2}+\Delta_{BN}\right)\,k^4\,u^2(k,\Lambda)}{(k^2+m^2)^{3/2}\,\left(\sqrt{k^2+m^2}+\Delta_{BN}\right)^2},$$

$$I_{\eta'}(m_\pi,\Lambda) = -\int_0^\infty dk\,\frac{k^4}{(k^2+m_\pi^2)^{\frac{5}{2}}}u^2(k,\Lambda)\,, \quad (5)$$

where Δ_{BN} is the relevant baryon mass splitting and the function $u(k,\Lambda)$ is the dipole-vertex regulator. The coefficients, χ, denote the known model-independent coefficients of the non-analytic terms for π and K mesons in QQCD [11,12].

Figure 3 illustrates a fit of FRR, quenched chiral perturbation theory (χPT) to our fermion lattice results (solid curve), where only the discrete momenta allowed in the finite volume of the lattice are summed in evaluating the chiral loop integrals. The long-dashed curve that also runs through the lattice results corresponds to replacing the discrete momentum sum by the infinite-volume, continuous momentum integral. For all but the lightest quark mass, finite volume effects are negligible.

Having determined the analytic coefficients $a_{0,2,4}^\Lambda$ for a particular choice of Λ, one can correct the chiral properties of the pion-cloud contribution from QQCD to full QCD [9, 13] by changing the coefficients of the loop integrals, $\chi_{\eta'}$, $\chi_{\pi B}$, χ_{KB} of (3), to their full QCD counter parts [11, 12]. Valence quark contributions in full QCD are indicated by the long-dash-dot curve in Fig. 3 (i.e. sea-quark loop charges are zero) and the full u-quark sector including the u-sea-quark loop contributions are indicated by the short-dash-dot curve for $\Lambda = 0.8$ GeV. Figures 4, 5 and 6 show similar results for the u quark in n, Σ^+, and Ξ^0 respectively.

The importance of correcting for both finite-volume and quenching artifacts is illustrated in Figs. 7 and 8, where the one standard deviation agreement between the chirally corrected lattice QCD simulation results and the experimentally measured baryon magnetic moments is highlighted.

4 Systematic errors

4.1 Regulator dependence

It is important to investigate systematic errors associated with the regulator-mass dependence of FRR χPT . The extrapolated results of finite-volume quenched chiral effective field theory should be insensitive to the choice of regulator parameter. When working to sufficient order in

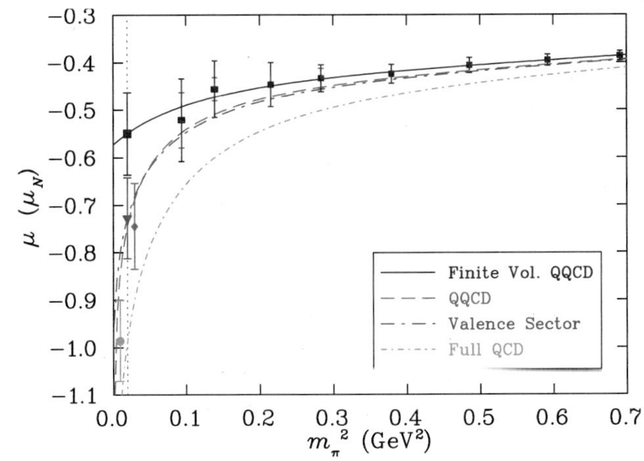

Fig. 4. The contribution of the u quark (with unit charge) to the magnetic moment of the neutron. *Curves* and *symbols* are as described in Fig. 3

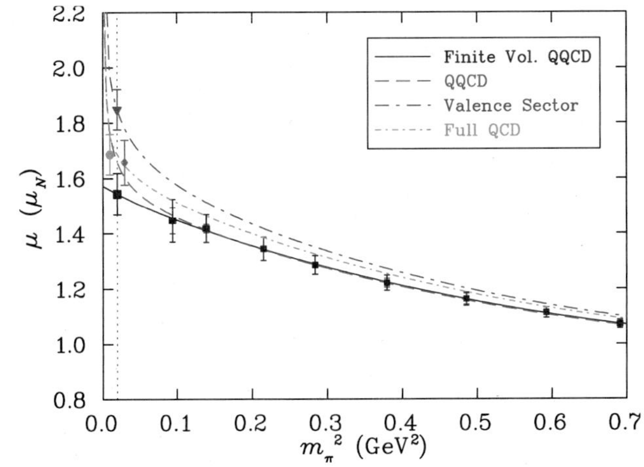

Fig. 5. The contribution of a single u quark (with unit charge) to the magnetic moment of Σ^+. *Curves* and *symbols* are as for Fig. 3

the chiral expansion, changes in the regulation of loop integrals should be absorbed by changes in the unrenormalized coefficients, $a_{0,2,4}^{\Lambda}$, in a manner which preserves the invariant renormalized coefficients. The latter are reflected in Fig. 9 which illustrates the insensitivity of quenched baryon magnetic moments in a finite-volume to the regulator parameter Λ. This systematic error is small relative to the statistical error.

Since the finite-volume and quenching corrections are applied only to the loop integral contributions, the final results are Λ dependent. In this case, the regulator of the loop integral has become a model for the axial-vector form factor of the nucleon, describing the coupling of pions to a core described by the analytic terms of the FRR expansion. This approach describes the relation between quenched and full QCD N and Δ mass simulation results as a function of m_π very accurately [13]. Figures 10 and 11 display results for positive and negative baryon magnetic moments in full QCD respectively. One-standard deviation agreement is achieved for $0.7 \leq \Lambda \leq 0.9$.

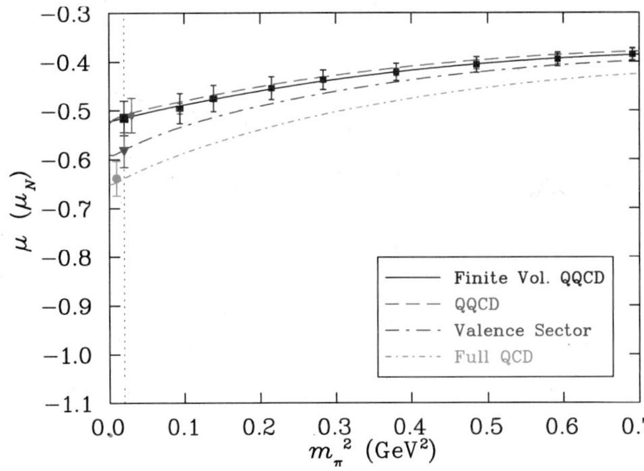

Fig. 6. The contribution of the u quark (with unit charge) to the magnetic moment of the Ξ^0 hyperon. *Curves* and *symbols* are as described in Fig. 3

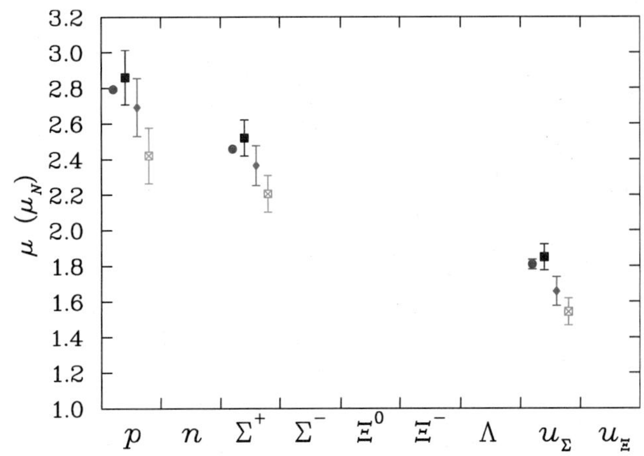

Fig. 7. The one standard deviation agreement between the chirally corrected lattice QCD simulation results (*square symbols*) and the experimentally measured baryon magnetic moments (*circular symbols*) having positive values. Finite-volume quenched results (*crossed boxes*) and infinite-volume quenched results (*diamonds*) are also illustrated to highlight the importance of correcting for both finite-volume and quenching artifacts

Because the strangeness magnetic moment of (1) and (2) depends only on ratios of magnetic moments, most of this Λ dependence cancels in the final ratios, as illustrated by the close clustering of points in Fig. 2.

4.2 Role of the decuplet in χPT

It is often argued that next-to-leading-order non-analytic (NLNA) contributions from the baryon decuplet are essential in describing the mass dependence of nucleon magnetic moments. While the decuplet baryon contributions are not necessarily small, we find the non-analytic curvature induced by these contributions is sufficiently subtle that it may be accurately absorbed by the analytic terms

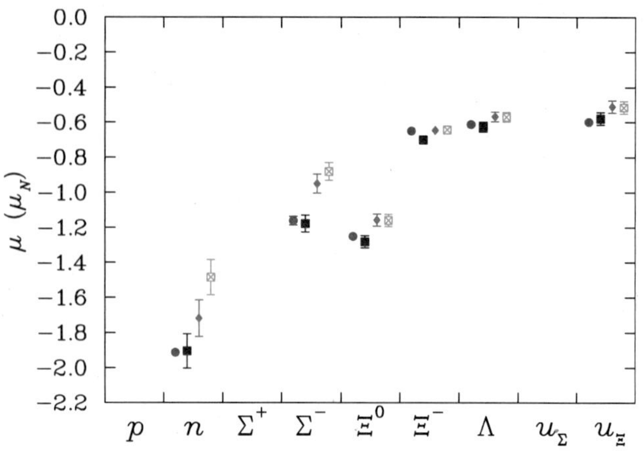

Fig. 8. The one standard deviation agreement between the chirally corrected lattice QCD simulation results and the experimentally measured baryon magnetic moments having negative values. Symbols are as in Fig. 7

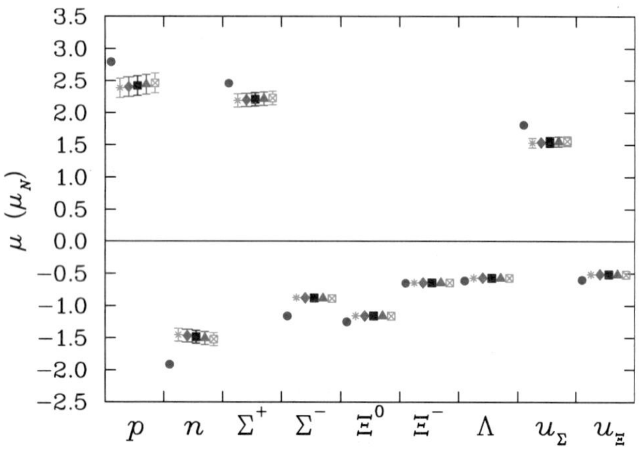

Fig. 9. The FRR χPT regulator mass dependence of finite-volume quenched chiral effective field theory. Experimental measurements are illustrated at left for each baryon for reference. Results for $\Lambda = 0.6$, 0.7, 0.8, 0.9 and 1.0 GeV are illustrated from left to right for each baryon. The small systematic dependence on Λ relative to the statistical *error bars* illustrated indicate the order of the chiral expansion is adequate for this analysis

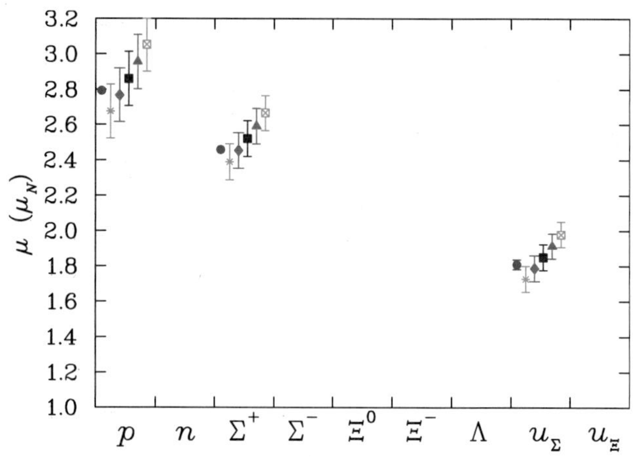

Fig. 10. The dependence of positive octet-baryon magnetic moments on the parameter $\Lambda = 0.6$, 0.7, 0.8, 0.9 and 1.0 GeV, governing the size of pion-cloud corrections associated with the finite-volume of the lattice and artifacts of the quenched approximation. Experimental measurements, illustrated at left by the *filled circle* for each baryon, indicate that optimal corrections are obtained for $0.6 \leq \Lambda \leq 0.9$ GeV

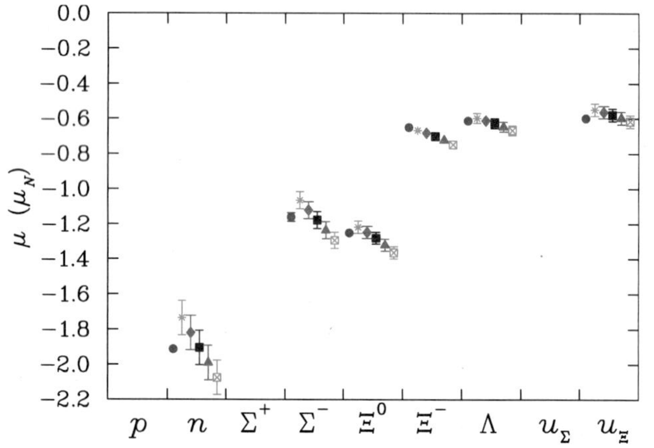

Fig. 11. The dependence of negative octet-baryon magnetic moments on the parameter $\Lambda = 0.6$, 0.7, 0.8, 0.9 and 1.0 GeV, governing the size of pion-cloud corrections associated with the finite-volume of the lattice and artifacts of the quenched approximation. Experimental measurements, illustrated at left by the *filled circle* for each baryon, indicate that optimal corrections are obtained for $0.7 \leq \Lambda \leq 0.9$ GeV

of the chiral expansion. Figure 12 illustrates the insensitivity of finite-volume quenched chiral effective field theory to NLNA decuplet-baryon contributions.

Figure 13 confirms that the NLNA decuplet-baryon contributions are indeed large in some cases and as such are important in correcting the artifacts of the quenched approximation. However, other more highly-excited baryon resonances have small couplings to the ground-state baryon octet relative to that for the decuplet and provide negligible corrections.

4.3 Scale dependence

Setting the scale in quenched QCD simulations is somewhat problematic. Different observables lead to different lattice spacings, a. If one is explicitly correcting the one-loop pion-cloud contributions to hadronic observables, as we are here, then clearly one must set the scale using an observable insensitive to chiral physics. This excludes observables such as the rho-meson mass, nucleon mass, or the pion decay constant commonly used in the literature to hide the artifacts of the quenched approximation. On the other hand, the heavy-quark phenomenology of the static-quark potential provides an optimal case. In particular, the Sommer parameter, r_0, is ideal as it sets the scale by equating the force between two static quarks in QQCD and full QCD at a precise separation of $r_0 = 0.49$ fm.

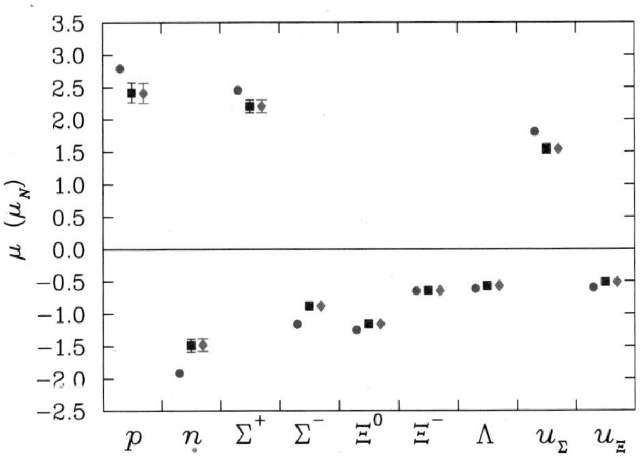

Fig. 12. The insensitivity of finite-volume quenched chiral effective field theory to NLNA decuplet-baryon contributions. Results including decuplet intermediate states in chiral effective field theory (*squares*) are compared with results excluding the decuplet (*diamonds*). Experimental measurements (*circles*), are illustrated at left for each baryon for reference

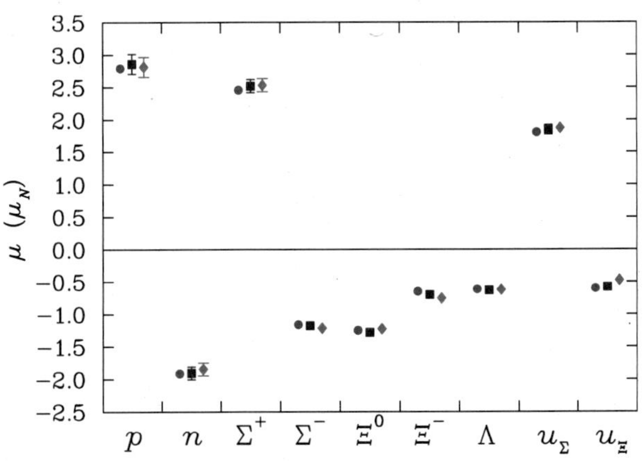

Fig. 13. The dependence of octet-baryon magnetic moments on the inclusion of NLNA decuplet-baryon contributions in the process of correcting the pion cloud of quenched chiral effective field theory. Preferred results including decuplet intermediate states in chiral effective field theory (*squares*) are compared with results excluding the decuplet (*diamonds*). Experimental measurements (*circles*), are illustrated at left for each baryon for reference

The string tension might also be used, but screening of the potential in full QCD makes this measure poorly defined. However, we consider it here as a measure of the systematic error encountered in setting the scale of the lattice QCD results. Figure 14 illustrates this scale dependence on baryon magnetic moments and Fig. 15 illustrates the rather minor impact this systematic uncertainty has on the valence-quark moment ratios vital to determining the sign of G_M^s.

Collecting the variation of the valence-quark moment ratios from variations in Λ (which maintain one-standard-deviation agreement with experiment), variations in setting the lattice scale and statistical errors determined by a

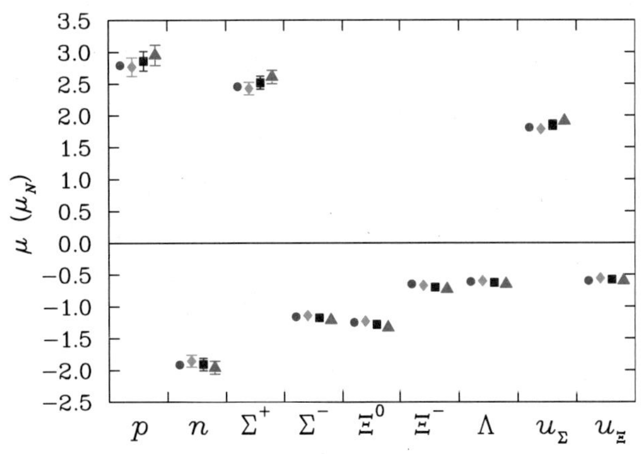

Fig. 14. The dependence of octet-baryon magnetic moments on the matching criteria for determining the lattice spacing, a. While $a_{r_0} = 0.128$ fm set by r_0 (*square symbols*) is the preferred method for determining the scale, results for $a_\sigma = 0.134$ fm set by the string tension (*triangles*) and a third spacing of 0.122 fm (*diamonds*) are also illustrated. Experimental measurements (*filled circles*), are illustrated at left for each baryon

third-order single-elimination jackknife analysis, one finds

$$\frac{u^p}{u^\Sigma} = 1.092 \pm 0.030 \text{ and } \frac{u^n}{u^\Xi} = 1.254 \pm 0.124 \,. \qquad (6)$$

Using the experimental magnetic moments, one observes that (2)

$$G_M^s = \left(\frac{{}^\ell R_d^s}{1 - {}^\ell R_d^s} \right) \left[-1.033 - \frac{u^n}{u^\Xi} (-0.599) \right] , \qquad (7)$$

is least sensitive to variation in the valence-quark moment ratio, and hence provides the most precise determination for G_M^s. Figure 16 plots G_M^s as a function of ${}^\ell R_d^s$ with standard error limits provided by (6).

5 Estimating ${}^\ell R_d^s \equiv G_M^s / {}^\ell G_M^d$

The symmetry of the three-point correlation function [14] describing the sea-quark loop contributions to the nucleon, depicted in the right-hand side of Fig. 1, ensure that the chiral expansion for this quantity is identical for all three quark flavours, up to simple charge factors. For the d- or s-quark loop contributions, the only difference that can arise is whether one evaluates the chiral expansion at the pion or kaon mass. The leading non-analytic contribution to the chiral expansion involves two pseudoscalar meson propagators, and therefore one expects contributions to ${}^\ell R_d^s$ in the ratio $m_\pi^4/m_K^4 \sim 0.1$.

To be more precise, one can use the same successful (single-parameter) model, previously used to correct the quenched simulation results to full QCD, as highlighted in Figs. 7 and 8, to provide an estimate for ${}^\ell R_d^s$. Evaluating the loop integrals with $\Lambda = 0.8 \pm 0.2$ GeV yields ${}^\ell R_d^s = 0.139$ with $0.096 \le {}^\ell R_d^s \le 0.181$. This uncertainty dominates the final uncertainty in G_M^s.

Table 1. Sources of uncertainty and their contribution to the strangeness magnetic moment of the nucleon, G_M^s, in units of nuclear magnetons, μ_N. Uncertainties are documented for G_M^s obtained from the valence-quark ratio u^p/u^Σ in (1), from the valence-quark ratio u^n/u^Ξ in (2) and from a statistically weighted (SW) average of these two determinations

Uncertainty Source	Parameter Range	u^p/u^Σ, (1) $G_M^s = -0.045$	u^n/u^Ξ, (2) $G_M^s = -0.046$	SW Average $G_M^s = -0.046$
Statistical Errors		0.016	0.009	0.008
Chiral corrections	$0.7 \leq \Lambda \leq 0.9$ GeV	0.001	0.002	0.002
Scale Determination	$0.122 \leq a \leq 0.134$ fm	0.001	0.002	0.002
${}^\ell R_d^s$ Determination	$0.096 \leq {}^\ell R_d^s \leq 0.181$	0.016	0.017	0.017
Total Uncertainty		0.023	0.019	0.019

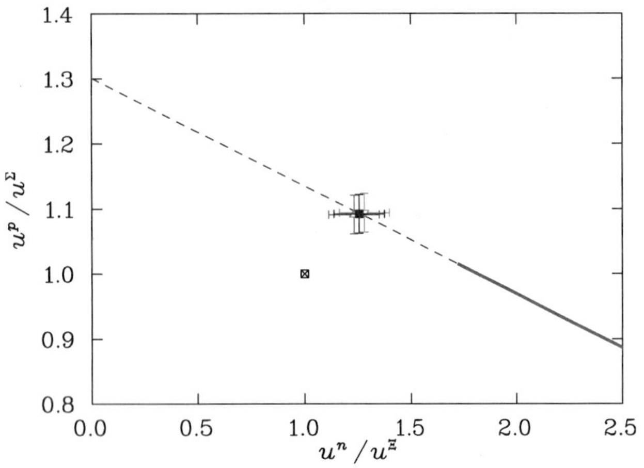

Fig. 15. The charge symmetry constraint line (*dashed* $G_M^s(0) < 0$, *solid* $G_M^s(0) > 0$) on the ratios u^p/u^Σ and u^n/u^Ξ. The dependence of the ratios from chirally-corrected quenched lattice QCD on the scale parameter, $a = 0.134$, 0.128, and 0.122 fm, is illustrated by the cluster of points with a decreasing from left to right

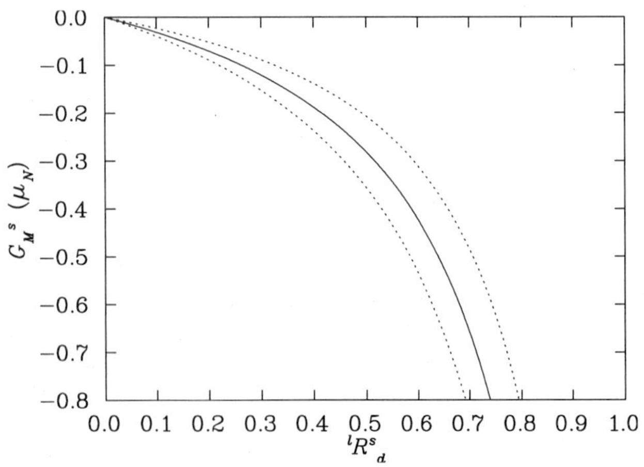

Fig. 16. The dependence of G_M^s on the strange to light sea-quark loop ratio ${}^\ell R_d^s$. Standard error limits have there origin in the systematic error summary of 6

Table 1 summarizes the sources of uncertainty and their contributions to the final determination

$$G_M^s = -0.046 \pm 0.019 \ \mu_N , \tag{8}$$

for the strange quark contribution to the magnetic moment of the nucleon.

Acknowledgements. We thank the Australian Partnership for Advanced Computing (APAC) for generous grants of supercomputer time which have enabled this project. Support from the South Australian Partnership for Advanced Computing (SAPAC) and the National Facility for Lattice Gauge Theory is also gratefully acknowledged. DBL thanks Jefferson Lab for their kind hospitality where the majority of this research was performed. This work is supported by the Australian Research Council and by DOE contract DE-AC05-84ER40150 under which SURA operates Jefferson Lab.

References

1. D.B. Leinweber et al.: arXiv:hep-lat/0406002
2. G.A. Miller, B.M. Nefkens, I. Slaus: Phys. Rept. **194**, 1 (1990)
3. J.M. Zanotti et al.: Phys. Rev. D **65**, 074507 (2002) [hep-lat/0110216]
4. D.B. Leinweber et al.: Eur. Phys. J. A **18**, 247 (2003) [nucl-th/0211014]
5. J.M. Zanotti et al.: hep-lat/0405015
6. J.M. Zanotti et al.: hep-lat/0405026
7. R.D. Young et al.: Prog. Part. Nucl. Phys. **50**, 399 (2003) [hep-lat/0212031]
8. D.B. Leinweber, A.W. Thomas, R.D. Young: Phys. Rev. Lett. **92**, 242002 (2004) [hep-lat/0302020]
9. R.D. Young, D.B. Leinweber, A.W. Thomas: arXiv:hep-lat/0406001
10. J.F. Donoghue, B.R. Holstein, B. Borasoy: Phys. Rev. D **59**, 036002 (1999)
11. D.B. Leinweber: Phys. Rev. D **69**, 014005 (2004) [hep-lat/0211017]
12. M.J. Savage: Nucl. Phys. A **700**, 359 (2002) [nucl-th/0107038]
13. R.D. Young et al.: Phys. Rev. D **66**, 094507 (2002) [arXiv:hep-lat/0205017]
14. D.B. Leinweber: Phys. Rev. D **53**, 5115 (1996) hep-ph/9512319

Eur Phys J A (2005) **24**, s2, 85–88
DOI: 10.1140/epjad/s2005-04-018-x

EPJ A direct
electronic only

Current status of parton charge symmetry

J.T. Londergan

Dept. of Physics, Indiana University, Bloomington, IN, 47404, USA

Received: 15 October 2004 / Published Online: 8 February 2005

Abstract. We review constraints on parton charge symmetry from various experiments. Recently charge symmetry violation (CSV) has been included in a global fit to high energy data. We show that CSV compatible with all high energy data would be able to remove completely the NuTeV anomaly.

PACS. 11.30.Hv Flavor Symmetries – 13.15.+g Neutrino Interactions – 13.60.Hb Total and inclusive cross-sections, including deep inelastic structure functions

1 Experimental limits on parton CSV

Charge symmetry is a restricted form of isospin invariance involving a rotation of 180° about the "2" axis in isospin space. For parton distributions, charge symmetry involves interchanging up and down quarks while simultaneously interchanging protons and neutrons. In nuclear physics, charge symmetry is generally obeyed at the level of a fraction of a percent [1,2]. Charge symmetry violation (CSV) in parton distribution functions (PDFs) arises from two sources; from the difference $\delta m \equiv m_d - m_u$ between down and up current quark masses, and from electromagnetic (EM) effects. Since charge symmetry is so well satisfied at lower energies, it is natural to assume that it holds for parton distributions. At present, there is no direct experimental evidence of substantial violation of charge symmetry in parton distribution functions (PDFs).

In a recent paper [3], we have reviewed experimental and theoretical estimates for parton CSV, and have discussed potential corrections to the extraction of the Weinberg angle in neutrino deep inelastic scattering (DIS). We summarize our arguments here. The most stringent upper limits on parton CSV come from comparing the structure function $\overline{F_2}^W$, the average of ν and $\bar{\nu}$ charged current reactions, and the structure function F_2^γ for charged lepton DIS, on isoscalar targets (N_0). In leading order (LO), F_2^γ depends on the squared charges of the quarks, while F_2^W depends on the quark weak charges. Assuming charge symmetry gives a simple relation between the structure functions, defined as the "charge ratio" $R_c(x, Q^2)$. To lowest order in the (presumably small) CSV terms

$$
\begin{aligned}
R_c(x) &\equiv \frac{F_2^{\gamma N_0}(x) + x\left(s(x) + \bar{s}(x) - c(x) - \bar{c}(x)\right)/6}{5\overline{F_2}^{W N_0}(x)/18} \\
&\approx 1 + \frac{3x\left(\delta u(x) + \delta\bar{u}(x) - \delta d(x) - \delta\bar{d}(x)\right)}{10\,Q(x)}
\end{aligned}
$$

Send offprint requests to: tlonderg@indiana.edu

$$Q(x) = \sum_{j=[u,d,s,c]} x\left[q_j(x) + \bar{q}_j(x)\right] \qquad (1)$$

1 introduces the CSV parton distributions,

$$\delta u(x) = u^p(x) - d^n(x); \;\; \delta d(x) = d^p(x) - u^n(x), \qquad (2)$$

with analogous relations for antiquarks. Deviation of $R_c(x)$ from unity would indicate a CSV contribution.

The most precise neutrino measurements were obtained by the CCFR group [4], who extracted the F_2 structure function for ν-Fe and $\bar{\nu}$-Fe reactions. Muon DIS measurements were obtained by the NMC group [5,6], who measured F_2 structure functions for muon interactions on deuterium at muon energies $E_\mu = 90$ and 280 GeV. Taking into account many corrections (relative normalizations; heavy quark threshold effects; nuclear effects; corrections for excess neutrons in iron; contributions from s and c quarks), CCFR obtained results consistent with unity at about the $2-3\%$ level, in the range $0.1 \le x \le 0.4$. From (1), this gives an upper limit to parton CSV effects in the $6-9\%$ range. At smaller x, R_c appeared to deviate significantly from unity. However, upon re-analysis [7] the ratio agrees with unity at the $2-3\%$ level down to $x \sim 0.03$, as significant effects were found from NLO treatment of charm mass corrections, and separation of the F_2 and F_3 structure functions in ν DIS.

Other limits on parton CSV can come from measurements of $W^\pm$ asymmetry in a $p-\bar{p}$ collider. Since u quarks carry more momentum than d quarks, the direction of the W^+ and p tend to be aligned, as do the W^- and $\bar{p}$. Measurement of the W charge asymmetry is thus quite sensitive to the proton's u and d distributions. Conversely, charged lepton DIS on an isoscalar target tends to be more sensitive to u^n than to d^p, as it is more heavily weighted due to the squared quark charge. Comparison of, say, the CDF W charge asymmetry [8] and NMC $\mu - D$ DIS can constrain some aspects of parton CSV.

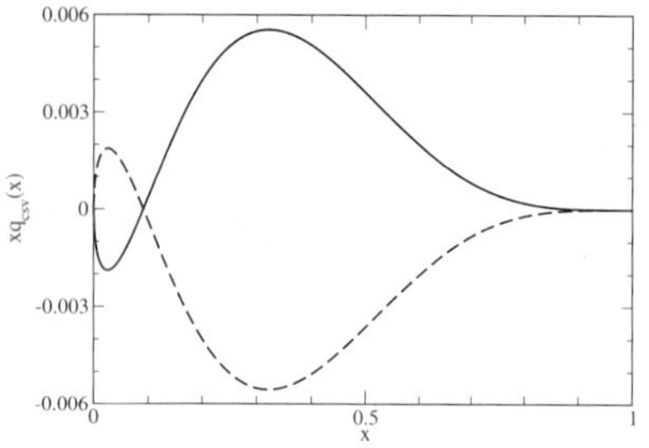

Fig. 1. The valence quark CSV function from [9], corresponding to best fit value $\kappa = -0.2$ defined in 3. *Solid curve*: $x\delta d_{\rm v}(x)$; *dashed curve*: $x\delta u_{\rm v}(x)$

2 Phenomenology and theory of parton CSV

Because CSV effects are typically very small at nuclear physics energy scales, all previous phenomenological PDFs have assumed the validity of parton charge symmetry. However, Martin, Roberts, Stirling and Thorne (MRST) [9] have recently studied the uncertainties in parton distributions arising from a number of factors, including isospin violation. MRST chose a specific model for valence quark charge symmetry violating PDFs:

$$\delta u_{\rm v}(x) = -\delta d_{\rm v}(x) = \kappa(1-x)^4 x^{-0.5}(x - .0909). \quad (3)$$

At both small and large x the MRST CSV PDFs have the same form as the valence distributions. The first moment of the MRST valence CSV function is zero; this must be the case since, *e.g.*, the integral $\langle \delta u_{\rm v}\rangle$ is just the total number of valence up quarks in the proton minus the number of down quarks in the neutron. The second moment of this function represents the CSV momentum asymmetry; $\delta U_{\rm v} \equiv \langle x\delta u_{\rm v}(x)\rangle$ is the difference in total momentum carried by $u^p_{\rm v}$ and $d^n_{\rm v}$. The MRST valence CSV distributions require that $\delta u_{\rm v}$ and $\delta d_{\rm v}$ have opposite signs at large x, in agreement with theoretical predictions. This condition also insures that valence quarks in the proton and neutron carry an equal amount of total momentum (this is strictly true only at the starting scale, since the momentum asymmetry is not constant under QCD evolution; however MRST find that it does not change very much over a fairly wide Q^2 range). The overall coefficient κ was varied in a global fit to a wide range of high energy data.

The value $\kappa = -0.2$ minimised χ^2. Their χ^2 had a shallow minimum with the 90% confidence level obtained for the range $-0.8 \le \kappa \le +0.65$. In Fig. 1 we plot the valence quark CSV PDFs corresponding to the MRST best fit value $\kappa = -0.2$. Within the 90% confidence region for the global fit, the valence quark CSV PDFs could be either four times as large as in Fig. 1, or it could be three times as big with the opposite sign. CSV distributions with this shape, and for κ within this range, will not disagree seriously with any of the high energy data used to extract quark and gluon PDFs.

The MRST group also searched for CSV in the sea quark sector. Again, they chose a specific form for sea quark CSV, dependent on a single parameter, *i.e.*,

$$\bar{u}^n(x) = \bar{d}^p(x)\,[1+\delta]$$

$$\bar{d}^n(x) = \bar{u}^p(x)\,[1-\delta] \quad (4)$$

Somewhat surprisingly, evidence for sea quark CSV in the global fit was substantially stronger than for valence quark CSV. The best fit was obtained for $\delta = 0.08$, corresponding to an 8% violation of charge symmetry in the nucleon sea. This is considerably larger than theoretical estimates of sea quark CSV [10]. The χ^2 for this value is substantially better than with no CSV, primarily because of improvement in fits to the NMC $\mu - D$ DIS data [5,6] and to the E605 Drell-Yan data [11], when sea quark CSV is included. The MRST best-fit values will necessarily give reasonable agreement with the charge ratio of (1), since both the CCFR ν X-sections and NMC muon DIS are included in the global fit. The MRST group also includes the CDF W charge asymmetry measurements [8], so that the MRST global fit PDFs including CSV are compatible with all data sets that are most sensitive to charge symmetry violating effects.

The MRST phenomenological CSV distributions agree rather well with two earlier predictions using quark models. The Adelaide group [12] developed a method for calculating twist-two valence PDFs from quark model wavefunctions. Unlike earlier calculations, this model guaranteed correct support for the PDFs. Rodionov et al. [13] extended this model to calculate valence quark CSV. Sather [14] approximated the dependence of valence quark PDFs on the quark and nucleon masses, and obtained analytic approximations relating valence quark CSV to derivatives of the valence PDFs. Although there are several differences between the models of Sather and Rodionov, their predictions of valence quark CSV are quite similar. In Fig. 2, we show the theoretical valence quark CSV prediction of Rodionov. The solid curve is $x\delta u_{\rm v}(x)$, while the dot-dashed curve is $x\delta d_{\rm v}(x)$, both evolved to $Q^2 = 10$ GeV^2. Qualitatively, the results of Rodionov *et al.* are very similar to the best-fit phenomenological CSV distribution of MRST, shown in Fig. 1. The sign and relative magnitude of both $\delta d_{\rm v}(x)$ and $\delta u_{\rm v}(x)$ are quite similar in both phenomenology and theory. The second moments of the CSV PDFs (which give the total momentum asymmetry between, *e.g.*, $u^p_{\rm v}$ and $d^n_{\rm v}$) of the MRST and Rodionov distributions are equal to within 10%.

3 Parton CSV and the NuTeV anomaly

In 1973, Paschos and Wolfenstein [15] suggested that the ratio of neutral-current (NC) and charge-changing (CC) neutrino cross sections on isoscalar targets could provide an independent measurement of the Weinberg angle ($\sin^2\theta_W$). The Paschos-Wolfenstein (PW) ratio R^- is given by

$$R^- \equiv \frac{\langle\sigma_{NC}^{\nu N_0}\rangle - \langle\sigma_{NC}^{\bar{\nu} N_0}\rangle}{\rho_0^2\left(\langle\sigma_{CC}^{\nu N_0}\rangle - \langle\sigma_{CC}^{\bar{\nu} N_0}\rangle\right)} = \frac{1}{2} - \sin^2\theta_W = \frac{R^\nu - R^{\bar{\nu}}}{1 - r\,R^\nu}. \quad (5)$$

In 5, $\langle\sigma_{NC}^{\nu N_0}\rangle$ is the NC inclusive total cross section for neutrinos on an isoscalar target. The quantity ρ_0 is one in the Standard Model.

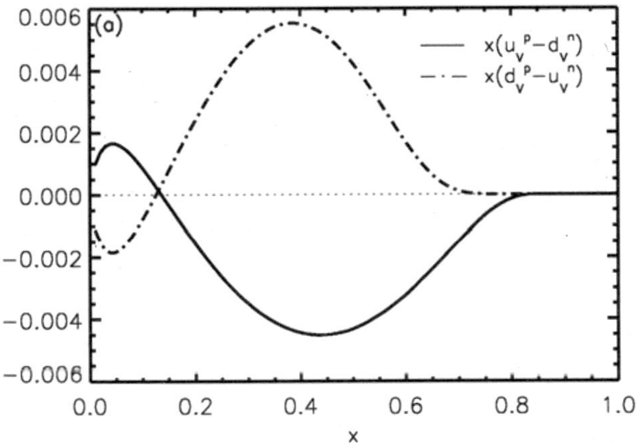

Fig. 2. Valence quark CSV contributions, $x\delta q_{\rm v}(x)$ vs. x. *Solid line*: $x\delta u_{\rm v}$; *dash-dot line*: $x\delta d_{\rm v}$. Calculated by Rodionov et al. [13] using MIT bag model wavefunctions, evolved to $Q^2 = 10$ GeV^2

The NuTeV group has measured NC and CC ν and $\bar{\nu}$ cross sections on iron [16] *[for more details, see talk by K. McFarland in these proceedings]*. They obtained the NC/CC ratios $R^\nu = 0.3916 \pm 0.0007$ and $R^{\overline{\nu}} = 0.4050 \pm 0.0016$, from which they extracted $\sin^2\theta_W = 0.2277 \pm 0.0013$ $(stat) \pm 0.0009$ $(syst)$. This value is three standard deviations above the measured value for the Weinberg angle obtained from electroweak (EW) processes near the Z pole, $\sin^2\theta_W = 0.2227 \pm 0.00037$ [17]. Such an effect can be interpreted as a 1.2% decrease in the left-handed coupling of light valence quarks to the weak neutral current. Davidson et al. [18] have examined possible contributions from "new physics" beyond the Standard Model. It is extremely difficult to find new particles that fit NuTeV without violating other experimental constraints, as several observables are constrained by very precise measurements, at or near the 0.1% level, in experiments near the Z pole [17]. Even modest success in removing the NuTeV anomaly, while leaving all other measurements within 1σ, can be achieved only with new particles whose masses, numbers, and couplings are very finely tuned – so-called "designer particles." Because of the serious difficulties in explaining the NuTeV result with particles outside the Standard Model, attention has focused on QCD effects within the Standard Model.

There are many QCD effects that must be taken into account [19]. For example, there is a correction due to the excess neutrons in Fe (iron is not an isoscalar target). Although the correction is large, it has been taken into account by NuTeV and should be well under control. Radiative corrections, which affect only CC reactions, are also substantial. These effects were calculated with a standard radiative correction model [20]. Recently Diener et al. have re-calculated the radiative corrections [21]. However, their calculation has yet to be incorporated into a reanalysis of the NuTeV data. The NuTeV group has also corrected for nuclear effects on the PDFs. There is still some uncertainty in the magnitude of such corrections; in particular, at present it is generally assumed that nuclear effects are identical for ν and charged-lepton DIS. Hirai et al. are currently calculating nuclear effects in neutrino reactions [22].

Finally, there is a possible contribution from a strange quark momentum asymmetry. The production of opposite-sign dimuons in ν and $\bar{\nu}$ reactions allows a separate extraction of s and $\bar{s}$ PDFs [23,24]. A difference in total momentum carried by s and $\bar{s}$ would affect the NuTeV result. Currently this has been analyzed by two groups; the CTEQ group extracts the strange PDFs in a global fit to high energy data [25], while the NuTeV group has analyzed the dimuon production cross Sects. [26,27]. At the moment the two results appear to disagree. The CTEQ analysis favors a positive momentum asymmetry $S_{\rm v} = \langle x(s - \bar{s})\rangle$, which would remove roughly 1/3 of the NuTeV anomaly, while the NuTeV analysis is consistent with $S_{\rm v}$ either zero or slightly negative. The two groups are currently collaborating on the analyses, although they both agree that strange quark effects alone cannot remove the anomaly.

Here we will review the effects of isospin violation on the NuTeV anomaly. The correction to the PW ratio arising from isospin violation in the PDFs has the form

$$\Delta R^-_{CSV} \approx \left[1 - \frac{7}{3}\sin^2\theta_W\right] \frac{\delta U_{\rm v} - \delta D_{\rm v}}{2(U_{\rm v} + D_{\rm v})}. \qquad (6)$$

Only valence quarks contribute to (6), and the correction depends on the second moment of valence PDFs, where $Q_{\rm v} \equiv \langle x(q - \bar{q})\rangle$. The numerator of (6) is equal to the momentum asymmetry between up quarks and down quarks in an isoscalar nucleus, *i.e.*, $U^p_{\rm v} + U^n_{\rm v} - (D^p_{\rm v} + D^n_{\rm v})$. However, estimates based on the PW ratio do not accurately predict contributions to the NuTeV result. The NuTeV group measures the NC/CC ratios R^ν and $R^{\overline{\nu}}$. Since these ratios have different cuts and acceptance corrections, one cannot simply combine them as in (5). To obtain the magnitude of a given effect on the NuTeV result for the Weinberg angle, it is necessary to fold that effect with functionals generated by NuTeV [27]. Thus, sea quark CSV makes a correction to the NuTeV extraction of the Weinberg angle, although it is much smaller than that from valence quark CSV. Using the best-fit MRST values for sea quark and valence quark CSV, would remove roughly 1/3 of the NuTeV anomaly. The value $\kappa = -0.6$, within the 90% confidence limit found by MRST, would completely remove the NuTeV anomaly, while the value $\kappa = +0.6$ would double the discrepancy. The MRST results show that isospin violating PDFs are able to completely remove the NuTeV anomaly in the Weinberg angle, or to make it twice as large, without serious disagreement with any of the data used to extract quark and gluon PDFs.

The model CSV predictions that we discussed earlier suggest that isospin violating corrections would tend to decrease the NuTeV anomaly for the Weinberg angle. Both the Rodionov [13] and Sather [14] theoretical models would remove about 1/3 of the NuTeV anomaly. There are other models that predict substantially smaller CSV effects on the NuTeV result [27,28,29], but all theoretical predictions are well within the phenomenological limits established by MRST. The magnitude of CSV effects allowed by the MRST fit makes isospin violation one of the only viable explanations for the NuTeV anomalous value for the Weinberg angle.

If CSV effects are sufficiently large to remove the Weinberg angle anomaly, such effects should be visible in various other experiments. Several possible experiments to test parton CSV were reviewed by Londergan and Thomas [30]. We briefly review three such possibilities. The first would be a comparison of Drell-Yan (DY) reactions from charged pions interacting with an isoscalar target. Comparison of, say, π^+-D and π^--D DY reactions would be sensitive to the presence of parton CSV. A study of these DY reactions [31] predicted CSV effects of about 2% in magnitude. Another experiment that could detect CSV effects would be semi-inclusive deep inelastic scattering (SIDIS) on an isoscalar target. A study of semi-inclusive π^+ and π^- leptoproduction on deuterium [32] predicted measurable effects from CSV. However, the ability to measure CSV effects in SIDIS reactions requires accurate knowledge of "favored" and "non-favored" fragmentation functions. In the studies of both DY and SIDIS reactions, the CSV effects were three times smaller than those necessary to explain the NuTeV anomaly. If isospin violating effects are really the explanation of the NuTeV effect, both of these reactions should produce effects at the several percent level. This is currently under investigation. A third possible test of parton isospin violation would be the measurement of W asymmetries in high-energy $p-D$ reactions. This could be carried out at RHIC if deuteron beams were available [33]. We are currently investigating the feasibility of this reaction, and the asymmetries that would be allowed by MRST phenomenological fits including CSV.

In conclusion, despite recent progress in constraining parton isospin violation, experimental data still allows parton CSV terms at the several percent level. This has been demonstrated by the MRST global fit that incorporates isospin violation, although the form of the CSV terms was fixed in their global fit. It is clearly of great interest to investigate this issue experimentally, either to decrease the allowed upper limits on isospin violating PDFs, or to measure isospin violating effects that might explain the anomalous NuTeV value for the Weinberg angle.

Theoretical work cited here was carried out with A.W. Thomas. The author thanks W. Melnitchouk, K. McFarland, S. Kretzer, F. Olness, W-K Tung and R. Thorne for useful discussions.

References

1. G.A. Miller, B.M.K. Nefkens, I. Slaus: Phys. Rep. **194**, 1 (1990)
2. E.M. Henley, G.A. Miller in *Mesons in Nuclei*, eds. M. Rho, D.H. Wilkinson (North-Holland, Amsterdam 1979)
3. J.T. Londergan, A.W. Thomas, arXiv:hep-ph/0407247
4. CCFR Collaboration, W.G. Seligman et al.: Phys. Rev. Lett. **79**, 1213 (1997)
5. NMC Collaboration, P. Amaudruz et al.: Phys. Rev. Lett. **66**, 2712 (1991); Phys. Lett. B **295**, 159 (1992)
6. NMC Collaboration, M. Arneodo et al.: Nucl. Phys. B **483**, 3 (1997)
7. CCFR Collaboration, U.K. Yang et al.: Phys. Rev. Lett. **86**, 2742 (2001)
8. CDF Collaboration, F. Abe et al.: Phys. Rev. Lett. **81**, 5754 (1998)
9. MRST Collaboration, A.D. Martin et al.: arXiv:hep-ph/0308087
10. C.J. Benesh, J.T. Londergan: Phys. Rev. C **58**, 1218 (1998)
11. E605 Collaboration, G. Moreno et al.: Phys. Rev. D **43**, 2815 (1991)
12. A.I. Signal and A.W. Thomas: Phys. Lett. B **191**, 205 (1987); Phys. Rev. D **40**, 2832 (1989)
13. E. Rodionov, A.W. Thomas, J.T. Londergan: Mod. Phys. Lett. A **9**, 1799 (1994)
14. E. Sather: Phys. Lett. B **274**, 433 (1992)
15. E.A. Paschos, L. Wolfenstein: Phys. Rev. D **7**, 91 (1973)
16. NuTeV Collaboration, G.P. Zeller et al.: Phys. Rev. Lett. **88**, 091802 (2002)
17. D. Abbaneo et al.: arXiv:hep-ex/0112021
18. S. Davidson, S. Forte, P. Gambino, N. Rius, A. Strumia: JHEP **202**, 037 (2002)
19. J.T. Londergan: arXiv:hep-ph/0408243
20. D.Yu. Bardin, V.A. Dokuchaeva: report JINR-E2- 86-260, unpublished
21. K-P.O. Diener, S. Dittmaier, W. Hollik: Phys. Rev. D **65**, 073005 (2004)
22. M. Hirai, S. Kumano, T-H. Nagai: arXiv:hep-ph/0408023
23. CCFR Collaboration, A.O. Bazarko et al.: Z. Phys. C **65**, 189 (1995)
24. NuTeV Collaboration, M. Goncharov et al.: Phys. Rev. D **64**, 112006 (2001)
25. S. Kretzer: arXiv:hep-ph/0405221; S. Kretzer et al.: arXiv:hep-ph/0312322; F. Olness et al.: arXiv:hep-ph/0312323
26. K. McFarland: these procedings
27. NuTeV Collaboration, G.P. Zeller et al.: Phys. Rev. D **65**, 111103 (2002)
28. C.J. Benesh, T. Goldman: Phys. Rev. C **55**, 441 (1997)
29. F.G. Cao, A.I. Signal: Phys. Lett. B **559**, 229 (2003)
30. J.T. Londergan, A.W. Thomas: in *Progress in Particle and Nuclear Physics*, Volume 41, p. 49, ed. A. Faessler: (Elsevier Science, Amsterdam, 1998)
31. J.T. Londergan, G.T. Garvey, G.Q. Liu, E.N. Rodionov, A. W. Thomas: Phys. Lett. B **340**, 115 (1994)
32. J.T. Londergan, A. Pang, A.W. Thomas: Phys. Rev. D **54**, 3154 (1996)
33. C. Boros, J.T. Londergan, A.W. Thomas: Phys. Rev. D **59**, 074021 (1999)

Eur Phys J A (2005) **24**, s2, 89–92
DOI: 10.1140/epjad/s2005-04-019-9

EPJ A direct
electronic only

Pion–nucleon interaction and the strangeness content of the nucleon

M.E. Sainio

[1] Helsinki Institute of Physics, P.O. Box 64, 00014 University of Helsinki, Finland
[2] Department of Physical Sciences, University of Helsinki, Finland

Received: 15 October 2004 / Published Online: 8 February 2005

Abstract. A brief review of the pion-nucleon sigma-term is given.

PACS. 13.75.Gx Pion-baryon interactions – 14.20.Dh Protons and neutrons

1 Introduction

Sigma-terms are proportional to the matrix elements

$$\langle A|m_q\bar{q}q|A\rangle \;\; ; q = u, d, s \;\; ; \;\; A = \pi, K, N$$

of scalar quark currents in the framework of QCD. These matrix elements are of interest, because they are related
- to the mass spectrum,
- to scattering amplitudes through Ward identities,
- to the strangeness content of A,
- to quark mass ratios.

In the following the status of the πN system is considered and the implications to the strangeness content in the nucleon are outlined.

2 The πN sigma-term

The pion-nucleon sigma-term is a measure of explicit chiral symmetry breaking in QCD and it is defined as

$$\sigma = \frac{\hat{m}}{2m}\langle p|\bar{u}u + \bar{d}d|p\rangle, \qquad \hat{m} = \frac{1}{2}(m_u + m_d), \tag{1}$$

which is the $t = 0$ value of the nucleon scalar form factor $\sigma(t)$

$$\bar{u}'\sigma(t)u = \hat{m}\,\langle p'|\bar{u}u + \bar{d}d|p\rangle, \;\; t = (p' - p)^2, \tag{2}$$

i.e. $\sigma = \sigma(t = 0)$. The nucleon mass is denoted by m.

The strangeness content of the proton can be defined as

$$y = \frac{2\,\langle p|\bar{s}s|p\rangle}{\langle p|\bar{u}u + \bar{d}d|p\rangle} \tag{3}$$

(the OZI rule would imply y=0).

Algebraically σ can be written in the form

$$\sigma = \frac{\hat{m}}{2m}\frac{\langle p|\bar{u}u + \bar{d}d - 2\bar{s}s|p\rangle}{1 - y}, \tag{4}$$

where the numerator is proportional to the octet breaking piece in the hamiltonian. To first order in SU(3) breaking we have now

$$\sigma \simeq \frac{\hat{m}}{m_s - \hat{m}}\frac{m_\Xi + m_\Sigma - 2m_N}{1 - y} \simeq \frac{26 \text{ MeV}}{1 - y}, \tag{5}$$

where the quark mass ratio takes the value

$$\frac{m_s}{\hat{m}} = 2\,\frac{M_K^2}{M_\pi^2} - 1 \simeq 25 \tag{6}$$

in terms of the kaon and pion masses. Chiral perturbation theory (ChPT) allows us to determine the combination

$$\hat{\sigma} = \sigma(1 - y) \tag{7}$$

from the baryon spectrum.

For $\hat{\sigma}$ we have:

- 26 MeV (leading order)
- 35±5 MeV, $\mathcal{O}(m_q^{3/2})$ [1]
- 36±7 MeV, $\mathcal{O}(m_q^2)$ [2]
- 33±3 MeV, $\mathcal{O}(m_q^2)$ [3],

where the difference in the last two determinations is the regularization method used, dimensional regularization [2] or cut-off [3].

With the help of the Feynman-Hellmann theorem the sigma-term can be extracted from the nucleon mass

$$\sigma = \hat{m}\frac{\partial m}{\partial \hat{m}}. \tag{8}$$

Equivalently, employing $M^2 = 2\hat{m}B$,

$$\sigma = M^2\frac{\partial m}{\partial M^2}, \tag{9}$$

where B is the scalar vacuum condensate. The quark mass expansion of the nucleon mass [4]

$$m = m_0 + k_1 M^2 + k_2 M^3 + k_3 M^4 \ln \frac{M^2}{m_0^2} + k_4 M^4 + \mathcal{O}(M^5) \tag{10}$$

yields for σ

$$\sigma = k_1 M^2 + \frac{3}{2} k_2 M^3 + k_3 M^4 \{2 \ln \frac{M^2}{m_0^2} + 1\} + 2k_4 M^4 + \mathcal{O}(M^5). \tag{11}$$

Here the factors k_i contain the low-energy constants appearing in the respective chiral order. Numerically

$$\sigma = (75 - 23 - 7 + 0) \text{ MeV} = 45 \text{ MeV},$$

where the $\mathcal{O}(M^2)$ term, k_1, has been fixed by taking σ to have the value 45 MeV [5].

3 The πN amplitude

To relate the sigma-term discussion to the scattering information the standard representation for the πN amplitude is adopted

$$T_{\pi N} = \bar{u}'[A(\nu,t) + \frac{1}{2}\gamma^\mu (q+q')_\mu B(\nu,t)]u. \tag{12}$$

The definition of the crossing variable is

$$\nu = \frac{s-u}{4m} = \omega + \frac{t}{4m}, \tag{13}$$

where ω is the pion laboratory energy. The amplitude D is defined as

$$D(\nu,t) = A(\nu,t) + \nu B(\nu,t) \tag{14}$$

and its imaginary part can be related to the total cross section through the optical theorem, $\text{Im}\, D(\omega, t=0) = k_{\text{lab}}\, \sigma$. The isoscalar (+) and isovector (−) amplitudes $D^\pm$ can be written in terms of the amplitudes in the physical channels as

$$D^\pm = \frac{1}{2}(D_{\pi^- p} \pm D_{\pi^+ p}). \tag{15}$$

4 A low-energy theorem

Chiral symmetry allows us to write at the Cheng-Dashen point, i.e. at $(\nu = 0, t = 2M_\pi^2)$,

$$\Sigma \doteq F_\pi^2 \bar{D}^+(\nu = 0, t = 2M_\pi^2) = \sigma(2M_\pi^2) + \Delta_R, \tag{16}$$

where F_π is the pion decay constant, $\bar{D}^+$ is the isoscalar D-amplitude with the pseudovector Born term subtracted and Δ_R is the remainder, which is formally of the order M_π^4 [6]. To one-loop in ChPT ($\mathcal{O}(q^3)$) [7]

$$\Delta_R = 0.35 \text{ MeV}. \tag{17}$$

One-loop in HBChPT ($\mathcal{O}(q^4)$) gives the upper limit [8]

$$\Delta_R \simeq 2 \text{ MeV} \tag{18}$$

and here it is notable that no logarithmic contribution to order M_π^4 appears. This allows us to write

$$\Sigma \simeq \sigma(2M_\pi^2). \tag{19}$$

What remains to be fixed in order to determine the σ is the form factor difference

$$\Delta_\sigma = \sigma(2M_\pi^2) - \sigma(0). \tag{20}$$

ChPT to one loop gives [7]

$$\Delta_\sigma \simeq 5 \text{ MeV}. \tag{21}$$

Dispersion analysis yields [9]

$$\Delta_\sigma = 15.2 \pm 0.4 \text{ MeV}. \tag{22}$$

Becher and Leutwyler obtain [4]

$$\Delta_\sigma = 14.0 \text{ MeV} + 2M^4 \bar{e}_2, \tag{23}$$

where $\bar{e}_2$ is a renormalized coupling constant appearing in the $\mathcal{L}_{\text{N}}^{(4)}$ lagrangian. The constant $\bar{e}_2$ is expected to be small [4].

5 The Σ-term

Inside the Mandelstam triangle it is convenient to employ the subthreshold expansion [10], where $\bar{D}^+$ is expanded in powers of ν^2 and t

$$\bar{D}^+ = d_{00}^+ + d_{10}^+\nu^2 + d_{01}^+ t + d_{20}^+\nu^4 + d_{11}^+\nu^2 t + \cdots \tag{24}$$

The curvature term Δ_D is defined as

$$\Sigma = F_\pi^2(d_{00}^+ + 2M_\pi^2 d_{01}^+) + \Delta_D \equiv \Sigma_d + \Delta_D \tag{25}$$

and it is dominated by the $\pi\pi$ cut giving the result [9]

$$\Delta_D = 11.9 \pm 0.6 \text{ MeV}. \tag{26}$$

The linear part Σ_d is a sensitive quantity due to the cancellation of the d_{00}^+ and d_{01}^+ pieces in

$$\Sigma_d(\text{A}) = (-91.3 + 138.8) \text{ MeV} \simeq 48 \text{ MeV} \tag{27}$$

$$\Sigma_d(\text{B}) = (-94.5 + 144.2) \text{ MeV} \simeq 50 \text{ MeV} \tag{28}$$

corresponding to the two solutions (A and B) discussed in [5]. These numbers lead to $\Sigma \simeq 60$ MeV, which is consistent with the old result of Koch [11] $\Sigma = 64 \pm 8$ MeV based on hyperbolic dispersion relations.

6 The strangeness content of the nucleon

Putting all these pieces together leads to a determination of the strangeness content of the proton

$$\frac{\hat{\sigma}}{1-y} = \Sigma - \Delta_R - \Delta_\sigma \tag{29}$$

and numerically with the solution A

$$\frac{35\text{ MeV}}{1-y} = (60 - 2 - 15)\text{ MeV}, \tag{30}$$

which gives $y \simeq 0.2$ with a sizeable error. This value of y corresponds to about 130 MeV in the proton mass being due to the strange sea.

7 Partial wave analysis

The analysis discussed in Sect. 5 was based on the KH80 solution of the Karlsruhe group [12]. The data basis used there contained mainly pre-meson-factory-era data and, therefore, it is of great interest to perform a new analysis with the new data in the spirit of the Karlsruhe group incorporating fixed-t constraints. This would hopefully help in fixing the value of Σ more accurately. In the forward direction it is feasible to solve the dispersion relations directly, but for $t < 0$ it is more practical to use the expansion method [10]. E.g., for the C^+ amplitude ($C = A + \nu/(1 - t/4m^2)B$)

$$C^+(\nu,t) = C_N^+(\nu,t) + H(Z,t)\sum_{n=0}^{N} c_n^+ Z^n,$$

where $C_N^+(\nu,t)$ is the Born term, the function H is adjusted to the asymptotic behaviour of the amplitude and Z is the conformal mapping

$$Z(\nu^2,t) = \frac{\alpha - \sqrt{\nu_{th}^2 - \nu^2}}{\alpha + \sqrt{\nu_{th}^2 - \nu^2}}, \tag{31}$$

where $\nu_{th} = \mathrm{M}_\pi + t/4m$ and α is a real parameter. The convergence and smoothing is taken care of by a convergence test function

$$\chi_T^2 = \lambda \sum_{n=0}^{N} c_n^2 (n+1)^3,$$

which is one component in the χ^2 expression to be minimized. Other contributions include χ^2_{DATA} and χ^2_{PW}, where the latter is calculated from the previous iteration of the partial wave solution.

To demonstrate the working of the expansion method at $t = 0$ with N=40, Figs. (1) and (2) display Re $C^+(\omega, t = 0)$ and $\sigma^T_{\pi^+p}$. For the real part of the C^+-amplitude there are three data points at low energy as input and they fix the subtraction constant appearing in the dispersion relation for the C^+.

The VPI/GWU group has recently published a partial wave analysis [13], which does employ fixed-t constraints. The publication does not, however, give a value for the Σ,

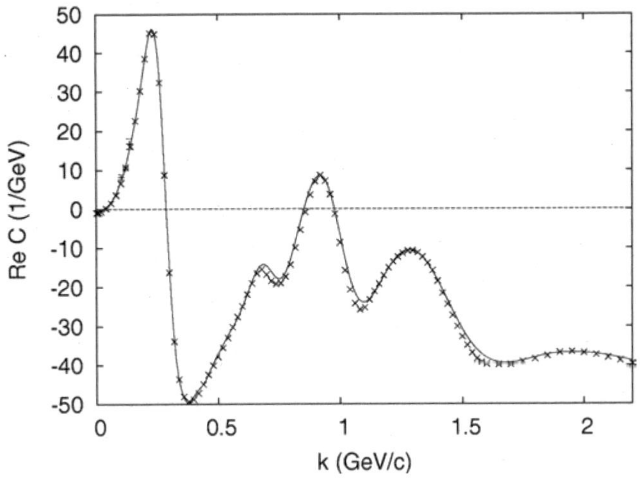

Fig. 1. The real part of the C^+-amplitude. The *crosses* refer to the tabulated values in [10]

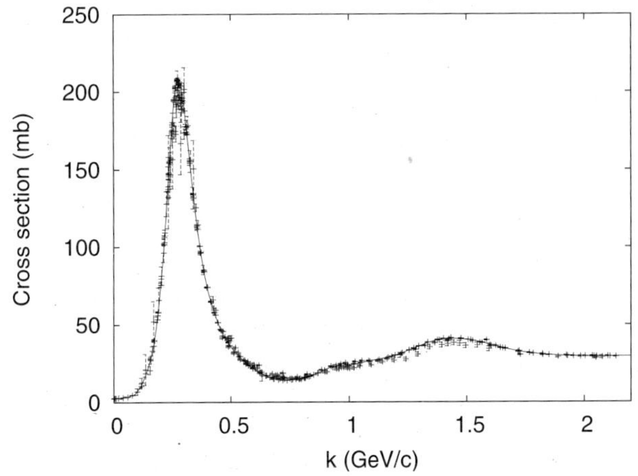

Fig. 2. The π^+p total cross section

but indications are [14] that the number could be 20-30 % larger than the numbers quoted above.

8 The relation $\Sigma \leftrightarrow$ threshold

The issue of relating the Σ to the values of threshold parameters is an old one [15]. In general, Σ can be expressed in terms of the threshold parameters [16]

$$\Sigma = F_\pi^2[L(a_l,\tau) + (1 + \frac{M_\pi}{m})\tau J^+] + \delta, \tag{32}$$

where $L(a_l,\tau)$ is a linear combination of the threshold parameters a_l, τ is a free parameter to single out individual scattering lengths and J^+ is the integral over the isoscalar combination of the total cross section. The remainder, δ, contains contributions from the Born term, the Δ and the loop corrections. The approach of Altarelli et al. [15] corresponds to choosing $\tau = -1$, but without loops. However, at present, one has to rely on dispersion methods to extract the threshold parameters anyway, so the value of any such formula is limited.

9 Lattice results

ChPT permits a study of the quark mass dependence of the nucleon mass. This makes it possible to have a connection to the lattice data, where, currently, only unphysically high quark masses can be dealt with. New accurate data from the CP-PACS, JLQCD and QCDSF-UKQCD collaborations (dynamical quarks, two flavours) give [17]

$$\sigma = 49 \pm 3 \text{ MeV} \tag{33}$$

to $\mathcal{O}(q^4)$ in ChPT. Another approach including the leading nonanalytic and next-to-leading nonanalytic behaviour yields [18]

$$\sigma = 35 - 73 \text{ MeV}. \tag{34}$$

10 Conclusions

The challenge at present seems to be in determining Σ. That involves a number of questions

– one has to deal with conflicting sets of data,

– one has to rely on the Tromborg [19] formalism for the electromagnetic corrections even though there are indications [20] that further improvements in this sector should be incorporated,

– the extrapolation from the low-energy region to the Cheng-Dashen point could, to some extent, be sensitive to the d-waves, which otherwise cannot be fixed with the low-energy scattering information [21].

The new direction with the lattice calculations is gradually getting very interesting as far as the sigma-term is concerned. However, further improvements, i.e. smaller m_q-values, will still be needed.

Acknowledgements. I wish to thank P. Piirola for providing the figures and A.M. Green for useful comments on the manuscript. Support from the Academy of Finland grant 54038 and the EU grant HPRN-CT-2002-00311, EURIDICE, is acknowledged.

References

1. J. Gasser, H. Leutwyler: Phys. Reports **87**, 77 (1982)
2. B. Borasoy, U.-G. Meißner: Ann. Phys. **254**, 192 (1997)
3. B. Borasoy: Eur. Phys. J. C **8**, 121 (1999)
4. T. Becher, H. Leutwyler: Eur. Phys. J. C **9**, 643 (1999); JHEP **6**, 017 (2001)
5. J. Gasser, H. Leutwyler, M.E. Sainio: Phys. Lett. B **253**, 252 (1991)
6. L.S. Brown, W.J. Pardee, R.D. Peccei: Phys. Rev. D **4**, 2801 (1971)
7. J. Gasser, M.E. Sainio, A. Švarc: Nucl. Phys. B **307**, 779 (1988)
8. V. Bernard, N. Kaiser, U.-G. Meißner: Phys. Lett. B **389**, 144 (1996)
9. J. Gasser, H. Leutwyler, M.E. Sainio: Phys. Lett. B **253**, 260 (1991)
10. G. Höhler: Pion-Nucleon Scattering, Landolt-Börnstein Vol.I/9b2 (Springer, Berlin, 1983)
11. R. Koch: Z. Phys. C **15**, 161 (1982)
12. R. Koch, E. Pietarinen: Nucl. Phys. A **336**, 331 (1980)
13. R.A. Arndt, W.J. Briscoe, I.I. Strakovsky, R.L. Workman, M.M. Pavan: Phys. Rev. C **69**, 035213 (2004)
14. M.M. Pavan, R.A. Arndt, I.I. Strakovsky, R.L. Workman: πN Newsletter **16**, 110 (2002)
15. G. Altarelli, N. Cabibbo, L. Maiani: Phys. Lett. B **35**, 415 (1971); Nucl. Phys. B **34**, 621 (1971)
16. J. Gasser: Proc. 2nd Int. Workshop on πN Physics, Los Alamos 1987, ed. W.R. Gibbs and B.M.K. Nefkens: Los Alamos report LA-11184-C 266 (1987)
17. M. Procura, T.R. Hemmert, W. Weise: Phys. Rev. D **69**, 034505 (2004)
18. D.B. Leinweber, A.W. Thomas, R.D. Young: hep-lat/0302020 (2003)
19. B. Tromborg, S. Waldenstrøm, I. Øverbø: Phys. Rev. D **15**, 725 (1977)
20. N. Fettes, U.-G. Meißner: Nucl. Phys. A **693**, 693 (2001)
21. J. Stahov: hep-ph/0206041 (2002)

Eur Phys J A (2005) **24**, s2, 93–96
DOI: 10.1140/epjad/s2005-04-020-4

EPJ A direct
electronic only

Strange form factors of the nucleon in the chiral quark-soliton model

A. Silva[1 a], D. Urbano[1,2], H.-C. Kim[3], and K. Goeke[4]

[1] Centro de Física Computacional, Departamento de Física da Universidade de Coimbra, P-3004-516 Coimbra, Portugal
[2] Faculdade de Engenharia da Universidade do Porto, R. Dr. Roberto Frias s/n, P-4200-465 Porto, Portugal
[3] Department of Physics, Pusan National University, 609-735 Pusan, Republic of Korea
[4] Institut für Theoretische Physik II, Ruhr-Universität Bochum, D-44780 Bochum, Germany

Received: 15 October 2004 / Published Online: 8 February 2005

Abstract. The results for the strange electromagnetic and axial form factors in the chiral quark-soliton model are reviewed. The roles of a new quantization method and of the meson asymptotic behaviour are discussed. Predictions for the A4 and G0 experiments are presented.

PACS. 12.40.-y Other models for strong interactions – 14.20.Dh Protons and neutrons – 13.40.Gp Electromagnetic form factors

1 Introduction

The question of the strange content of the nucleon is a very important one in the understanding of hadron structure, which eventually is to be settled by experiment. Indeed, on one hand there is no obvious theoretical reason why strange $s\bar{s}$ pairs should not contribute to the nucleon properties. On the other hand the theoretical mechanism and the actual contribution of strange quarks are not yet understood as well.

Early results about the strange quark bilinears in the nucleon included the analysis of the sigma term [1] and polarized DIS. These first results indicated strange quark nonvanishing contributions to the nucleon's mass and spin, respectively. The strange vector currents associated with electromagnetic form factors were discovered to be accessible by means of weak neutral current experiments [2,3, 4]. For reviews see [5,6].

Due to its importance, the question of the strange form factors has been addressed before in the context of the chiral quark-soliton model (CQSM) [7,8]. This work reviews the CQSM results [9,10] for the strange electromagnetic form factors and presents also the strange axial form factor. These form factors are obtained using a new quantization method [11] and studied phenomenologically in terms of the effects of the asymptotic behavior of the meson fields.

[a] Financial support from the organization of PAVI04 and from the Center for Computational Physics (CFC) for the participation in PAVI04 is kindly acknowledged. The work of A.S. has been partially supported by the grant PRAXIS XXI BD/15681/98.

2 The chiral quark–soliton model (CQSM)

The chiral quark-soliton model (CQSM) has been applied successfully to the description of many baryon observables, both in flavours SU(2) and SU(3). See [12,13] for reviews. The Lagrangian of the CQSM reads,

$$\mathcal{L} = \bar{\psi}(x)\left(i\,\partial\!\!\!/ - m - MU^{\gamma_5}\right)\psi(x) \qquad (1)$$

where $m = \mathrm{diag}(m_u, m_d, m_s)$ is the current quark mass matrix and M the constituent quark mass. M is the only free parameter in the model. The main property of the model in relation with hadron physics is chiral symmetry and its spontaneous breaking. The auxiliary chiral fields $U^{\gamma_5} = e^{i\gamma^5 \pi\cdot\tau}$ in (1), may be interpreted as physical meson fields and, thus, the Lagrangian (1) corresponds to a theory of constituent quarks interacting through the exchange of mesons.

However, the field description embodied in (1) is not renormalizable. A regularization method is therefore necessary to completely define the model. This work uses the proper-time regularization, which introduces one cut-off parameter into the model. The remaining parameters of the model are the current quark masses. In this work isospin breaking effects have been neglected and therefore the current mass parameters are the equal current masses $(\overline{m})$ of the u and d quarks and the current mass (m_s) of the strange quarks. In order to fit these parameters, one calculates the pion and kaon masses as well as the pion decay constant in terms of the parameters of the model. Equating the model results for these quantities with their experimental values allows to fix the parameters of the model: the current nonstrange quark masses, the strange quark mass and the cut-off come out around 8 MeV, 164 MeV and 700 MeV, respectively.

3 The model baryon state

In broad terms, the construction of the model baryon states necessary to compute baryon matrix elements are obtained in a projection after variation procedure, with the variation referring to the mean field solution.

Taking the large N_c limit of the correlation function of two baryon currents and neglecting the strange quark mass lead to an effective action from which the mean field solution is obtained. At this level the effective action is proportional to the energy of the mean field, $S_{\mathrm{eff}}[U] \sim \mathcal{M}[U]$, which is a functional of the chiral fields U: $\mathcal{M}[U] = N_c\epsilon_{\mathrm{V}}[U] + \epsilon_{\mathrm{S}}[U]$, with ϵ_{V} the energy of the valence level, occupied by Nc quarks, and ϵ_{S} the regularized energy of the Dirac sea. Both result from the solution of the Dirac one-particle problem $\epsilon_n[U] = \langle n|\, h(U)\, |n\rangle$ with

$$h(U) = -i\gamma^0\gamma^i\partial_i + \gamma^0 M U^{\gamma_5} + \overline{m}\gamma^0 \,. \qquad (2)$$

In order to solve such problem one further restricts the mesons to the hedgehog shape, $\pi \cdot \tau = P(r)\,\hat{n} \cdot \tau$, with a profile function $P(r)$ vanishing at large distances. The minimization of the action corresponds to a profile function $P_c(r)$ which represents the mean field soliton. The formalism in flavour SU(3) is built upon the embedding of the SU(2) hedgehog into an SU(3) matrix [14]:

$$U = e^{iP(r)\sum_{a=1}^{8} n^a\lambda^a} \rightarrow \begin{pmatrix} e^{iP(r)\sum_{i=1}^{3} n^i\tau^i} & 0 \\ 0 & 1 \end{pmatrix}. \qquad (3)$$

3.1 Quantization

The quantization of the mean field is achieved in the context of collective coordinates. The collective coordinates are the orientations of the soliton in configuration as well as in flavour spaces (related by the hedgehog) and the position of the center of mass. These coordinates are most suited to describe the rotational zero modes, i.e. unconstrained large amplitude motion, which should be treated exactly. The model calculation neglects modes normal to the zero modes, which are subjected to restoring forces and hence suppressed.

The time dependent solutions are constructed on the basis of the restriction to the zero modes by the ansatz[1]

$$U(\boldsymbol{x}) \rightarrow A(t) U_c(\boldsymbol{x}) A^\dagger(t) \qquad (4)$$

with the collective coordinates contained in A.

In the laboratory system the Dirac operator becomes

$$D(U) = A\left[D(U_c) + A^\dagger\dot{A} + \gamma^0 A^\dagger \delta m A\right] A^\dagger \,, \qquad (5)$$

which clearly shows the two expansion parameters used in this work: the angular velocity $A^\dagger\dot{A}(t) = i\Omega_E^\alpha\lambda^\alpha/2$ and the strange mass related parameter δm, discussed below, in Sect. 3.2, in connection with the asymptotic meson behaviour effects.

[1] Not showing translational zero modes explicitly.

The collective coordinate Hamiltonian is

$$H = \mathcal{M}(U_c) + H^{\mathrm{sc}}_{\mathrm{coll}} + H^{\mathrm{sb}}_{\mathrm{coll}} \,, \qquad (6)$$

i.e. it is composed by a flavour symmetry conserving (sc) and a symmetry breaking (sb) pieces. While the first is treated exactly, the second is treated perturbatively. The wave functions of the symmetry conserving part are the Wigner SU(3) matrices. For the octet

$$|B, 8\rangle^{\mathrm{sc}} \rightarrow \Psi^{(8)\mathrm{sc}}_{\nu\nu'}(A) = \sqrt{8}\,(-)^{Y'/2+J_3}\, D^{(8)*}_{\nu\nu'}(A) \,. \qquad (7)$$

with $\nu = (Y, T, T_3)$ and $\nu' = (Y', J, J_3)$ standing for the quantum numbers of the baryon state ($Y' = -1$). The symmetry breaking part of the Hamiltonian leads to a representation mixing, in first order perturbation theory, from which the baryon wave function $|B, 8\rangle^{\mathrm{sc}}$ picks terms from higher order representations [14].

3.2 Asymptotics of the meson fields and symmetry conserving quantization

The question of the asymptotics of the meson fields originates in the fact that the embedding (3) forces all the meson fields to have a common asymptotic behavior. This is not satisfactory in the SU(3) case due to the large mass difference between pions and kaons. In order to have some information on how large these effects may be, one exploits in this work the fact that there is no prescription to fix the multiples of the unity mass matrices in (2) and in the symmetry breaking piece δm of (5):

$$\gamma^0 D(U_c) + A^\dagger \delta m A = -i\gamma^i\partial_i + M U^{\gamma^5} + \overline{m} + A^\dagger \delta m A \quad (8)$$

where $\delta m = M_1 + M_8\lambda^8$, with $M_1 = (m_s - \overline{m})/3$ and $M_8 = (\overline{m} - m_s)/\sqrt{3}$. In this case one may obtain larger mass asymptotics of the meson fields, by, in (8), increasing $\overline{m}$ at the expense of a lower M_1. The pion asymptotics will be denoted in the following by $\mu \rightarrow \pi(\sim 140$ MeV) and the kaon asymptotics by $\mu \rightarrow K(\sim 490)$ MeV).

The symmetry conserving quantization [11] used in this work avoids the problem which is encountered in this model by using the ansatz (4) to compute the relation between isospin (T) and spin (S) operators. While the result from (4) reads

$$T_\alpha = -D^{(n)}_{\alpha\beta}(A)\, J_\beta - \sqrt{3} D^{(n)}_{\alpha 8}(A)\, \frac{I'_2}{I_2} \,, \qquad (9)$$

the correct relation implies $I'_2 = 0$. The symmetry conserving quantization identifies the terms like I'_2 in the expressions for observables on the basis of the quark model limit of the CQSM.

4 Strange electromagnetic form factors

The knowledge of the form factors of the octet vector currents is enough to provide the form factors for each flavour:

$$V^0_\mu = \bar{u}\gamma_\mu u + \bar{d}\gamma_\mu d + \bar{s}\gamma_\mu s \,, \qquad (10)$$

$$V^3_\mu = \bar{u}\gamma_\mu u - \bar{d}\gamma_\mu d \,, \qquad (11)$$

$$\sqrt{3} V^8_\mu = \bar{u}\gamma_\mu u + \bar{d}\gamma_\mu d - 2\bar{s}\gamma_\mu s \,. \qquad (12)$$

Table 1. Strange magnetic moment (in n.m.) and electric and magnetic radii (in fm^2). The constituent quark mass is 420 MeV and the strange quark mass 180 MeV. π and K stand for the two asymptotic descriptions of the mesons as discussed in Sect. 3.2.

μ	π	K
$\langle r^2\rangle_E^s$	−0.220	−0.095
μ_s	0.074	0.115
$\langle r^2\rangle_M^s$	0.303	0.631

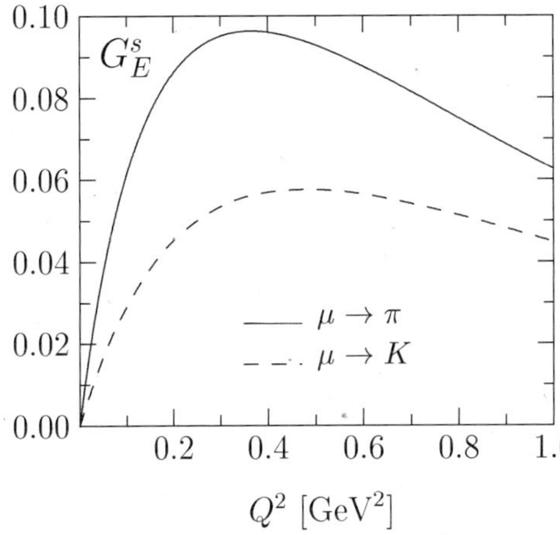

Fig. 1. Strange electric form factor of the nucleon. Conventions and model parameters as in Table 1.

In this work the form factors for these currents were first obtained for the two asymptotic behaviours. The effects of the asymptotics was then studied at the level of baryon form factors by combining flavour form factors with pion asymptotics for nonstrange quarks with kaon asymptotics for strange form factors, i.e. by taking

$$G^B_{E,M}(Q^2) = G^{uB(\pi)}_{E,M}(Q^2) + G^{dB(\pi)}_{E,M}(Q^2) + G^{sB(K)}_{E,M}(Q^2)\,. \tag{13}$$

The most significant consequences of this ansatz for the baryon octet electromagnetic form factors is the improvement in the overall description of magnetic form factors and the neutron electric form factor. This supports the expectation that the strange form factors are better described in terms of kaon ($\mu \to K$) rather than in pion ($\mu \to \pi$) asymptotics. The results for ($\mu \to \pi$) are nevertheless always presented.

The strange magnetic moment and the radii are presented in Table 1 and the strange electric and magnetic form factors in the CQSM are shown in Fig. 1 and Fig. 2, respectively. A particular aspect of these results is the positive value for the strange magnetic moment, which seems to be at variance with many theoretical calculations based in different approaches. The CQSM seems nevertheless to be able to accommodate higher values of this quantity, up to 0.41 n.m., as is found in the "model independent analysis" of [15]. This pinpoints the SU(3) structure of the rotational and mass corrections as the reason for such

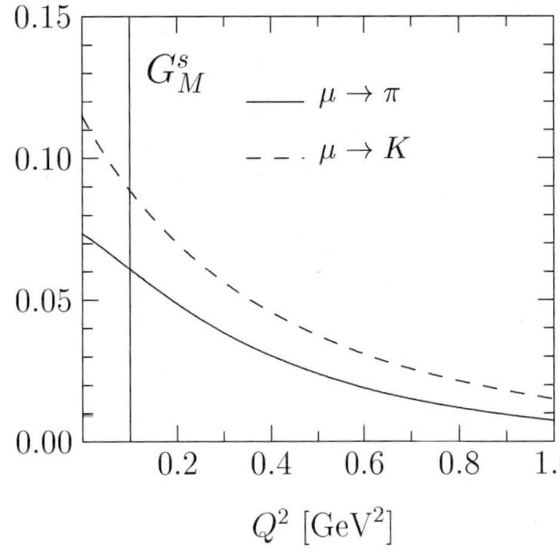

Fig. 2. Strange magnetic form factor in nuclear magnetons. Conventions and model parameters as in Table 1.

positive value, similarly to what is observed in other models by maintaining just the SU(3) structure [16].

4.1 The SAMPLE, HAPPEX, A4, and G0 experiments

The experimental value for the magnetic form factor obtained by the SAMPLE collaboration [17] is

$$G^s_M(Q^2 = 0.1) = +0.37 \pm 0.20 \pm 0.26 \pm 0.07 \text{ (n.m.)}\,. \tag{14}$$

The CQSM results of this work underestimates this experimental result, still falling however within the error bar, shown in Fig. 2 at $Q^2 = 1$. The HAPPEX result [18]

$$(G^s_E + 0.392 G^s_M)(Q^2 = 0.477) = 0.014 \pm 0.030\,, \tag{15}$$

is, on the contrary, overestimated in the CQSM, with the nearest result (kaon asymptotics) being 0.073.

Using the necessary form factors, one may study the combination $G^s_E(Q^2) + \beta(Q^2,\theta) G^s_M(Q^2)$, with $\beta(Q^2,\theta) = \tau G^{p\gamma}_M / \epsilon G^{p\gamma}_E$, $\tau = Q^2/(4M_N^2)$, $\epsilon^{-1} = 1 + 2(1+\tau)\tan^2(\theta/2)$, for the values of θ and Q^2 used in current or near future experiments. Figure 3 compares the model predictions with the recent forward data,

$$(G^s_E + 0.225 G^s_M)(Q^2 = 0.23) = 0.039 \pm 0.034 \tag{16}$$

$$(G^s_E + 0.106 G^s_M)(Q^2 = 0.10) = 0.074 \pm 0.036\,, \tag{17}$$

with (16) from [19] and (17) from [20], by the A4 Collaboration and gives predictions for the backward angle. Figure 4 gives predictions for the G0 experiment.

5 Strange axial form factor

The axial currents of interest are expressed in term of flavour components in the same way as (12). The axial form factors for the octet axial currents have also been computed in the model with the same accuracy of the

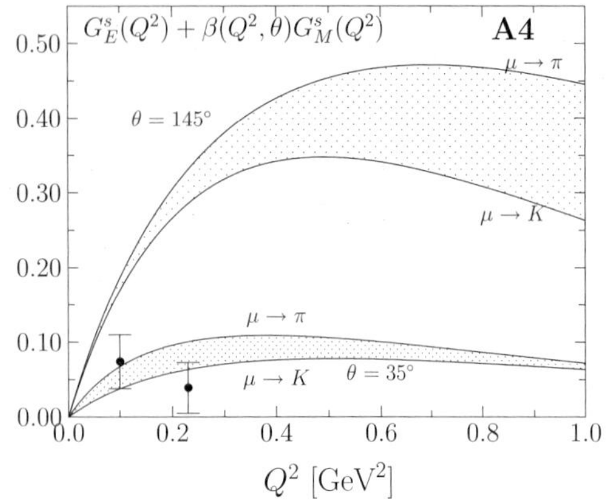

Fig. 3. Predictions of $G_E^s(Q^2)+\beta(Q^2,\theta)G_M^s(Q^2)$ for the A4 experiment. Conventions and model parameters as in Table 1.

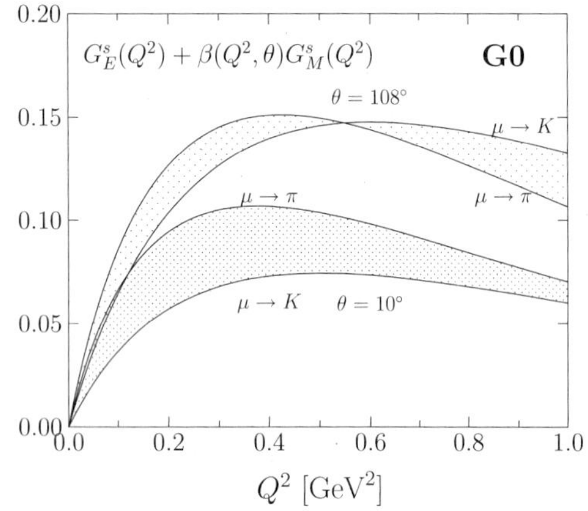

Fig. 4. Predictions for the G0 experiment. Conventions and model parameters as in Table 1.

Table 2. Δs and strange axial r.m.s. Conventions and model parameters as in Table 1.

	π	K
Δs	-0.086	-0.075
$\langle r_s^2\rangle^{1/2}$	0.554	0.172

electromagnetic ones. The axial constants and axial dipole mass reproduce reasonably well the available experimental data. The results for $\Delta s = G_A^s(Q^2=0)$ and the strange axial r.m.s. are shown in Table 2 and the strange axial form factor is presented in Fig. 5. The CQSM results for Δs seem to fall below a recent analysis of DIS data, which yields $\Delta s = -0.14 \pm 0.03$ [21].

6 Conclusions

The CQSM results for the strange form factors are not ruled out by the presently available experimental data. The results are stable from the point of view of the model parameters, with higher order corrections in N_c and δm

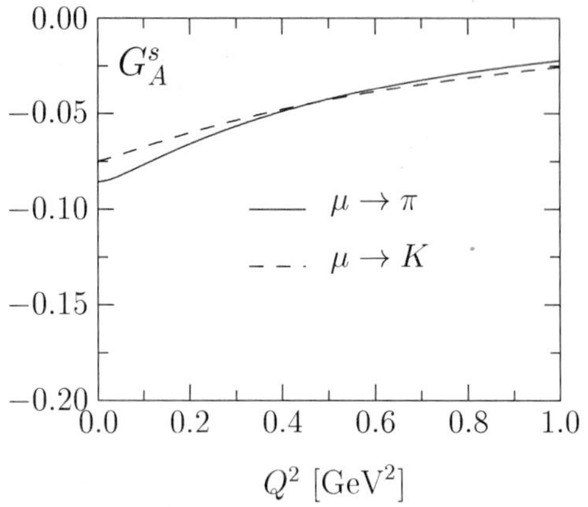

Fig. 5. Strange axial form factor. Conventions and model parameters as in Table 1.

expected below 15 %. The phenomenological observed effects related to the asymptotic behaviour of the meson fields requires nevertheless a more rigorous treatment.

References

1. J. Gasser, H. Leutwyler, M.E. Sainio: Phys. Lett. B **253** (1991) 252
2. D.B. Kaplan, A. Manohar: Nucl. Phys. B **310** (1988) 527
3. D.H. Beck: Phys. Rev. D **39** (1989) 3248
4. M.J. Musolf et al.: Phys. Rept. **239** (1994) 1
5. W.M. Alberico, S.M. Bilenky, C. Maieron: Phys. Rept. **358** (2002) 227 [arXiv:hep-ph/0102269]
6. D.H. Beck, R.D. McKeown: Ann. Rev. Nucl. Part. Sci. **51** (2001) 189 [arXiv:hep-ph/0102334]
7. H.C. Kim, T. Watabe, K. Goeke: Nucl. Phys. A **616** (1997) 606 [arXiv:hep-ph/9606440].
8. H. Weigel, A. Abada, R. Alkofer, H. Reinhardt: Phys. Lett. B **353** (1995) 20 [arXiv:hep-ph/9503241]
9. A. Silva, H.C. Kim, K. Goeke: Phys. Rev. D **65** (2002) 014016 [Erratum-ibid. D **66** (2002) 039902] [arXiv:hep-ph/0107185]
10. A. Silva, H.C. Kim, K. Goeke: arXiv:hep-ph/0210189, to appear in EPJA
11. M. Praszalowicz, T. Watabe, K. Goeke: Nucl. Phys. A **647** (1999) 49 [arXiv:hep-ph/9806431]
12. C.V. Christov et al.: Prog. Part. Nucl. Phys. **37** (1996) 91 [arXiv:hep-ph/9604441]
13. D. Diakonov, V.Y. Petrov: arXiv:hep-ph/0009006
14. A. Blotz et al.: Nucl. Phys. A **555** (1993) 765
15. H.C. Kim, M. Praszalowicz, M.V. Polyakov, K. Goeke: Phys. Rev. D **58** (1998) 114027 [arXiv:hep-ph/9801295]
16. S.T. Hong, B.Y. Park, D.P. Min: Phys. Lett. B **414** (1997) 229 [arXiv:nucl-th/9706008]
17. D.T. Spayde et al. (SAMPLE Collaboration): Phys. Lett. B **583** (2004) 79 [arXiv:nucl-ex/0312016]
18. K.A. Aniol et al. (HAPPEX Collaboration): Phys. Rev. C **69** (2004) 065501 [arXiv:nucl-ex/0402004]
19. F.E. Maas et al. (A4 Collaboration): Phys. Rev. Lett. **93** (2004) 022002 [arXiv:nucl-ex/0401019]
20. F.E. Maas (for the A4 Collaboration): these proceedings
21. B.W. Filippone, X.D. Ji: Adv. Nucl. Phys. **26** (2001) 1 [arXiv:hep-ph/0101224]

Eur Phys J A (2005) **24**, s2, 97–100
DOI: 10.1140/epjad/s2005-04-021-3

EPJ A direct
electronic only

Strange form factors and Chiral Perturbation Theory

Bastian Kubis[a]

Institut für theoretische Physik, Universität Bern, Sidlerstrasse 5, CH-3012 Bern, Switzerland

Received: 15 October 2004 / Published Online: 8 February 2005

Abstract. We review the contributions of Chiral Perturbation Theory to the theoretical understanding or not-quite-yet-understanding of the nucleon matrix elements of the strange vector current.

PACS. 12.39.Fe Chiral Lagrangians – 14.20.Dh Protons and neutrons

1 Introduction

Chiral Perturbation Theory (ChPT), as the low-energy effective field theory of the Standard Model, ought to be predestined to describe the response of the nucleon to the strange vector current: it incorporates all symmetry constraints of the fundamental theory and contains the degrees of freedom relevant at low energies, Goldstone bosons (pions and kaons) as well as matter fields (nucleons). In Sect. 2, we shall point out some specific aspects of ChPT that are essential for the following discussion. We shall reiterate in Sect. 3 why symmetry considerations alone are not sufficient for ChPT to be predictive for the leading moments of the strange form factors, the strange magnetic moment μ_s and the strange electric radius $\langle r^2_{E,s}\rangle$. A low-energy theorem for the strange *magnetic* radius $\langle r^2_{M,s}\rangle$ is presented in Sect. 4, its usefulness and limitations are discussed. Some alternative regularization schemes for loop diagrams have been suggested in order to achieve improved convergence behavior of the chiral series, we shall say a few words on these in Sect. 5. Finally, a conclusion is given in Sect. 6.

2 Some aspects of Chiral Perturbation Theory

2.1 Goldstone bosons and counterterms

ChPT [1] systematically exploits the far-reaching consequences of chiral symmetry: it dictates the appearance of (pseudo) Goldstone bosons (π, K, η) and tightly constrains their interaction with each other as well as with external currents and matter fields. As a consequence, the Goldstone boson dynamics is completely calculable, while the influence of heavier states can be parameterized, at low energies, by polynomials. The polynomial coefficients (called low-energy constants) are not numerically known *a priori*, but are still far from arbitrary, as they potentially interrelate many different observables. A particularly famous example is the quark mass expansion of the pion mass:

$$M_\pi^2 = 2B\hat{m} - \frac{\bar{\ell}_3}{32\pi^2 F_\pi^2}\left[2B\hat{m}\right]^2 + \mathcal{O}\left(\hat{m}^3\right) \quad , \qquad (1)$$

where $\hat{m} = (m_u + m_d)/2$. The first term on the right constitutes the well-known Gell-Mann–Oakes–Renner relation. The correction of order $\hat{m}^2$ is proportional to the low-energy constant $\bar{\ell}_3$ that can be determined from $\pi\pi$ scattering. Given such independent experimental information, the relative size of the contributions to M_π^2 linear and quadratic in the quark masses is known [2].

2.2 Power counting

In the Goldstone boson sector of ChPT, Lorentz invariance dictates that only even powers of momenta appear in the effective Lagrangian. The typical low-energy expansion parameters therefore are $M_\pi^2/\Lambda_\chi^2 \approx 0.02$ (for chiral SU(2)) or $M_K^2/\Lambda_\chi^2 \approx 0.2$ (for chiral SU(3)), where the chiral symmetry breaking scale is $\Lambda_\chi = 4\pi F_\pi \approx 1160$ MeV. In contrast, due to the appearance of spin, there are also odd powers of momenta in the effective pion-nucleon Lagrangian, such that the convergence order-by-order is markedly slower, typical expansion parameters are $M_\pi/m_N \approx 0.15$ in SU(2) or $M_K/m_N \approx 0.5$ for SU(3). Obviously, there is no way we can expect ChPT to work as well for the baryon sector with strangeness as it does for, say, $\pi\pi$ scattering.

It is important to understand the "generic" chiral orders of the lowest moments in the nucleon vector form factors, i.e. the orders at which polynomial contributions occur. The low-energy expansion of the strange Sachs form factors is given by

$$G_{E,s}\left(Q^2\right) = \underbrace{Q_s}_{\mathcal{O}(p)} - \underbrace{\frac{1}{6}\langle r^2_{E,s}\rangle\, Q^2}_{\mathcal{O}(p^3)} + \ldots \ ,$$

[a] *Present address:* HISKP (Th), Universität Bonn, Nussallee 14-16, D-53115 Bonn, Germany

$$G_{M,s}(Q^2) = \underbrace{\mu_s}_{\mathcal{O}(p^2)} - \underbrace{\frac{1}{6}\langle r^2_{M,s}\rangle Q^2}_{\mathcal{O}(p^4)} + \ldots ,$$

where one has $Q_s = 0$ due to gauge invariance for the strange charge. We note in particular that polynomial contributions to the radius terms $\langle r^2_{E,s}\rangle$, $\langle r^2_{M,s}\rangle$ appear at leading and subleading one-loop order, respectively.

3 Why ChPT cannot predict μ_s and $\langle r^2_{E,s}\rangle$

The reason why it is impossible to predict μ_s and $\langle r^2_{E,s}\rangle$ from first principles in ChPT was identified several years ago [3]. There are three independent diagonal vector currents in SU(3),

$$J^{(i)}_\mu = \bar{q}\frac{\lambda^i}{2}\gamma_\mu q , \qquad i = 3, 8, 0, \tag{2}$$

where $\lambda^{3/8}$ are the usual Gell-Mann matrices and $\lambda^0 = \sqrt{2/3}\,\mathrm{diag}(1,1,1)$. The three are proportional to the isovector and isoscalar electromagnetic currents and the baryon number current, respectively. The electromagnetic and the strangeness currents are linear combinations of these,

$$J^{\mathrm{EM}}_\mu = J^{(3)}_\mu + \frac{1}{\sqrt{3}}J^{(8)}_\mu , \quad J^s_\mu = \sqrt{\frac{2}{3}}J^{(0)}_\mu - \frac{2}{\sqrt{3}}J^{(8)}_\mu , \tag{3}$$

i.e. the response to one component of the strangeness current, the baryon number, is going to be completely independent of what we know from electromagnetic probes. In the effective Lagrangian language, this means that wherever matrix elements of the electromagnetic current depend on low-energy constants, there will be a *new, independent* constant for the strangeness current. As an example, consider the leading terms contributing to the magnetic moments,

$$\mathcal{L}^{(2)} = \frac{b_6^{D/F}}{8m_N}\langle \bar{B}\sigma^{\mu\nu}\left[F^+_{\mu\nu}, B\right]_\pm\rangle + \frac{b_6^0}{8m_N}\langle \bar{B}\sigma^{\mu\nu}B\rangle\langle F^+_{\mu\nu}\rangle . \tag{4}$$

The constants $b_6^{D/F}$ can be fitted alternatively to the magnetic moments of proton and neutron or to all octet moments, but b_6^0 only appears in the strange magnetic moment. The same pattern emerges for all low-energy constants, therefore instead of predicting the strange vector form factors, ChPT can only adjust its constants in order to reproduce experimental findings.

4 The strange magnetic radius

Besides being a quantity of interest in its own right, the strange magnetic radius $\langle r^2_{M,s}\rangle$ is also of high importance for the experimental determination of the strange magnetic *moment* μ_s for the following reason: Experimental measurements of $G_{M,s}(Q^2)$ always have to be performed at finite, non-vanishing Q^2, therefore one needs to extrapolate to $G_{M,s}(0) = \mu_s$ [4].

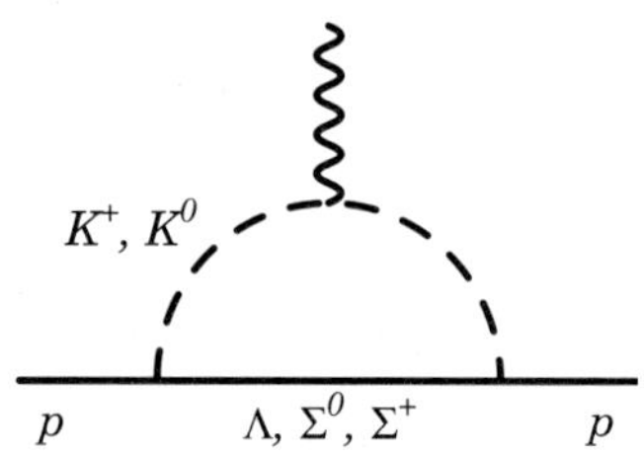

Fig. 1. Diagram generating the leading contribution to $\langle r^2_{M,s}\rangle$

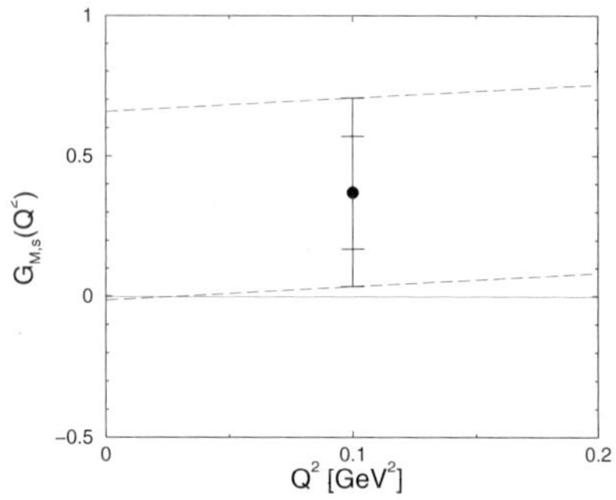

Fig. 2. Extrapolation of the SAMPLE value $G_{M,s}(Q^2)$ to μ_s

4.1 A low-energy theorem for $\langle r^2_{M,s}\rangle$

After the pessimistic finding of the last section, how can there possibly be any low-energy theorem for any strange vector form factor? The answer is, through leading non-analytic loop effects. The diagram of order $\mathcal{O}(p^3)$ displayed in Fig. 1 generates a contribution to the strange magnetic radius according to [4]

$$\langle r^2_{M,s}\rangle^{(3)} = -\frac{5D^2 - 6DF + 9F^2}{48\pi F_K^2}\frac{m_N}{M_K} + \mathcal{O}(M_K^0) . \tag{5}$$

As a low-energy constant can only contribute to $\langle r^2_{M,s}\rangle$ at the next order, generating a term of $\mathcal{O}(M_K^0)$, the term in (5) is (at least formally) dominant. All the masses and coupling constants in (5) are known, hence we have a parameter-free prediction for $\langle r^2_{M,s}\rangle$,

$$\langle r^2_{M,s}\rangle = -0.115\ \mathrm{fm}^2 . \tag{6}$$

We can use the leading chiral prediction for $\langle r^2_{M,s}\rangle$ to extrapolate from the SAMPLE result [5] $G_{M,s}(0.1\,\mathrm{GeV}^2) = 0.37 \pm 0.20 \pm 0.26 \pm 0.07$ to the strange magnetic moment,

$$\mu_s = 0.32 \pm 0.20 \pm 0.26 \pm 0.07 . \tag{7}$$

This extrapolation is visualized in Fig. 2. Furthermore, in [7] this was combined with a second experimental result, the HAPPEX measurement [6] for $G_{E,s} + 0.39 G_{M,s}$, in order to fix both unknowns in the $\mathcal{O}(p^3)$ representation of the strange form factors (for μ_s and $\langle r^2_{E,s}\rangle$) and predict also the Q^2–dependence of the strange electric form factor, see Fig. 3. We note from that figure that even over a large Q^2 range, the form factor looks essentially linear

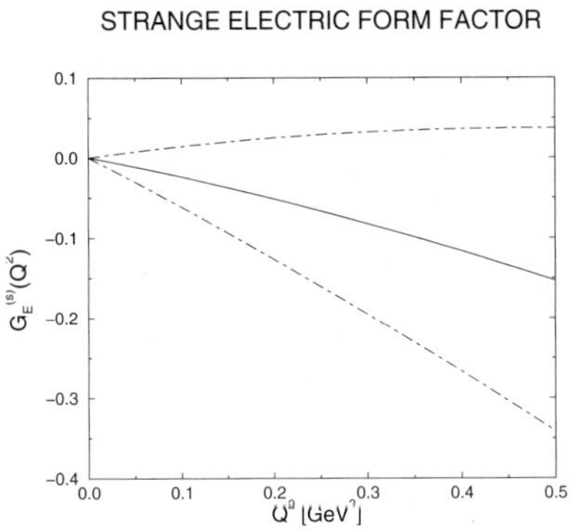

Fig. 3. Q^2–dependence of $G_{E,s}$ at $\mathcal{O}(p^3)$. The figure is taken from [7]

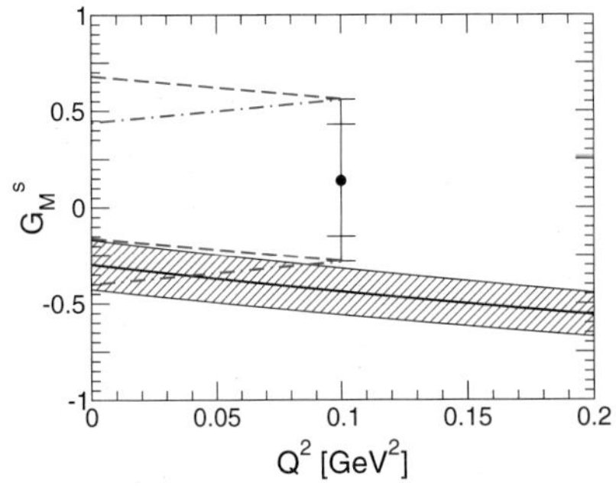

Fig. 4. Extrapolation of the SAMPLE value $G_{M,s}(Q^2)$ to μ_s, using the $\mathcal{O}(p^4)$ value for $\langle r^2_{M,s}\rangle$. The *grey band* indicates a dispersion-theoretical prediction. The figure is taken from [8]

and displays only very little curvature. The reason for this is that the closest infrared singularity in this representation is the $K\bar{K}$ cut, which has a threshold far away from the physical region, and that counterterms providing curvature in the polynomial part are suppressed to two-loop order.

4.2 Stability of the low-energy theorem

Even though the low-energy theorem (5) is strictly a *theorem*, it is an important problem to study its stability when subject to higher-order corrections. A calculation to $\mathcal{O}(p^4)$ has been performed [8] (see also [9]), from which we only quote the numerical result,

$$\langle r^2_{M,s}\rangle^{(4)} = -\big(0.04 + 0.3\, b^r_s\big)\ \text{fm}^2 \ , \tag{8}$$

where the low-energy constant b^r_s is expected to be of order 1. We note that the corrections to the central value as compared to (6) are sizeable, and that the unknown constant induces a theoretical uncertainty in the extraction of μ_s from experiment. The corresponding plot is shown in Fig. 4 (taken from [8]).

5 Variants of loop regularization

We have seen that the low-energy theorem for $\langle r^2_{M,s}\rangle$ is given by a leading-order kaon-loop effect, but that the next-to-leading order corrections are sizeable. We therefore want to discuss here two alternative schemes to evaluate loops that may resum these corrections more efficiently, or may give a fairer estimate of these contributions altogether.

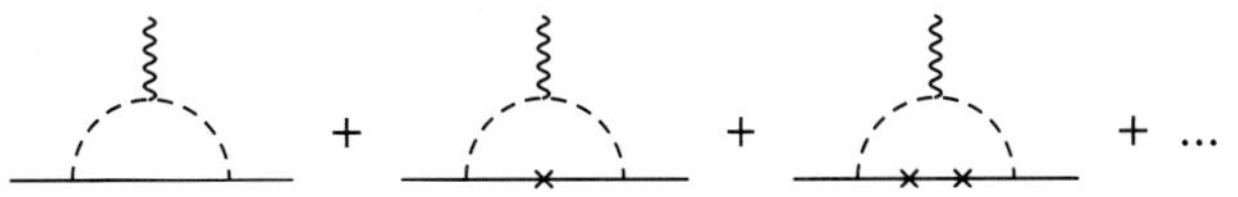

Fig. 5. Resummation of relativistic corrections in the infrared regularization scheme. The crosses denote $1/m_N$ insertions in the nucleon propagator

5.1 Infrared regularization

All loop results discussed so far have been calculated in what is called Heavy Baryon Chiral Perturbation Theory. In this formalism, the nucleons are treated as heavy, non-relativistic fields, for which the relativistic corrections suppressed by powers of $1/m_N$ can be calculated systematically. This method has the advantage that loop diagrams always follow naive power counting rules and that leading loop calculations are technically simple, but the downside that the analytic structure is sometimes distorted even in the low-energy region, and that in practice $1/m_N$ corrections are awkward to calculate.

One alternative scheme that has been suggested to overcome these shortcomings is "infrared regularization" [10], which is a variant of dimensional regularization that preserves power counting also in relativistic baryon ChPT. In this scheme, all the $1/m_N$ corrections are automatically resummed as depicted schematically in Fig. 5.

It has been shown in [11] that in some cases particularly sensitive to relativistic "recoil effects" (in [11]: the neutron electric form factor), this relativistic resummation leads to a much improved convergence behavior. It is therefore interesting to see what happens to the strange magnetic radius in this scheme, even beyond $\mathcal{O}(p^4)$. We find the following *partial* corrections of $\mathcal{O}(p^5)$ [9]:

$$\begin{aligned}\Delta\langle r^2_{M,s}\rangle^{(5)} &= -\frac{35(5D^2-6DF+9F^2)}{384\pi F_K^2}\frac{M_K}{m_N} + \mathcal{O}\left(\frac{\Delta m_{\Lambda/\Sigma}}{M_K}\right) \\ &= +0.053\ \text{fm}^2 \ ,\end{aligned} \tag{9}$$

where the additional terms proportional to $\Delta m_{\Lambda/\Sigma} = m_{\Lambda/\Sigma} - m_N$ have been included in the numerical value. Although these expressions are not complete (there will also be two-loop contributions), they are non-analytic in the quark masses and therefore not modified by counterterms. Numerically, they turn out to be large, about half the size of the leading $\mathcal{O}(p^3)$ value. This supports once more the conclusion that, unfortunately, the low-energy theorem for $\langle r^2_{M,s}\rangle$ at $\mathcal{O}(p^3)$ is numerically not very reliable.

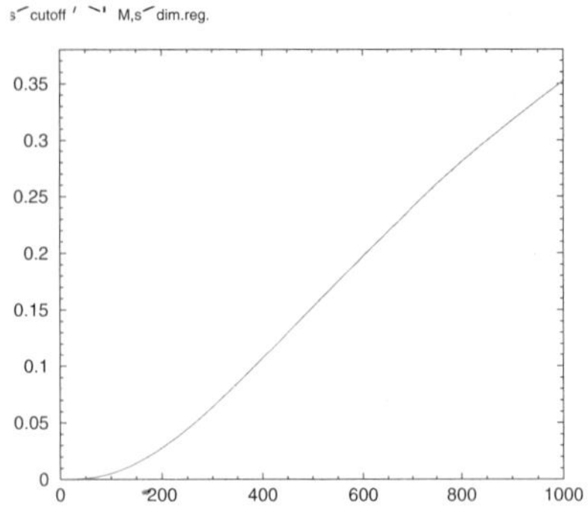

Fig. 6. Cutoff dependence of $\langle r^2_{M,s}\rangle$

5.2 Cutoff regularization

Various attempts at cutoff regularization have been undertaken in the context of ChPT. We do not try to give an overview of these, but instead concentrate on one specific approach [12] in which $\langle r^2_{M,s}\rangle$ has explicitly been considered [13]. The idea is that the nucleon has an intrinsic size, and that ChPT can only reliably predict the π or K fluctuations that are long-ranged on this scale. Dimensional regularization does not separate these different ranges of momenta in the loop integration, such that it might be more efficient to employ a cutoff in order to reduce unrealistically strong short-distance effects.

The method in [12] amounts to the use of a KNN form factor instead of a constant vertex. In [13], a dipole form has been chosen, which leads to the analytic result

$$\langle r^2_{M,s}\rangle_{\rm cutoff} = \langle r^2_{M,s}\rangle_{\rm dim.reg.} \times X\big(\tfrac{\Lambda}{M_K}\big) \;, \qquad (10)$$
$$X(x) = \frac{x^3(x^2+\frac{14}{5}x+1)}{(x+1)^5} \;.$$

This cutoff dependence is displayed in Fig. 6. It is obvious that for cutoffs of the order of $\Lambda = 400\ldots600$ MeV, $\langle r^2_{M,s}\rangle_{\rm cutoff}$ is sizeably reduced compared to the dimensional regularization result.

Despite the host of plausible arguments in favor of cutoff schemes, there are also some drawbacks:

1. The cutoff is an additional parameter that can only, at best, be estimated.
2. It presents a deviation from the strict effective Lagrangian approach in ChPT, as a consequence of which gauge invariance usually has to be cured "by hand", and chiral symmetry is by no means guaranteed.[1]
3. The cutoff upsets the analytic structure of the form factors: it produces additional unphysical poles and/or cuts. This means in particular that is is unclear how a marriage with dispersion relations (see e.g. [8] and references therein) might be achieved.
4. Finally, the reduction of the numerical value of $\langle r^2_{M,s}\rangle$ must not be confused with the inclusion of higher order corrections as done in the infrared regularization scheme. The problem of large $1/m_N$ corrections in chiral SU(3) is by no means solved here.

[1] This problem has recently been addressed in [14].

6 Summary and conclusions

The matrix elements of the strange vector current seem to remain elusive quantities for a description in Chiral Perturbation Theory. They are problematic as Goldstone boson dynamics is not overly dominant here, and in most cases, unknown low-energy constants appear at leading order. We have pointed out that there is an exception, the strange magnetic moment for which a parameter-free low-energy theorem exists. However, higher order corrections are found to be sizeable, such that the low-energy theorem is very unstable once these are taken into account. Also cutoff calculations indicate a smaller absolute size of $\langle r^2_{M,s}\rangle$. The role of ChPT remains however to aid to interrelate more data, as they become available [15].

Acknowledgements. I would like to thank the organizers of PAVI 04 for the great opportunity to participate in this workshop, and for the wonderful organization of the whole event. I am grateful to T.R. Hemmert and U.-G. Meißner for the fruitful collaboration that originally introduced me to this subject, and to various colleagues for interesting discussions that deepened my understanding of the field, in particular H.-W. Hammer, M.J. Ramsey-Musolf, J.F. Donoghue, B.R. Holstein, T. Huber, and A. Roß. This work was supported in part by RTN, BBW-Contract No. 01.0357, and EC-Contract HPRN–CT2002–00311 (EURIDICE).

References

1. J. Gasser, H. Leutwyler: Annals Phys. **158**, 142 (1984); J. Gasser, H. Leutwyler: Nucl. Phys. B **250**, 465 (1985)
2. G. Colangelo, J. Gasser, H. Leutwyler: Phys. Rev. Lett. **86**, 5008 (2001)
3. M.J. Musolf, H. Ito: Phys. Rev. C **55**, 3066 (1997)
4. T.R. Hemmert, U.-G. Meißner, S. Steininger: Phys. Lett. B **437**, 184 (1998)
5. D.T. Spayde et al., [SAMPLE Collaboration]: Phys. Lett. B **583**, 79 (2004)
6. K.A. Aniol et al., [HAPPEX Collaboration]: Phys. Rev. Lett. **82**, 1096 (1999); Phys. Lett. B **509**, 211 (2001)
7. T.R. Hemmert, B. Kubis, U.-G. Meißner: Phys. Rev. C **60**, 045501 (1999)
8. H.W. Hammer, S.J. Puglia, M.J. Ramsey-Musolf, S.L. Zhu: Phys. Lett. B **562**, 208 (2003)
9. B. Kubis: Thesis, *Berichte des FZ Jülich*, Jül-4007 (2002)
10. T. Becher, H. Leutwyler: Eur. Phys. J. C **9**, 643 (1999)
11. B. Kubis, U.-G. Meißner: Nucl. Phys. A **679**, 698 (2001)
12. J.F. Donoghue, B.R. Holstein, B. Borasoy: Phys. Rev. D **59**, 036002 (1999)
13. T. Huber: *Master thesis* (Univ. of Massachusetts, 2002); A. Roß: *Master thesis* (Univ. of Massachusetts, 2003)
14. D. Djukanovic, M.R. Schindler, J. Gegelia, S. Scherer: (2004), arXiv:hep-ph/0407170
15. F.E. Maas et al. [A4 Collaboration]: Phys. Rev. Lett. **93**, 022002 (2004)

Eur Phys J A (2005) **24**, s2, 101–102
DOI: 10.1140/epjad/s2005-04-022-2

EPJ A direct
electronic only

Time-like compton scattering and the Bethe–Heitler process

Stephen R. Cotanch[a]

Department of Physics, North Carolina State University, Raleigh NC 27695-8202, USA

Received: 15 October 2004 / Published Online: 8 February 2005

Abstract. Arguments are presented for $p(\gamma, e^+e^-)p$ measurements to obtain new information about the off-shell time-like nucleon form factors, especially in the ϕ meson region governing the ϕN couplings $g^{V,T}_{\phi NN}$. Theoretical predictions based upon a Quantum Hadrodynamic model and vector meson dominance are highlighted for both the proton form factor and the time-like Compton scattering cross section. The Bethe-Heitler process is also calculated but is only important at low momentum transfer $|t|$ permitting a novel high $|t|$ ϕ enhancement in the Compton cross section related to nucleon strangeness to emerge.

PACS. 12.40.Nn Regge theory – 12.40.Vv Vector-meson dominance – 13.40.Gp Electromagnetic form factors – 13.40.Hq Electromagnetic decays

1 Introduction

While Compton scattering using real, $p(\gamma,\gamma)p$, and even virtual (space-like), $p(e,e'\gamma)p$, photons has been measured, the time-like virtual Compton scattering [TVCS] process, $p(\gamma, e^+e^-)p$, has received little attention. However such measurements can provide new information about the time-like proton form factor in the unobserved region $0 \leq q^2 \leq 4M_p^2$ (all measurements utilized $e^+e^- \leftrightarrow N\bar{N}$ where $q^2 \geq 4M_p^2$) since the three-body final state has an essentially unrestricted virtual photon mass $q^2 \geq 4M_e^2 \sim 0$. Previous theoretical studies using vector meson dominance [VMD] have predicted that the $p(\pi^-, e^+e^-)n$ [1,2] and $p(\gamma, e^+e^-)p$ [3] cross sections both exhibit a dramatic, dual peaked resonant signature for time-like virtual photon four-momentum spanning the vector meson masses ($q^2 \approx M_V^2$ for $V = \rho, \omega, \phi$). This paper extends the work of [3] by calculating the competing Bethe-Heitler [BH] process, $\gamma p \to \gamma\gamma_v p \to e^+e^-p$, and documenting that it is only important for small $|t|$, significantly below the interesting high $|t|$ region where the s and u channel processes embodying the VMD resonant signature dominate.

2 Hidden strangeness

In addition to mesons, other eigenstates of the QCD Hamiltonian also contain hidden strangeness. One clear example is the ground state vacuum with non-zero quark condensates $\langle O|\bar{s}s|O\rangle \simeq \langle O|\bar{u}u|O\rangle \simeq \Lambda^{QCD}$. Of current intense interest and debate is nucleon hidden strangeness which is completely specified by the $n = 1, 2 \ldots 16$ matrix elements, $\langle N|\bar{s}\Gamma_n s|N\rangle$, involving the Lorentz bilinear covariants. This study addresses the vector and tensor elements, γ_μ and $\sigma_{\mu\nu}$, corresponding to the coupling constants $g^V_{\phi NN}$ and $g^T_{\phi NN}$, respectively, since the ϕ is predominantly $s\bar{s}$. These ϕN coupling constants govern the Quantum Hadrodynamical [QHD] Lagrangian

$$\mathcal{L}_{\phi NN} = g^V_{\phi NN}\bar{N}\gamma_\mu N\phi^\mu + g^T_{\phi NN}\bar{N}\sigma_{\mu\nu}N[\nabla^\mu\phi^\nu - \nabla^\nu\phi^\mu]$$

used for both the VMD proton form factor and TVCS cross section predictions discussed below. If there is no or insignificant nucleon strangeness then ϕN coupling should be suppressed due to the dominant $s\bar{s}$ structure of the ϕ and the OZI rule. However, significant OZI violations have been observed in inelastic μp and elastic νp scattering, $p\bar{p}$ annihilation experiments and measurements of the πN sigma term, which collectively suggest appreciable strangeness in the proton. Evidence for nucleon strangeness is further discussed and reviewed in [4]. Related, a previous analysis [3] of space-like neutron electric form factor data and high $|t|$ ϕ photoproduction data yielded $g^V_{\phi NN} = 1.3$, $g^T_{\phi NN} = 2.3$. Accurate TVCS measurements at high $|t|$ will permit extraction of these couplings which quantify the degree of nucleon strangeness.

3 Predictions for TVCS

As detailed in [3] the nucleon form factors were calculated using a generalized VMD model. A good description of the baryon octet form factors was obtained, especially the sensitive space-like neutron electric form factor and the proton EM form factors in both the space-like and time-like regions as depicted in Fig. 1. Note the resonant peaks in the unmeasured time-like vector meson region, in particular the ϕ peak which scales with the ϕN coupling.

[a] *E-mail address:* cotanch@ncsu.edu

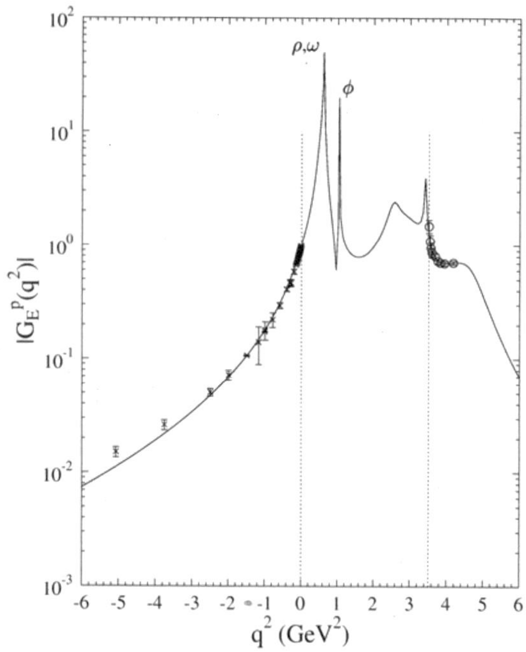

Fig. 1. Data and VMD (absolute value) for the proton electric form factor

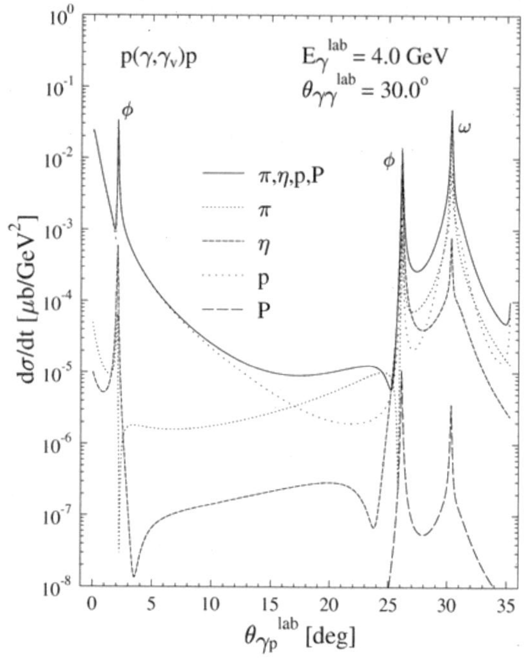

Fig. 2. VMD prediction for TVCS, $p(\gamma, \gamma_v)p$. The smaller angle ϕ peak quantifies the proton's strangeness

The form factor peaks are reflected in the TVCS cross section displayed in Fig. 2. A novel, dual peak profile arises from the quadratic relation between q^2 and the recoil proton lab angle. The smaller angle ϕ peak corresponds to high $|t|$ and is dominated by the u channel proton propagator with $g_{\phi NN}$ coupling (sparse dotted curve). The magnitude of this peak represents the degree of the proton's strangeness. The other two peaks near 30° entail lower $|t|$ and involve ϕ and ω coupling, respectively, to π (dense dotted curve) and η (short dashed curve) t channel exchange. These two peaks represent the expected ϕ production background since they are based upon established results. This dual peak ϕ signature follows from only VMD and should occur in other dynamic models. Therefore VMD predicts that a measurement of the high $|t|$ TVCS cross section ratio $R = \sigma(q^2 = M_\phi^2)/\sigma(q^2 = M_\omega^2)$ is proportional to $g^2_{\phi NN}/g^2_{\omega NN}$. This has been numerically confirmed in this model giving $R = 0.14\ f$ (where f is a kinematic quantity of order unity) which is an order of magnitude larger than the OZI prediction [4], $R = tan^2\ \delta\ f = 0.0042\ f$, where $\delta = 3.7^o$ is the deviation from the ideal quark flavor mixing angle in the ϕ. TVCS measurements would therefore appear to be an excellent probe of the proton's strangeness content.

4 The Bethe–Heitler process

Because the BH and TVCS processes compete it is necessary to assess their relative magnitudes. Even if the amplitudes are comparable it is still possible to extract the TVCS amplitude by measuring the charge asymmetry $(\sigma(e^+e^-) - \sigma(e^-e^+))$ since in TVCS the e^+e^- pair has C-parity -1 (single photon production) while for the theoretically known BH process they have C = 1 (two photon production). For the kinematics listed in Fig. 2, the calculated BH cross section dominates the TVCS cross section for $|t| \leq 0.01$ GeV^2, is comparable for $|t|$ up to 0.04 GeV^2 and is an order of magnitude smaller for $|t| > 0.06$ GeV^2. Hence the charge asymmetry measurement will only be necessary for small $|t|$ where meson and pomeron (long dashed curve in Fig. 2) exchange dominate the TVCS process. To extract the ϕN couplings at high $|t|$, the TVCS cross section is sufficiently large for direct measurement without competition from the BH process.

5 Conclusion

With GeV electron facilities, such as Jlab, TVCS experiments appear quite feasible, providing an opportunity to obtain the unknown nucleon on and off-shell time-like form factors. If VMD is valid the ϕN couplings can then be extracted which in turn permits a direct assessment of nucleon hidden strangeness.

R. A. Williams and C. W. Kao are acknowledged. This work was supported by DOE grant DE-FG02-97ER41048.

References

1. R.A. Williams, S.R. Cotanch: Phys. Rev. Lett. **77**, 1008 (1996)
2. S.R. Cotanch, R.A. Williams: Nucl. Phys. A **631**, 478 (1998)
3. S.R. Cotanch, R.A. Williams: Phys. Lett. B **549**, 85 (2002)
4. J. Ellis, M. Karliner, D.E. Kharzeev, M.G. Sapozhnikov: Phys. Lett. B **353**, 319 (1995)

Eur Phys J A (2005) **24**, s2, 103–103
DOI: 10.1140/epjad/s2005-04-023-1

EPJ A direct
electronic only

Corrections to the nuclear axial vector coupling in a nuclear medium

G.W. Carter and E.M. Henley

Department of Physics,Box 351560, University of Washington, Seattle, WA 98195-1560, USA

Received: 15 October 2004 / Published Online: 8 February 2005

Abstract. The temporal component of the weak axial vector current in nuclei is increased due to meson exchange currents. We consider further corrections from pions and the sigma mean-field.

PACS. 23.40.-s Beta decay; double beta decay; electron and muon capture – 23.40.Bw Weak interaction and lepton (including neutrino) aspects

The weak axial current j^5_μ is not conserved in a nuclear medium or nucleus; the space component is reduced whereas the time component, g^0_A increases by $50-100\%$ [1].

Already in the 1970's it was shown [2] that nuclear exchange currents are responsible for the major part of the increase of g^{0*}_A. A simplistic explanation is that the pion adds an extra γ^5 which makes the space component of $O(p/M)$, whereas the time component becomes of $O(1)$. Pion exchange currents have been examined by several authors [3]. Some aspects of the exclusion principle have been included in pion exchange calculations, e.g., via use of the shell model. However, the theoretical enhancement falls about 10% short of the measurements [4]. There is, to the same order, the one nucleon pion loop contribution. It differs from the free nucleon counterpart by the effect of the Pauli exclusion principle. It is this effect we have examined in order to see whether it might contribute the missing $\sim 10\%$ of g^{0*}_A.

We use the chiral Lagrangian of Carter, et al. [5], treating the σ field in mean field theory.

$$L_1 = -\bar{N}g\left[(\sigma + i\tau\cdot\pi\gamma_5) + \frac{1}{2}g'\gamma^\mu\tau\cdot\mathbf{a}_\mu\gamma_5\right]N\,, \quad (1)$$

$$L_2 = \frac{1}{2}\frac{D}{\sigma_0^2}\bar{N}\,[\gamma^\mu\tau\cdot[\pi\times\Delta_\mu\pi + \gamma_5(\sigma\Delta_\mu\,\pi - \pi\Delta_\mu\sigma)]]\,N\,, \quad (2)$$

$$\Delta_\mu\sigma = \partial_\mu\sigma + g'\mathbf{a}_\mu\cdot\pi\,, \quad (3)$$

$$\Delta_\mu\pi \approx \partial_\mu\pi - g'\sigma\mathbf{a}_\mu\,, \quad (4)$$

where g' is the coupling to the chiral vector mesons ρ and A_1, and bold characters indicate isospin vectors.

To start with, the nucleon has a momentum $\leq k_F$. Since we need the propagators in the presence of the nucleus, they have to be above the Fermi sea (e.g. $|\mathbf{p}| > k_F$).

The effect of the in-medium sigma field is to reduce the nucleon mass, $M \to M^* \sim 0.8M$. We assume that the pion coupling to the nucleon has a dipole form factor with a cutoff $\Lambda = 0.9$ or 1.1 GeV. The constant D is fixed to fit $g_A = 1.26$ for the free nucleon,

$$g_A \approx \left(1 + D\frac{\bar{\sigma}^2}{\sigma_0^2}\right)\,, \quad (5)$$

where σ_0 is the vacuum expectation value of the σ field, $\sigma_0 = 102$ MeV, and $\bar{\sigma}$ is the mean field result at finite density.

We fnd that $g^*_A = g_A(1+\delta)$, with $\delta = -.13(-.11)$ for $\Lambda = 1.1(0.9)$ GeV. Thus, the omitted effect is indeed of the order of magnitude required (10-15%), but with the wrong sign. These corrections only serve to increase the discrepancy between theory and experiment.

The authors thank Mary Alberg for helpful comments. This work was supported by the U. S. Department of Energy grant DE-FG02-97ER4014.

References

1. See, e.g., D.H. Wildenthal, M. S. Curtin, B. A. Brown: Phys. Rev. C **28**, 1343 (1983)
2. L. Kubodera, J. Delorme, M. Rho: Phys. Rev. Lett. **40**, 755 (1978)
3. See e.g., T-S. Park, H. Jung, D-P. Min: Phys. Lett. B **409**, 26 (1997)
4. E.K. Warburton: Phys. Rev. Lett. **66**, 1823 (1991); Phys. Rev. C **44**, 233 (1991)
5. G.W. Carter, P.J. Ellis, S. Rudaz: Nucl. Phys. A **603**, 367 (1996), [Erratum-ibid., **608**, 514] (1996)

Eur Phys J A (2005) **24**, s2, 105–105
DOI: 10.1140/epjad/s2005-04-024-0

EPJ A direct
electronic only

Strangeness-conserving effective weak chiral Lagrangian

Hee-Jung Lee[1], Chang Ho Hyun[2], Chang-Hwan Lee[3], and Hyun-Chul Kim[3]

[1] Departament de Fisica Teòrica, Universitat de València E-46100 Burjassot (València), Spain
[2] Institute of Basic Science, Sungkyunkwan University, Suwon 440-746, Republic of Korea
[3] Department of Physics and NuRI, Pusan National University, Busan 609-735, Republic of Korea

Received: 15 October 2004 / Published Online: 8 February 2005

Abstract. We consider the strangeness–conserving effective weak chiral Lagrangian based on the nonlocal chiral quark model from the instanton vacuum. We incorporate the effect of the strong interaction by the gluon into the effective Lagrangian. The effect of the Wilson coefficients on the weak pion–nucleon coupling constant is discussed briefly.

PACS. 12.40.-y, 14.20.Dh Effective chiral Lagrangian, Wilson coefficients, weak pion-nucleon coupling constant

In this work, we briefly illustrate the $\Delta S = 0$ effective weak chiral Lagrangian from the instanton vacuum, which is essential to study the weak interaction of hadrons at low-energy regions. Since parity violation (PV) can provide high–precision tests of the electro–weak standard model (SM) [1], a great deal of attention has been paid to PV in the SM. It is known that there is discrepancy in the weak charge between measurements in atomic physics and the prediction from the SM [2]. Moreover, it is well known that there is still disagreement theoretically as well as experimentally [3] in determining the weak pion-nucleon coupling constant h_π^1.

Recently, Meissner et al. [4] studied h_π^1 within the SU(3) Skyrme model, based on the effective current–current interaction which is equivalent to a factorization scheme at the leading order in N_c. However, processes such as nonleptonic weak processes defy any explanation from the factorization. Therefore, we first derive the $\Delta S = 0$ effective weak chiral Lagrangian, incorporating the effective Hamiltonian [5] within the nonlocal chiral quark model from the instanton vacuum. Using the derivative expansion, we obtain the effective weak Lagrangian to order $\mathcal{O}(p^2)$ and to the next–to–leading order (NLO) in N_c. We will use this derived effective weak chiral Lagrangian as a starting point for investigating h_π^1.

We include two effects from the strong interaction in the effective Lagrangian : The first one is the QCD vacuum effect which is implemented in the chiral quark model from the instanton vacuum, and the other is the perturbative gluon effect (the strong enhancement effect) which is encoded in the Wilson coefficients [5]. The terms at the leading order of N_c in the effective Lagrangian are expressed in terms of the current–current interactions which correspond to the factorization scheme. It can be easily shown that they are identical to those in [4] when the perturbative gluon effect is turned off. On the other hand, the NLO terms from the non–factorization scheme have a more complicated form. The explicit form of the $\Delta S = 0$ effective weak Lagrangian with the Wilson coefficients to the NLO can be found in [6].

The role of the Wilson coefficients can be investigated by calculating h_π^1 from the effective weak chiral Lagrangian. As done in [4], we employ the chiral soliton with the zero-mode quantization. The pion which couples to the nucleon can be introduced from the meson fluctuation around the soliton field. When the perturbative gluon effect is turned off and NLO terms are not considered, h_π^1 turns out to be the same as that in [4]. On the other hand, if the strong enhancement effect is taken into account at the leading order in N_c, a rough estimation of the effect shows that h_π^1 is enhanced by 20 %, compared to that without the Wilson coefficients. However, it should be noted that if we restrict ourselves to SU(2), h_π^1 vanishes anyway. The investigation in the SU(3) flavor space is under progress.

References

1. M.A. Bouchiat, C.C. Bouchiat: Phys. Lett. B **48**, 111 (1974)
2. C.S. Wood et al.: Science **275**, 1759 (1997); S.C. Bennett, C.E. Wieman: Phys. Rev. Lett. **82**, 2484 (1999)
3. W.S. Wilburn, J.D. Bowman: Phys. Rev. C **57**, 3425 (1998)
4. U.G. Meissner, H. Weigel: Phys. Lett.B **447**, 1 (1999)
5. B. Desplanques, J.F. Donoghue, B.R. Holstein: Annals Phys. **124**, 449 (1980)
6. H.-J. Lee et al.: hep-ph/0405217

IV Experimental techniques in PV electron scattering

IV-1 Beam asymmetry

Eur Phys J A (2005) **24**, s2, 109–114
DOI: 10.1140/epjad/s2005-04-025-y

EPJ A direct
electronic only

Overview of laser systematics

Gordon D. Cates, Jr.

Department of Physics, University of Virginia, Charlottesville, Virginia, USA

Received: 15 January 2005 / Published Online: 8 February 2005

Abstract. This paper discusses systematic effects in parity experiments that originate from the laser and optics system that are used in a polarized electron source. Covered are both the sources of systematics, as well as strategies for their minimization.

PACS. 29.25.Bx Electron sources – 29.27.Hj Polarized beams – 42.25.Ja Polarization – 42.25.Lc Birefringence – 11.30.Er Charge conjugation, parity, time reversal, and other discrete symmetries

1 Introduction

In general, experiments that study parity violation in electron scattering utilize a polarized electron source that is based on photo-emission from various types of gallium arsenide (GaAs) crystals. The photo-emission is induced using circularly polarized light from a laser. Because the helicity of the electron beam is determined by the polarization state of the laser light, the electron polarization can be reversed or "flipped" quickly and in a quasi-random manner by using an electro-optical device such as a Pockels cell to circularly polarize the light. Parity experiments typically measure tiny helicity-dependent asymmetries in the scattering of polarized electrons off unpolarized targets. The asymmetries themselves might range from 0.1 parts per million (ppm) to 100 ppm, and it is sometimes necessary to control helicity-correlated systematic effects at the level of parts per billion. In experiments where electronic crosstalk is sufficiently under control, the helicity-correlated changes in the parameters of the electron beam generally originate in helicity-correlated changes in the light used to induce photo-emission. Laser systematics are thus critical to achieving increasingly accurate measurements of parity violation in electron scattering.

Ever since the first pioneering experiment that observed parity violation in electron scattering [1], the understanding of laser-based systematics has grown. It is now possible to catalog some of the dominant effects, including how they can be diagnosed, and in some cases corrected. Indeed, the improvement of our understanding of laser systematics has played a critical role in making increasingly sensitive parity experiments possible.

This paper will examine some of the laser systematics that were dominant during several experiments with which the author was involved [2–5], as well as discussing strategies for their minimization.

2 The various types of systematics

In the absence of any effort to control them, the largest systematic in a parity experiment will generally be helicity-correlated asymmetries in the charge delivered to the target. While it is certainly important to measure and correct for "charge asymmetries" this can only be done up to a point. Beam current measuring devices will always have nonlinearities at some level. Even if one had a perfect device for measuring beam current, there is still the possibility that through interaction with the accelerator, such as beam loading, charge asymmetries could be translated into other helicity-correlated effects. Fortunately, it is reasonably straightforward to "balance" the charge delivered to the target over the course of an experiment at the level of a few hundred parts per billion (ppb) resulting in systematic uncertainties in the parity violating asymmetry of a few ppb.

If charge asymmetries are the "zeroth-order" effects, the first-order effects are then helicity-correlated differences in the beam position. Since charge asymmetries are reasonably straightforward to control, these "position differences" end up being a more troublesome problem. Many recent parity experiments have had significant contributions to their systematic errors from helicity-correlated position differences [4,5].

Even if charge and position differences are reduced to negligible levels it is still possible to be troubled by higher-order effects. For instance the spot size of the laser can systematically change while position and charge are held fairly constant. One issue that needs to be considered is whether the elimination of lower-order effects simply results in the increase of higher-order effects. This makes it desirable to have diagnostics for higher-order effects even if there is no obvious way to control them.

3 The sources of systematics

3.1 Charge asymmetries

Charge asymmetries result when the average current associated with one helicity state is different from the average current associated with the other helicity state. The dominant mechanisms associated with this effect are well understood, and have been described in some detail for both simple [6] and more complex [7] optics setups. The asymmetries stem from the fact that when making circularly polarized light, there are always small admixtures of linear polarization which cause a small degree of ellipticity. When the helicity of the light is flipped, it is often the case that the major axis of the polarization ellipse will rotate by 90°. Since most optics systems have many elements (for instance mirrors) that transport one linear polarization better than another (a property we will refer to as a transport asymmetry), flipping the helicity can cause a change in the efficiency with which the light delivered to the cathode. Historically, this type of effect has sometimes been referred to as the "PITA" effect, where PITA is an acronym standing for "polarization induced transport asymmetry" [6]. The PITA effect thus results from the fact that the optics system has an "analyzing power" with an accompanying analyzing-power axis.

In polarized electron sources, the optics transport system is not the only component with an analyzing power. Bulk GaAs has a theoretical maximum polarization of 50%, and values of 35-45% are typical. Recently it has become common to use modified GaAs crystals such as strained GaAs or super-lattice GaAs because in such crystals a degeneracy associated with the valence band is broken raising the theoretical maximum polarization to nearly 100%, with values of 70-82% being typical. The improved polarization of these photocathodes comes at a price. When irradiated with linearly polarized light, these photocathodes have a quantum efficiency (QE) that depends on the orientation of the light's polarization axis with respect to an axis that lies in the plane of the crystal's surface. The crystal itself thus has an analyzing power. Figure 1a illustrates the direction of the analyzing-power axis. The QE anisotropy associated with the analyzing-power axis can be as much as 15%.

The analyzing power of the crystal has essentially the same effect on beam current as does a transport asymmetry. When the crystal is illuminated with elliptically polarized light, the photo-emitted current depends critically on the position of the major axis with respect to the analyzing-power axis. If the polarization ellipses associated with the two helicity states are as indicated in Fig. 1b, a maximal charge asymmetry will result. If the polarization ellipses are oriented as indicated in Fig. 1c, a minimal charge asymmetry will result.

Whether a charge asymmetry results from a laser-beam transport asymmetry, or from an anisotropy in the photocathode's QE, it is straightforward, and useful, to characterize the effect quantitatively. For definiteness, we will assume that the device used to produce circular polarization is a Pockels cell, oriented so that its fast axis is at ±45° with respect to horizontal. We will further assume that prior to traveling through the Pockels cell, the light is linearly polarized in the horizontal direction. It is convenient to parameterize the phases introduced by the Pockels cell as

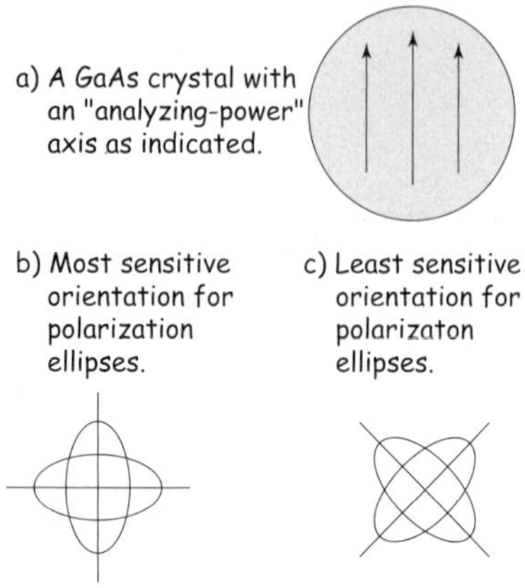

Fig. 1. a Illustrated is a GaAs crystal with a quantum efficiency that is sensitive to the orientation of linear polarization with respect to the indicated analyzing-power axis. **b** Polarization ellipses for nominally positive and negative helicity light resulting in maximum charge asymmetry. **c** Polarization ellipses for nominally positive and negative helicity light resulting in minimum charge asymmetry

$$\delta^R = -\left(\frac{\pi}{2} + \alpha_1\right) - \Delta_1 \quad (1)$$

$$\delta^L = +\left(\frac{\pi}{2} + \alpha_1\right) - \Delta_1 \ . \quad (2)$$

If $\Delta_1 = \alpha_1 = 0$, the phases introduced by the Pockels cell are $\pm\frac{\pi}{2}$, and in principle, the light will have perfect circular polarization. If either Δ_1 or α_1 are nonzero, however, elliptical polarization will result. I note in passing that we are using the subscript 1 for α and Δ to be consistent with the notation of [7]. Let us now assume that the light, after passing through the Pockels cell, passes through an asymmetric transport system, characterized by two orthogonal axes x' and y', where the x' axis makes an angle θ with respect to the horizontal. We will assume that light linearly polarized along the $x'(y')$ axis will be transported with a transmission coefficient $T_{x'}(T_{y'})$, and define the quantities $T = (T_{x'} + T_{y'})/2$, and $\varepsilon = T_{x'} - T_{y'}$. With these definitions, the charge or equivalently current asymmetry A_I can be written to first order as

$$A_I = \frac{I^R - I^L}{I^R - I^L} = \frac{\varepsilon}{T}\Delta_1 \cos 2\theta \quad (3)$$

where $I^R(I^L)$ are the electron beam intensities associated with the Pockels cell phases $\delta^R(\delta^L)$. The reason we chose to write the phases as we did in 1 and 2 is immediately apparent. The equation for A_I depends linearly on Δ_1, but not at all on α_1. The reason is as follows. If $\Delta_1 \neq 0$, polarization ellipses result whose major axes will rotate by 90° when the helicity is flipped. That is, if $\Delta_1 \neq 0$ we have a situation such as is illustrated in Fig. 1b. If $\alpha_1 \neq 0$, the polarization ellipses for the two helicity states will be

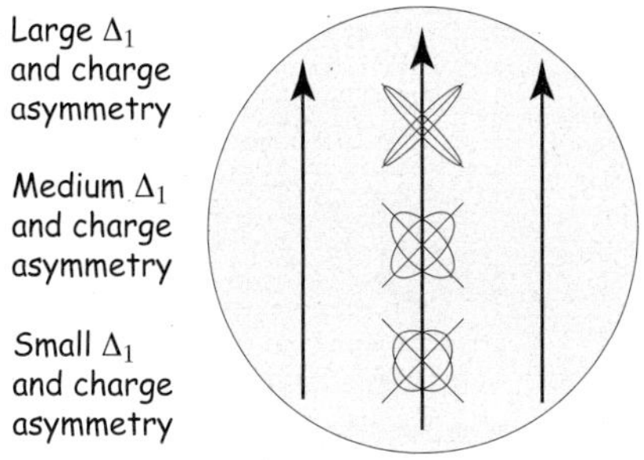

Fig. 2. Illustrated is a GaAs crystal being irradiated by light in which the residual linear polarization is varying from a maximum toward the top of the crystal to a minimum toward the bottom of the crystal

coincident with one another. Only the *direction* that the electric vector travels around the ellipse will change. We will discuss later how 3 can give us guidance in suppressing charge asymmetries.

One final comment regarding the phase Δ_1. It should be noticed that the *sign* of Δ_1 does not change when the phase on the Pockels cell is flipped. Thus, any object in the optics system, a vacuum window, a mirror, or even residual birefringence in the Pockels cell, can cause a non-zero value of Δ.

3.2 Position differences from phase gradients

In the previous section we discussed how a charge asymmetry can result from a non-zero value of the phase Δ_1. If we consider a laser beam spot illuminating a GaAs crystal, it can also be the case that the phase Δ_1, and hence the associated charge asymmetry, varies in some manner across the laser spot. Such a situation is illustrated in Fig. 2. If the charge asymmetry for the emitted electrons changes as we move from the top of the crystal to the bottom, the beam profiles for the two helicity states will have centroids that are shifted vertically with respect to one another. From the perspective of our beam position monitors, these shifts will be seen as helicity-correlated position differences. To first order, the position differences will be *proportional* to the gradient of the phase Δ_1, and *independent* of the average value of Δ_1.

3.3 Position differences from steering effects

Another source of helicity-correlated position differences is steering caused by the Pockels cell. The Pockels cell is alternately pulsed to positive and negative high voltage in order to introduce the phases given by 1 and 2. Empirically, it appears that this results in the Pockels cell behaving alternately as a diverging and converging lens. If a laser beam is sufficiently small in diameter and goes through the very center of the Pockels cell, the steering effects can be kept quite small. As one goes off center, however, some steering occurs as would also be the case with any lens. This effect, which is illustrated in Fig. 3 can be quite significant.

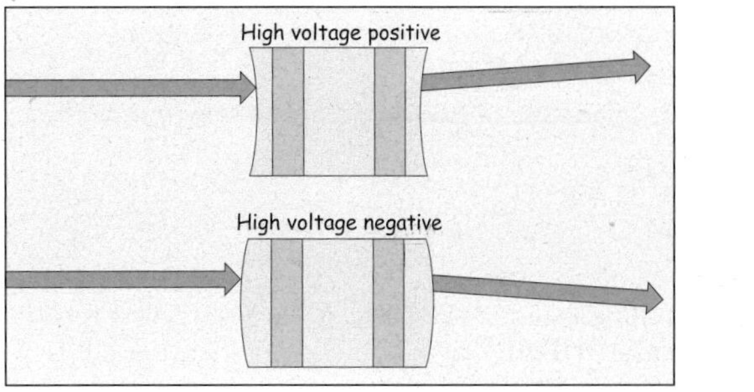

Fig. 3. Illustrated is steering due to a Pockels cell having lens-like properties when it is pulsed at high voltages

3.4 Position differences from gradients in the analyzing power of the cathode

The last class of effects we will mention arises from changes in the QE anisotropy as we move across the cathode. For instance, assume that the *direction* of the analyzing-power axis is constant, but that the *magnitude* of the anisotropy changes from 5% at the top of the cathode to 10% at the bottom. If the incident light is perfectly circularly polarized, there is no component of linear polarization and hence no charge asymmetry. If there is residual linear polarization, however, there could be a charge asymmetry whose size varies across the crystal, resulting in changes in the centroid of the electron beam's position much as was discussed in Sect. 3.1. The reader should note that this type of position difference should be proportional to the degree of ellipticity of the light, and hence Δ_1. This is in contrast to the position differences from phase gradients (Sect. 3.1) where to first order we would expect no dependence on Δ_1. This difference in response to Δ_1 is a useful diagnostic tool for distinguishing between phase gradients in the optics system and analyzing-power gradients on the cathode.

4 Controlling systematics

4.1 Charge asymmetries

There are at least three strategies for controlling charge asymmetries and 3 is useful for understanding two of them.

4.1.1 Phase adjustments

The phase Δ_1 can be controlled using the Pockels cell. The nominal voltage at which the Pockels cell is pulsed is ± 2.7 kV. If a fixed voltage is added to the voltage associated with each polarity, one can introduce an arbitrary Δ_1. For instance one could run at $+2,900$V and $-2,500$V. Notice that it is not the magnitude of each voltage that is changed, but rather its actual value. If instead we changed the magnitude of each voltage by a few hundred volts, we would be adjusting α_1 and not Δ_1, and no change to the charge asymmetry would occur.

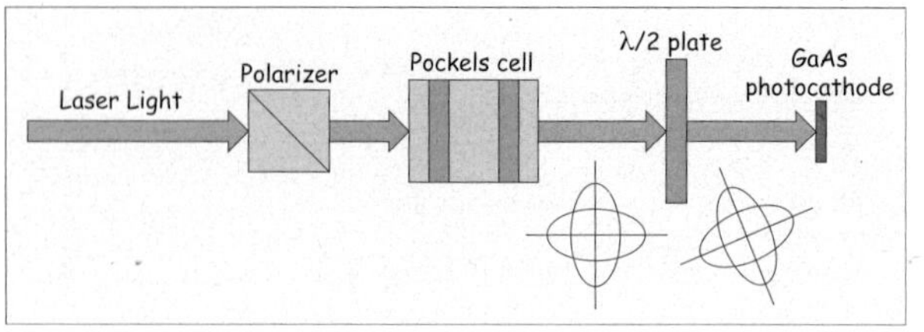

Fig. 4. Illustrated is the basic setup for using a rotating half-wave plate (RHWP) to rotate the polarization ellipses associated with the laser light

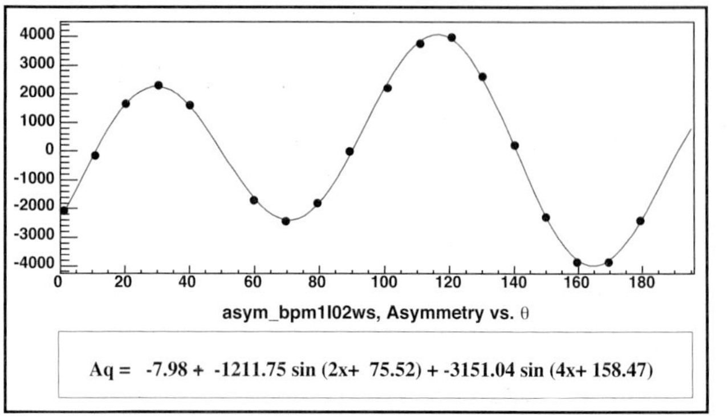

Fig. 5. Charge asymmetry (in ppm) is plotted as a function of the angle of the RHWP. Also shown are the results of a fit including a constant, a 2θ, and a 4θ component

4.1.2 Rotating half-wave plate

Another way of controlling charge asymmetries is through θ (the angle that appears in 3). When the light emerges from the Pockels cell, the major axes of the ellipses are typically either vertical or horizontal. This is determined by the orientation of the fast axis of the Pockels cell, which is usually at $\pm 45°$. By introducing a half-wave plate, however, we can rotate the orientation of the ellipses. The basic setup is illustrated in Fig. 4. The goal is to rotate the ellipses toward the orientation illustrated by Fig. 1c, where the major axes are at $\pm 45°$ with respect to the analyzing-power axis. This is equivalent to adjusting θ to 45°. In a practical situation it is actually desirable to retain a small amount of sensitivity to the analyzing power of the cathode. In this way the charge asymmetry induced by adjustments of Δ_1 can be used as a diagnostic for maximizing the circular polarization of the light that actually strikes the cathode. This helps with, among other things, the position differences due to analyzing-power gradients on the cathode.

The effect of the rotating half-wave plate is illustrated dramatically by the data shown in Fig. 5. The charge asymmetry is shown as a function of the rotation angle θ of a half-wave plate using a setup similar to that shown in Fig. 4 (this θ is distinct from the θ appearing in 3). From the discussion presented here, we should expect the charge asymmetry to vary sinusoidally with 4θ. This is because a 90° rotation of the half-wave plate will rotate the polarization ellipses by 180°, at which point the pattern should repeat. This is close to what we see in Fig. 5, but not exactly. An analysis of the data (shown on the figure) reveals both 4θ and 2θ components. This is because the half-wave plate itself will have imperfections and introduce a "Δ-like" phase, and the fast and slow axes associated with that phase rotates with the half-wave plate.

For parity experiments at JLab, a rotating half-wave plate is an essential part of controlling charge asymmetries. I note in passing that at SLAC, we accomplished essentially the same thing (rotating the ellipses to an arbitrary angle) using two Pockels cell. For the sake of brevity, however, I will not discuss that approach here.

4.1.3 IA cell

The final technique for controlling charge asymmetries is to use what has come to be called an "Intensity Asymmetry" or IA cell. The IA cell is a Pockels cell that is set up between two polarizers so that it can be used as part of an electro-optical shutter. The charge asymmetry can be carefully measured, and the IA cell pulsed to a slightly different voltage for each helicity to achieve balance. This is sort of a brute-force technique, in that it does not address the underlying problems that are causing the asymmetry. It is well suited for use in a feed-back mechanism, however, and it can be done very quickly. For this reason, the use of an IA cell has become standard at SLAC, JLab, and other labs. During HAPPEX II (an IA cell was not used during HAPPEX I) and during E158, however, great care was taken to reduce charge asymmetries to something on the order of 100 ppm independent of the function of the IA cell. At SLAC this was accomplished using a "double-feedback" method. The IA cell was used in a relatively fast feed-back loop to minimize charge asymmetries, and the size of the correction being applied by the IA cell became the error signal for a second feedback loop which corrected the value of Δ_1. For HAPPEX II Δ_1 was adjusted manually before turning on the IA-cell feedback loop, and the size of the IA cell correction was monitored "by hand" to determine when further adjustments to Δ_1 were necessary. Keeping charge asymmetries small before turning on the IA cell is a good way to help ensure that higher-order effects do not become a problem.

4.2 Controlling position asymmetries

As it turns out, it is relatively easy to control charge asymmetries. Position asymmetries, however, can present a real challenge to parity experiments. There are several techniques, however, that are useful.

4.2.1 Minimizing steering

To the extent that a Pockels cell behaves like a lens, the minimization of steering can be accomplished by centering the laser beam on the Pockels cell. The Pockels cell is translated in two dimensions while monitoring the position differences. The one difficulty comes from the fact that there is no obvious way to separate position differences due to steering from position differences due to other sources. Thus, the experimenter can fool themselves that they are centering the Pockels cell when actually they are compensating for several problems at once.

In addition to precise centering of the Pockels cell, another technique for reducing the effects of steering is imaging. By using lenses to create an image of the Pockels cell at the location of the photocathode, steering effects can hypothetically be eliminated. In this configuration, rays of light emanating from the same point on the Pockels cell will all be imaged to the same spot on the cathode, regardless of their exit angle leaving the Pockels cell. Between centering and imaging, steering effects can be greatly suppressed.

4.2.2 Minimizing the effects of phase gradients

One important source of phase gradients is the Pockels cell itself. A Pockels cell will often have a residual birefringence that varies across the aperture of the cell. It is straightforward, however, to construct a small setup on an optics table to characterize such gradients. We have found that if we communicate to the vendor that this is a specification about which we are concerned, they are capable of controlling it at some level. If we subsequently characterize the phase gradients of each Pockels cell and select the best one, the effects from the Pockels cell can be substantially reduced.

Another source of phase gradients is the vacuum window through which the laser beam enters the polarized electron source. This window is generally under stress, which causes induced birefringence. It is useful to explore ways of minimizing these effects, even if this means nothing more than carefully selecting the window that is used.

Once the best Pockels cell and vacuum window have been chosen, it may still be possible to reduce the effects of phase gradients further. The position differences due to phase gradients are independent of the average value of Δ_1, but they do depend on the orientation of the fast axis associated with the gradients to the analyzing-power axis of the system. If the phase gradients are dominated by the Pockels cell, the rotating half-wave plate can in principle be used to "dial away" the effect. If the phase gradients are dominated by the vacuum window, the rotating half-wave plate will have little or no effect.

4.2.3 Minimizing the effects of QE anisotropy gradients

One obvious way of minimizing the effects of QE anisotropy gradients is to choose a photocathode in which the anisotropies are small. Assuming this has been done, it is still possible to reduce the effects to a negligible size by ensuring that the light falling on the cathode is perfectly circularly polarized. It must be remembered, however, that even if light of arbitrary polarization state can be produced outside of the vacuum system, the vacuum window will have an effect, the exact nature of which is generally unknown. In general it is not trivial to completely eliminate the effect.

If a single Pockels cell and a rotating half-wave plate are being used, there is a way to eliminate the effect of QE anisotropy gradients in the limit where the vacuum window has no birefringence. As mentioned earlier, QE anisotropy gradients are proportional to Δ_1. With a large Pockels cell voltage offset (large Δ_1), the effect will be large, and the rotating half-wave plate can be used for zeroing. In a practical situation where the vacuum window does matter, this orientation may or may not be optimal to minimize all the sources of position differences.

4.2.4 Active feedback

It is always possible to employ a piezo-electric driven mirror to actively steer the laser beam in a helicity-correlated fashion to suppress position differences. Such techniques can be quite effective, but it should be remembered that higher order moments of the laser spot will still be varying in a fully helicity-correlated manner. It is thus prudent to minimize the sources of position differences as much as possible before using brute-force feedback for suppression.

4.2.5 Adiabatic damping

If the accelerator is appropriately tuned and free of XY coupling, helicity-correlated position differences are damped as $\sqrt{A/P}$, where A is a constant and P is the momentum. This is due to the adiabatic damping of phase space that takes place as the beam is accelerated. In practice, adiabatic damping has resulted in factors of 3-10 in suppression of position differences. Factors of 50 do not appear unrealistic, but have not yet been achieved.

5 Reversals

Once everything has been done to minimize laser and optics related systematics, the experimenter can still reduce the size of systematic errors by employing a range of "reversals" in which some action is taken that changes the sign of the physics asymmetry without changing the sign of certain helicity-correlated systematics. In this manner the effect of the systematic cancels out when computing the final physics asymmetry. In certain cases reversals can play a large role in taking a very troublesome systematic and suppressing it to a negligible level.

One such reversal utilizes an "insertable half-wave plate", not to be confused with the rotating half-wave plate discussed earlier. Inserting such a half-wave plate will flip the helicity of the light striking the cathode, ultimately resulting in a flip of the sign of the physics asymmetry. Certain systematics, however, will not flip sign, including steering effects and electronic cross talk. Helicity-correlated steering is one of the important sources of beam-position differences, making the the insertable half-wave plate an important part of the full setup.

Another reversal that was used during SLAC E158 was an "asymmetry inverter" [7], which is essentially a pair of two beam expanders which, together with the other optical elements, provides either a positive or negative magnification of equal magnitude between the the object point near the Pockels cell and the image point at the cathode. Both position and angle differences should in theory be eliminated by using an asymmetry inverter.

The final reversal we will mention involves the use of $g-2$ precession, the fact that when an electron beam at high energies traverses an angle θ the spins will precess by an amount $(g-2)\gamma\theta/2$, where γ is the Lorentz factor. If it is possible to run at two different energies such that the electron will arrive at the target with two different helicity states while other energy-dependent quantities are sufficiently well understood, one has a way of flipping the physics asymmetry that truly has nothing to do with the polarized electron source, and would thus leave source-related helicity-correlated systematics unchanged. Perhaps the largest drawback of this approach is that changing energy is time consuming, and hence cannot be done as frequently as one might want from the perspective of controlling systematics. It is very reassuring, however, to see a physics asymmetry flip sign during such an energy change.

6 Summary

There has been significant progress in the understanding of laser systematics since the early studies of parity nonconservation in electron scattering in the late 1970's. The control of such effects has made it possible to study progressively smaller asymmetries with increasing accuracy.

Certainly an area of notable progress has been an improved understanding of the origins of position differences. Whereas steering effects have been identified for some time [2], the identification of phase gradients and QE anisotropy gradients as a source of position differences has brought an important new perspective to the field. I try to summarize some of what we have discussed in Table 1. In the first row of this 2×2 matrix are a few of the things to which the effects of phase gradients are sensitive. The first column represents the sensitivities of the effects of phase gradients in the Pockels cell and the second column represents the sensitivities of the effects of phase gradients in the vacuum window. It is noted that in both cases, the position differences are independent of Δ_1. In the case of phase gradients from the Pockels cell, however, the rotating half-wave plate (RHWP) can be used to zero out the effect. In the second row of this 2×2 matrix are a few of the things to which the effects of QE anisotropy gradients are sensitive. The first column represents the interaction of QE anisotropy gradients with the residual birefringence of the Pockels cell. The second column represents the the interaction of QE anisotropy gradients with the birefringence of the vacuum window. The table indicates that both Δ_1 and the RHWP can be used to control the effects of the Pockels cell residual birefringence with the QE anisotropy gradients. The table entry under vacuum window in the second column is indicating that these effects are zeroed in the limit of perfect circular polarization, but no specifics are given regarding how to achieve that limit.

Table 1. The sensitivities of position differences

	Pockels Cell	Vacuum window
Phase gradients	ind. of Δ_1 sens. to RHWP	ind. of Δ_1 insens. to RHWP
QE anisotropy gradients	prop. to Δ_1 sens. to RHWP	sens. to degree of circ. pol.

Table 1 is useful for formulating strategies to minimize position differences. One possibility is to magnify the effects of QE anisotropy gradients due to the Pockels cell by using a large offset voltage (Δ_1). The RHWP can then be scanned, and should show four zero-crossings for the position differences. Δ_1 can then be set to a nominal zero, and the RHWP scan repeated. The experimenter can then set the RHWP to a position close to the zero from the first scan for which the position differences are minimal during the second scan. The effects of column one are then minimal, and the effects of "column two row two" are probably also fairly small. What is left is presumably mostly due to phase gradients in the vacuum window.

The discussion given here certainly does not represent a definitive analysis of all forms of laser systematics. I hope, however, that this paper provides a useful perspective as we advance forward in our understanding of these often subtle effects. Future parity experiments will make increasingly stringent demands on the control of laser systematics. There is every reason to remain optimistic that these challenges can be met.

References

1. C.Y. Prescott et al.: Phys. Lett. **77B**, 347 (1978)
2. P.A. Souder et al.: Phys. Rev. Lett. **65**, 694 (1990)
3. K. Aniol et al.: Phys. Rev. Lett. **82**, 1096 (1999)
4. K. Aniol et al.: Phys. Lett. B **509**, 211 (2001)
5. P.L. Anthony et al.: Phys. Rev. Lett. **92**, 181602-1 (2004)
6. G.D. Cates et al.: Nucl. Instr. and Meth. in Phys. Res. A **278**, 293 (1989)
7. W.B. Humensky et al.: Nucl. Instr. and Meth. in Phys. Res. A **521**, 261 (2004)

Eur Phys J A (2005) **24**, s2, 115–118
DOI: 10.1140/epjad/s2005-04-026-x

EPJ A direct
electronic only

Beam optics for electron scattering parity-violation experiments

Douglas H. Beck[1] and Mark L. Pitt[2]

[1] Nuclear Physics Laboratory and Loomis Laboratory of Physics, University of Illinois at Urbana-Champaign, 1110 West Green Street, Urbana, IL 61801, USA
[2] Department of Physics Virginia Polytechnic Institute and State University Blacksburg, VA 24061-0435, USA

Received: 15 November 2004 / Published Online: 8 February 2005

Abstract. Parity-violating electron scattering experiments at intermediate energies measure asymmetries in the $10^{-6} - 10^{-5}$ range and therefore require stringent control of false asymmetries. One of the primary sources of such asymmetries is the combined effect of helicity-correlated changes in a certain beam property, accompanied by a change in the detector response. Careful control of the beam, including the optical properties of the acceleration and transport system, is required in order to reduce these false asymmetries to a manageable level. Developments in beam optics associated with the HAPPEX and G0 experiments at Jefferson Lab are presented.

PACS. 29.27.Hj Polarized beams – 13.60.-r Photon and charged-lepton interactions with hadrons – 25.30.-c Elastic electron scattering

1 Introduction

Parity-violating electron scattering experiments measure the interference between the electromagnetic and neutral weak interactions with the target. As such, the experimental asymmetries are of order 10^{-5} with nucleon/nuclear targets at intermediate energies. The counting rates required to accumulate 10^{12} or more events are achieved with the combination of high intensity beams, thick targets and detectors with large solid angle acceptance as discussed extensively at this conference. Control of false asymmetries is achieved by careful control of the beam, starting on the laser Table [1] and extending throughout the accelerator to the target.

The most important aspect of beam control is maintaining a constant current independent of the beam helicity. This is achieved by measuring the electron beam current for each helicity state and feeding the resulting error signal back to control the laser intensity. Such systems have become fairly standard. The emphasis here is instead on the control of helicity-correlated position and angle differences to which the present and future experiments have a significant sensitivity. To set the scale, suppose the response of the detector yield (Y) to a change in beam position x is $(1/Y)(dY/dx) = 0.1\%/\mathrm{mm}$. In order for the asymmetry induced by such a sensitivity to be $A_f \leq 10^{-7}$, the helicity-correlated beam motion must be $\Delta x \leq 10^{-4}$ mm, or 100 nm. Control of beam position and angle at these scales is non-trivial. Helicity-correlated motion of the polarized laser beam in the electron source (used to produce electrons by the photoelectric effect) can easily be 1000 nm without careful attention.

Two different approaches were used by the G0 and HAPPEX experiments to reduce the helicity-correlated beam position and angle differences. In the G0 experiment, the position of the laser beam on the cathode was actively controlled to reduce beam position differences at the target with an automatic feedback system. To date, the HAPPEX experiment has essentially relied on careful setup of the optical polarization and transport system for the laser beam, together with "damping" of the position differences inherent in the acceleration of the beam. They have also recently introduced careful control of the optics near the target ("phase trombone"). The status of both techniques is discussed below.

2 General beam optics

Beams which are launched off the axis ("central orbit") of an optical system will oscillate about the axis with a motion known as betatron oscillation. The formalism of linear beam optics can be used to determine the transverse motion of beam particles in the vicinity of the nominal beam trajectory. The particle orbit is described by the quantities $[x, x', y, y']$ which are a function of the distance s along the central orbit. Here, x and y are the transverse displacements from the central orbit, while $x' \equiv dx/ds$ and $y' \equiv dy/ds$ are the inclination angles of the particle orbit relative to the central orbit. Under the assumptions of linear beam optics (small inclination angles and only constant or linearly increasing magnetic restoring forces) the solution of the equations of motion for the transverse

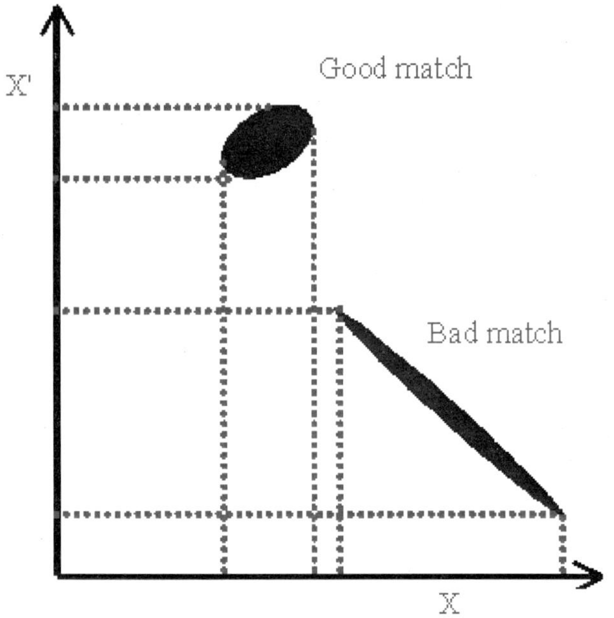

Fig. 1. A schematic illustration of the phase-space ellipses at the end of a well-matched and poorly-matched optical transport system. In the case of the badly matched system, the phase-space ellipse area is preserved, but the ellipse is distorted leading to a larger betatron amplitude than expected from the adiabatic damping factor

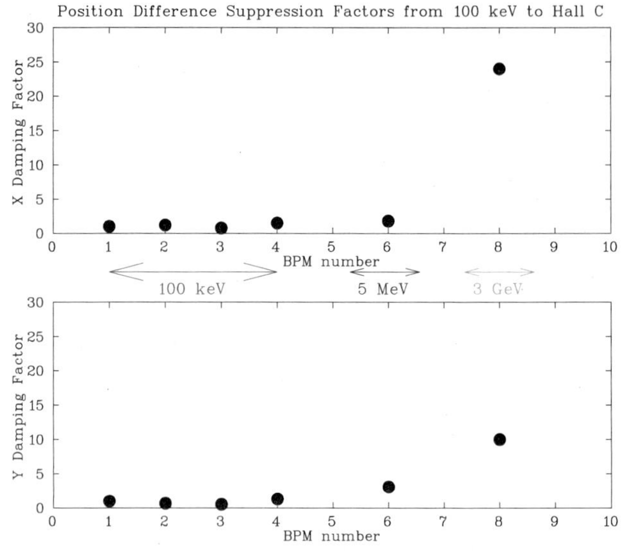

Fig. 2. Measured damping factors in the G0 experiment. The arrows below the graph show the kinetic energy of the electron beam at the various beam position monitors where the measurement are made. The damping factor is defined to be the helicity-correlated position difference observed at the first BPM (BPM number 1) divided by the helicity-correlated position difference observed at the BPM of interest. The expected damping in the injector (from 100 keV to 5 MeV kinetic energy) is not observed because of optics mismatches. Damping is observed in the accelerator (from 5 MeV to 3 GeV kinetic energy)

displacement in the case of no acceleration is given by [3]:

$$x(s) = \sqrt{\epsilon}\sqrt{\beta(s)}\,\cos\{\psi(s) + \phi\} \qquad (1)$$

where

$$\psi(s) = \int_0^s \frac{ds'}{\beta(s')} \qquad (2)$$

is the accumulated betatron phase, ϕ is the initial phase, ϵ is the emittance, and $\beta(s)$ is referred to as the beta function or amplitude function. So, the transverse position of the beam at a given point along the central orbit (e.g. at the target) depends on the initial position and angle of the beam as well as on the intervening optics (the integrated beta function along the orbit).

The solutions for x and x' can be combined in the form

$$\gamma(s)\, x^2(s) + 2\,\alpha(s)\, x(s)\, x'(s) + \beta(s)\, x'^2(s) = \epsilon \qquad (3)$$

which defines a phase-space ellipse in the $x - x'$ plane. The three parameters that characterize this ellipse (α, β and γ) are referred to as the "Twiss parameters". Note that these parameters are functions of the path length s along the central trajectory, so the phase-space ellipse can change shape as the particle moves along its orbit. The area of this phase-space ellipse is simply $\pi\epsilon$, where ϵ is the emittance. For the case of no acceleration, Liouville's theorem states that the area of the phase-space ellipse (and therefore the emittance) remains constant.

When there is acceleration, and the relative momentum changes are small over the scale of the optical element spacing, the expression for the transverse position is modified in a simple way

$$x(s) = \sqrt{\epsilon}\sqrt{\beta(s)}\sqrt{\frac{p_0}{p}}\,\cos\{\psi(s) + \phi\}\,, \qquad (4)$$

where p_0 and p are the initial and final beam particle momenta, respectively. This reduction in the amplitude of the betatron motion as the beam momentum is adiabatically increased is referred to as adiabatic damping. Therefore the acceleration of the machine gradually reduces the amplitude of the betatron oscillation, and the transverse displacement from the central orbit can be reduced further at a given point by controlling the overall accumulated betatron phase.

In practice, one works to achieve the minimum possible helicity-correlated position and angle differences in the beam in the injector by proper alignment and configuration of elements in the polarized injector laser path as described in reference [1]. One then expects additional reduction in the position and angle differences observed at the experimental target by the adiabatic damping factor - $\sqrt{p_0/p}$. The full expected adiabatic damping factor is often not achieved due to an optically mismatched beam transport system. In a perfectly matched system, the Twiss parameters after passing through each beamline element match the design parameters. As discussed in the next section, various types of imperfections in the transport system lead to a deviation from this ideal case.

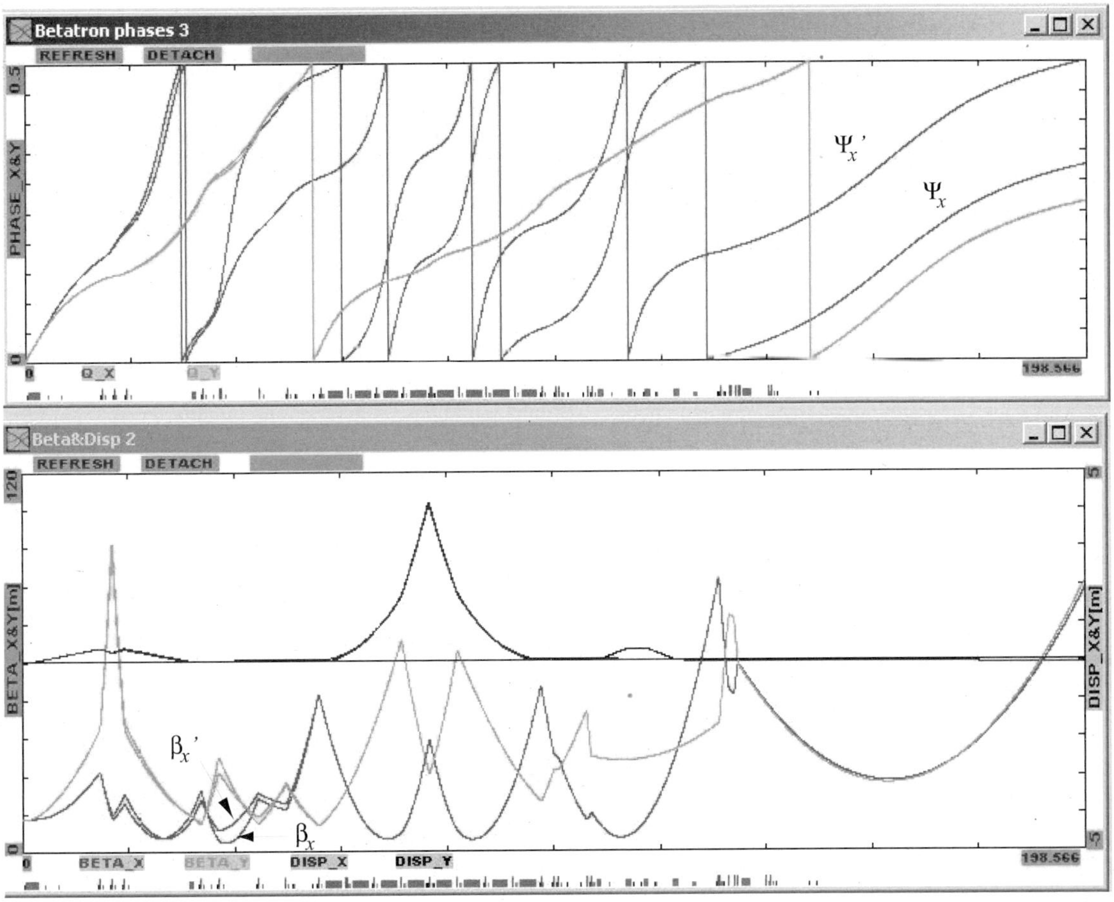

Fig. 3. Calculated effects of the phase trombone in the Hall A beam line at Jefferson Lab. The *upper panel* shows a difference in the phase of the β_x function of 60° (*upper curves at right:* Ψ_x *and* Ψ'_x) for slightly different tunings. The small difference in the tuning can be seen in the *lower panel* where the nearly identical pair of curves show β_x, β'_x for the two cases. The y functions are unaffected

The result of a mismatched transport setup are shown schematically in Fig. 1. At a given location, the phase-space ellipse preserves its area, but in a badly matched system the ellipse becomes distorted leading to a larger betatron amplitude ("orbit blow-up") than ideal adiabatic damping would predict.

3 Active feedback of beam position difference

In the G0 experiment, the relatively large bunch charge (running at 31 MHz pulse rate, necessitated by time-of-flight measurements, see [2]) required a non-standard tuning of the injector which made reduction of the helicity-correlated beam position differences with the standard damping difficult to achieve. Nominally, with 3 GeV incident energy, the damping from the injector to the target would be

$$\sqrt{\frac{3\ \text{GeV}}{355\ \text{keV}}} = 95 \qquad (5)$$

However the damping factors measured in the x and y directions were typically 25 and 10, respectively as shown in Fig. 2.

The precise cause of the reduced damping is not clear; however, there are several likely contributors. In order for the damping to be realized, the beam, characterized by its Twiss parameters, must have the envelope to which the subsequent optical elements are matched. In practice, correction elements are used to restore the envelope after sections of the beamline (e.g. linac, arc, etc.). Matching is particularly important in the injector where the relative acceleration is large, but is particularly difficult because of the focusing effects of the accelerating sections. The impact of this focusing is larger in the injector because of the lower energy of the beam; it is further complicated by the focusing components that mix the x and y phase spaces. Work continues to make improvements in this difficult tuning problem [4].

During the G0 experiment, the lack of damping was overcome with active feedback on the beam position at the target [5]. Helicity-correlated beam position difference measurements were made near the target and used to move the polarized source laser beam (using a piezo-electric actuator to move a reflecting mirror) in a helicity-correlated manner to null the error signal. In practice, this feedback was somewhat more difficult as the beam current and motions in the x and y directions were fully coupled, e.g. changing the y position of the laser beam at the source changed the x and y positions at the target, as well as the beam current. Successful operation of the feedback system required periodic calibration of this "response" matrix, as well as some optical adjustments to insure that the matrix was non-singular.

4 Phase trombone

In the most recent run of the HAPPEX experiments [6], beam position differences were mitigated using different techniques. As HAPPEX ran with the standard bunch charge (499 MHz pulse rate), the tuning of the injector was more standard and larger damping factors (though not the theoretical values) were routinely obtained. The position differences of the laser beam on the polarized source crystal were also reduced through a combination of a new, larger diameter Pockels cell (the electro-optic $\lambda/4$ plate that sets the laser beam polarization) and more careful optical alignment.

In the current run, a newly developed technique was also initiated. Starting with reduced position differences at the target, the HAPPEX group, in collaboration with the JLab Accelerator Division used a group of eight quadrupole magnets in the arc to adjust the beta function phase advance at the target (hence the name phase trombone) to trade off helicity-correlated position and angle differences (see Fig. 3). The basic idea is to change the phase advance periodically during the experiment to trade, e.g., a large position difference for a large angle difference or even to reverse the sign of a position difference to cancel the effect in an earlier part of the run. Technically, this amounts to rotating the phase space ellipse that describes the beam envelope. Development of this tool is also continuing.

5 Conclusion

Because the asymmetries in parity-violating experiments are small, careful control of the beam is required to reduce false asymmetries to acceptable levels. Therefore, in a real sense, the apparatus for these experiments involves the entire accelerator as an integral part. Among other accelerator challenges, controlling helicity-correlated beam positions and angles at the target at a level of a few nm and a few nrad, respectively, requires sophisticated control mechanisms. Achieving the natural (due to acceleration) damping of motion in the transverse planes is becoming easier as new sources of optical mismatches are identified. In the G0 experiment, active feedback was used with some success to reduce the position and angle differences, although changing beam conditions required periodic adjustments. The present HAPPEX run has seen the beginning of development of a new optical tool to allow position and angle differences to be traded off, reducing the overall effect on the experiment. Continued development of these techniques will be important to the success of future measurements of even smaller asymmetries.

We gratefully acknowledge contributions to this presentation from A. Bogacz, Y. Chao, K. Nakahara, and K. Paschke.

References

1. G. Cates: this workshop
2. P. Roos: this workshop
3. K. Wille: *The Physics of Particle Accelerators: An Introduction* (Oxford University Press, New York 2000) 80ff
4. Y. Chao et al.: proceedings of the 9th European Particle Accelerator Conference, 2004
5. K. Nakahara: this workshop
6. R. Holmes: this workshop

Eur Phys J A (2005) **24**, s2, 119–120
DOI: 10.1140/epjad/s2005-04-027-9

EPJ A direct
electronic only

G0 beam quality and multiple linear regression corrections

Kazutaka Nakahara

University of Illinois at Urbana-Champaign 1110 West Green St. Urbana, IL 61801, USA e-mail: nakahara@jlab.org

Received: 15 October 2004 / Published Online: 8 February 2005

Abstract. The G0 experiment measures parity-violating elastic e-p scattering asymmetries to probe the strange quark content of the nucleon. The goal is to measure the asymmetries with an overall uncertainty of 5% of the measured asymmetries which will be of order 10^{-5} to 10^{-6}. In order to achieve the above precision, systematic effects which can induce false asymmetries must be controlled. One such systematic error is helicity-correlated changes in beam parameters, which, coupled with the sensitivity of the G0 spectrometer to such beam variations, can induce false asymmetries. Beam parameters monitored for helicity-correlation are beam current, beam position, beam angle, and beam energy. A feedback loop was successfully used to reduce the helicity-correlation in beam current and beam position. The sensitivity of the G0 spectrometer to fluctuations in beam parameters has also been measured, and the false asymmetries have been determined to be of order 10^{-8}. This contribution will address the sensitivity of the G0 spectrometer to changes in beam conditions, the performance of the feedback loops as well as the resulting parity quality of the G0 beam, and the resulting false asymmetry.

PACS. 29.27.Hj Polarized beams – 13.60.-r Photon and charged-lepton interactions with hadrons – 25.30.-c Elastic electron scattering

1 Introduction

The physics process responsible for the asymmetry measured in the G0 experiment is the weak parity-violating amplitude in elastic scattering [1].

$$\mathtt{A} = (\sigma_+ - \sigma_-)/(\sigma_+ + \sigma_-) \tag{1}$$

where $\sigma_{+/-}$ are the elastic e-p cross sections in the positive/negative helicity states of the electron beam. Thus, the measured asymmetries are susceptible to helicity - correlated fluctuations in beam conditions. These fluctuations can alter the measured cross section in a manner unrelated to the physics process we wish to probe, and consequently appear as unwanted false asymmetries. As parity-violation experiments strive to achieve greater precision in the measurements of their asymmetries, the tolerance on helicity-correlated fluctuations in beam conditions have become more stringent. However, recent developments in improving beam quality for such experiments have yielded significant results in reducing false asymmetries coming from helicity-correlated beam fluctuations. These developments, together with the standard multiple linear regression analysis techniques, have rendered such false asymmetries to be a well-controllable systematic effect [2,3]. Although there have been many improvements made both in controlling the helicity-correlation at the source as well as through improved beam transport, this contribution will detail the methods specifically used by the G0 collaboration to reduce these false asymmetries.

2 Parity quality beam

Processes which can induce helicity-correlation in the beam may include effects such as an imperfect circular polarization of the laser, helicity-correlated laser motion at the cathode, and beam loading effects in the accelerating cavities [4]. In the G0 experiment the cumulative effect from all such processes was measured, and an active feedback device was used to null out the resulting helicity-correlated differences. An IA (intensity attenuator) cell and a PZT (piezo-electric transducer) mirror were used to modulate the intensity and position of the laser at the cathode in order to null out any charge asymmetry and position differences that exist within the beam. Existing correlations between charge, position, angle and energy made it possible to reduce all 6 helicity-correlated differences simply by feeding back on the above 3 parameters, although it was not guaranteed that the energy difference would be reduced. A halfwave plate was inserted/retracted at the source to reverse the circular polarization of the laser every few days to monitor any further systematics effects coming from polarization reversal.

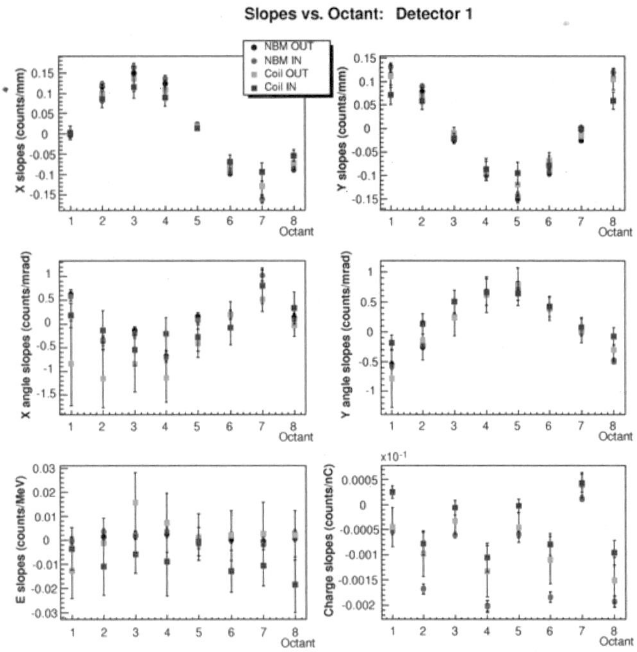

Fig. 1. Linear regression slopes determined from natural beam motion (NBM) and coil modulation show consistent results. The beam current slopes shows a large odd vs even octant dependence due to differing deadtimes from the 2 distinctly separate electronics that were used in those octants

3 Multiple linear regression

In order to determine how much false asymmetry results from the helicity-correlation in the beam, the sensitivity of the G0 spectrometer to changes in beam parameters was measured. The false asymmetry resulting from the helicity-correlated parameter differences is thus, [2]

$$A_{false} = \Sigma_i 1/Y * (dY/2dP_i) * \delta P_i \qquad (2)$$

where Y is the yield seen on the detectors, δP_i denotes the helicity-correlated differences in the six beam parameters, and dY/dP_i is the slope which characterizes the sensitivity of the spectrometer yield to fluctuations in the beam. The spectrometer is composed of 8 azimuthally symmetric octants, each with an array of 16 detectors which corresponds to different Q^2, so the false asymmetry contribution due to position and angular motion of the beam tends to cancel out when summed over all octants. In addition to determining the slopes with the natural motion of the beam, a set of steering coils were used upstream of the target to modulate the beam by large amounts in order to gain more dynamic range in determining the sensitivity of the spectrometer to beam position and angle. The two methods have shown consistent results throughout the run and the expected octant-dependence is seen. Figure 1 shows the determined sensitivity to changes in the 6 beam parameters for one particular detector element. The resulting false asymmetries coming from the slopes and the helicity-correlation in the 6 beam parameters are summarized on Table 1.

Table 1. The false asymmetries from the helicity-correlated differences as well as the sensitivity of the spectrometer

Beam Parameter	Helicity correlated difference (IN-OUT)	Slopes (octant dependent)	False asymmetry (octant-sum)
X	6 ± 4 nm	-1 to 1 %/mm	10^{-9}
Y	8 ± 4 nm	-1 to 0.8 %/mm	10^{-9}
θ_x	2 ± 0.3 nrad	-6 to 8 %/mrad	10^{-8}
θ_y	3 ± 0.5 nrad	-4 to 5 %/mrad	10^{-8}
E	58 ± 4 eV	-0.01 to 0.02 %/MeV	10^{-9}
I	-0.28 ± 0.28 ppm	$\sim 10^{-3}$ %/nC	10^{-8}

4 Conclusions

The parity beam feedback system for the G0 forward angle production was successfully implemented, and parity quality beam was achieved to the level where the resulting false asymmetry of order 0.01ppm is considered a negligible contribution to the 5% overall uncertainty that the G0 experiment aims for in the determination of its parity-violating asymmetries.

Acknowledgements. Thanks to Doug Beck, Mark Pitt, Matt Poelker, Joe Grames, Yu-Chu Chao, Reza Kazimi, and the Jefferson Lab HallC and Accelerator groups for their tireless effort and support.

References

1. Douglas H. Beck, Barry R. Holstein: International Journal of Modern Physics E (2001), p. 5–10
2. Damon Spayde: Thesis at the U. of Maryland, *Measurement of the Strange Magnetic Form Factor of the Proton using Elastic Electron Scattering*, (2001) p. 55 and p. 101–117
3. K.A. Aniol, et al.: Phys. Rev. C **69**, 065501 (2004), p. 4–6
4. P.A. Souder et al.: Phys. Rev. Lett. **65**, 694 (1990), p. 695–697

IV Experimental techniques in PV electron scattering

IV-2 Polarimetry

Eur Phys J A (2005) **24**, s2, 123–126
DOI: 10.1140/epjad/s2005-04-028-8

EPJ A direct
electronic only

Møller polarimetry with atomic hydrogen targets

E. Chudakov[1a] and V. Luppov[2b]

[1] Thomas Jefferson National Accelerator Facility, Newport News, VA23606, USA
[2] University of Michigan Spin Physics Center, Ann Arbor, MI 48109-2036, USA

Received: 15 October 2004 / Published Online: 8 February 2005

Abstract. A novel proposal of using polarized atomic hydrogen gas, stored in an ultra-cold magnetic trap, as the target for electron beam polarimetry based on Møller scattering is discussed. Such a target of practically 100% polarized electrons could provide a superb systematic accuracy of about 0.5% for beam polarization measurements. Feasibility studies for the CEBAF electron beam have been performed.

PACS. 07.60.Fs Polarimeters – 29.25.Pj Polarized targets – 67.65.+z Spin-polarized hydrogen and helium

1 Motivation

Precise electron beam polarimetry will become increasingly important for the next generation of parity violation experiments. The systematic errors (polarimetry excluded) and statistical errors of some of these experiments will become better that 0.5%. For example, the measurement of the neutron skin of the ^{208}Pb nucleus, proposed at Jefferson Lab [1], requires a 1% polarimetry accuracy for the 850 MeV, 50 μA polarized electron beam, and would benefit from a polarimetry accuracy of 0.5%.

Compton polarimetry, while accurate enough at the energies $>$ 4 GeV [2, 3] has difficulties at low energies $\sim$800 MeV and a 1% accuracy has not been achieved so far.

Møller polarimetry does not depend considerably on the beam energy, but the accuracy is limited by the choice of the polarized electron target. Ferromagnetic foils, used so far, provide electron polarization of about 8%, known either with an accuracy of about 2-3% (see, for example [4, 5, 6]) if the foil is magnetized along its surface in a a field of 10-30 mT, or with an accuracy $\sim$0.3%, if it is magnetized in a very strong field of $\sim$4 T [6]. There are other systematic errors, associated with ferromagnetic targets. A kinematics difference in scattering on the external and internal atom shells lead to a systematic error (the so-called Levchuk effect [7]). The target heating limits the beam current to 2-3μA, a factor of 10-30 below the typical currents needed for the experiments. Also, the dead time gives a systematic error.

With all this in mind it seems very attractive to use atomic hydrogen gas, held in an ultra-cold magnetic trap [8], as the source of 100% polarized electrons. Møller polarimetry with such a target would be free of the accuracy limitations discussed above. The target polarization would be close enough to 100% and there will be no need to measure it. There will be no Levchuk effect or noticeable dead time. Here, a feasibility study of such an option is presented.

2 Polarized atomic hydrogen target

2.1 Hydrogen atom in magnetic field

The magnetic field B_S and the hyperfine interaction split the ground state of hydrogen into four states with different energies. The low energy states are $|a\rangle = |\downarrow\Uparrow\rangle \cdot \cos\theta - |\uparrow\Downarrow\rangle \cdot \sin\theta$ and $|b\rangle = |\downarrow\Downarrow\rangle$, where the first and second (crossed) arrows in the brackets indicate the electron and proton spin projections on the magnetic field direction. As far as the electron spin is concerned, state $|b\rangle$ is pure, while state $|a\rangle$ is a superposition. The mixing angle θ depends on the magnetic field B_S and temperature T: $tan\ 2\theta \approx 0.05\ \mathrm{T}/B_S$. At B_S =8 T and T = 0.3 K the mixing factor is small: $\sin\theta \approx 0.003$. State $|b\rangle$ is 100% polarized. State $|a\rangle$ is polarized in the same direction as $|b\rangle$ and its polarization differs from unity by $\sim 10^{-5}$.

2.2 Storage cell

In a magnetic field gradient, a force $-\nabla(\mu_{\mathbf{H}}\mathbf{B})$, where μ_H is the atom's magnetic moment, separates the lower and the higher energy states. The lower energy states are pulled into the stronger field, while the higher energy states are repelled from the stronger field. The 0.3 K cylindrical storage cell, made usually of pure copper, is located

---

[a] This work was supported by the Southeastern Universities Research Association (SURA), which operates the Thomas Jefferson National Accelerator Facility for the United States Department of Energy under contract DE-AC05-84ER40150.

[b] Now with Janis Research Company, Wilmington, MA 01887-0696

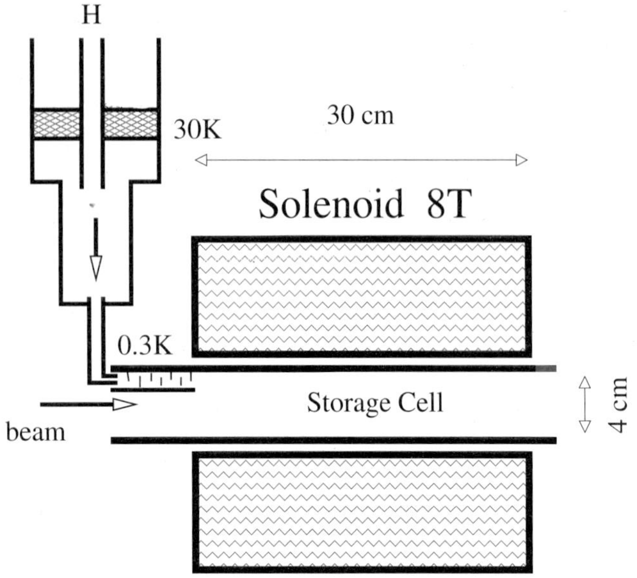

Fig. 1. A sketch of the storage cell

in the bore of a superconducting ~8 T solenoid. The polarized hydrogen, consisting of the low energy states, is confined along the cell axis by the magnetic field gradient, and laterally by the wall of the cell (Fig. 1).

At the point of statistical equilibrium, the state population, p follows the Boltzmann distribution:

$$p \propto \exp(\mu_e B/kT), \qquad (1)$$

where μ_e is the electron's magnetic moment ($\mu_H \approx \mu_e$) and $k = k_B$ is the Boltzmann constant. The cell is mainly populated with states $|a\rangle$ and $|b\rangle$, with an admixture of states $|c\rangle$ and $|d\rangle$ of $\exp(-2\mu_e B/kT) \approx 3 \cdot 10^{-16}$. In the absence of other processes, states $|a\rangle$ and $|b\rangle$ are populated nearly equally. The gas is practically 100% polarized, a small ($\sim 10^{-5}$) oppositely polarized contribution comes from the $|\uparrow\Downarrow\rangle$ component of state $|a\rangle$.

The atomic hydrogen density is limited mainly by the process of recombination into H_2 molecules (releasing ~4.5 eV). The recombination rate is higher at lower temperatures. In gas, recombination by collisions of two atoms is kinematically forbidden but it is allowed in collisions of three atoms. On the walls, which play the role of a third body, there is no kinematic limitation for two atom recombination. At moderate gas densities only the surface recombination matters. In case of polarized atoms, the cross section for recombination is strongly suppressed, because two hydrogen atoms in the triplet electron spin state have no bound states. This fact leads to the possibility of reaching relatively high gas densities for polarized atoms in the traps.

A way to reduce the surface recombination on the walls of the storage cell is coating them with a thin film (~50 nm) of superfluid ^{4}He. The helium film has a very small sticking coefficient[1] for hydrogen atoms. In contrast, hydrogen molecules in thermal equilibrium with the film are absorbed after a few collisions and are frozen in clusters on the metal surface of the trap [9].

The higher energy states are repelled from the storage cell by the magnetic field gradient and leave the cell. Outside of the helium-covered cell, the atoms promptly recombine on surfaces into hydrogen molecules which are either pumped away or are frozen on the walls. Some of the higher energy states recombine within the cell and the molecules eventually are either frozen on the helium-coated wall, or leave the cell by diffusion.

The cell is filled with atomic hydrogen from an RF dissociator. Hydrogen, at 80 K, passes through a Teflon[2] pipe to a nozzle, which is kept at ~30 K. From the nozzle hydrogen enters into a system of helium-coated baffles, where it is cooled down to ~0.3 K. At 30 K and above, the recombination is suppressed because of the high temperature, while at 0.3 K it is suppressed by helium coating. In the input flow, the atoms and molecules are mixed in comparable amounts, but most of the molecules are frozen out in the baffles and do not enter the cell.

The gas arrives at the region of a strong field gradient, which separates very efficiently the lower and higher atomic energy states, therefore a constant feeding of the cell does not affect the average electron polarization.

This technique was first successfully applied in 1980 [10], and later a density[3] as high as $3 \cdot 10^{17}$ atoms/cm^3 was achieved [8] in a small volume. So far, the storage cell itself has not been put in a high-intensity particle beam.

For the project being discussed a normal storage cell design can be used, with the beam passing along the solenoid axis (Fig. 1). The double walls of the cylindrical copper cell form a dilution refrigerator mixing chamber. The cell is connected to the beam pipe with no separating windows. The tentative cell parameters are (similar to a working cell [11]): solenoid maximum field of $B_S = 8$ T, solenoid length of $L_S = 30$ cm, cell internal radius of $r_\circ = 2$ cm, cell length of $L_C = 35$ cm and temperature of $T = 0.3$ K. The effective length of such a target is about 20 cm.

For the guideline, we will consider a gas density of $3 \cdot 10^{15}$ cm^{-3}, obtained experimentally [12], for a similar design.

2.3 Gas properties

Important parameters of the target gas are the diffusion speed. At 300 mK the RMS speed of the atoms is ~80 m/s. For these studies we used a calculated value [13] of the hydrogen atoms cross section $\sigma = 42.3 \cdot 10^{-16}$ cm^2, ignoring the difference between the spin triplet and singlet cross sections. This provided the mean free path $\ell = 0.57$ mm at density of $3 \cdot 10^{15}$ cm^{-3}.

[1] The sticking coefficient defines the atom's adsorption probability per a collision with a surface.

[2] Teflon has a relatively small sticking coefficient for hydrogen atoms.

[3] This parameter is called concentration, but we will use the word density in the text, since the mass of the gas is not important here.

The average time, τ_d for a "low field seeking" atom to travel to the edge of the cell, assuming its starting point is distributed according to the gas density, is[4]: $\tau_d \approx 0.7$ s. This is the cleaning time for an atom with opposite electron spin, should it emerge in the cell and if it does not recombine before. The escape time depends on the initial position of the atom, going from ~ 1 s at $z = 0$ to 0.1 s at $z = 8$ cm. The average wall collision time is about 0.5 ms.

2.4 Gas lifetime in the cell

For the moment we consider the gas behavior with no beam passing through it. Several processes lead to losses of hydrogen atoms from the cell: thermal escape through the magnetic field gradient, recombination in the volume of gas and recombination on the surface of the cell.

The volume recombination can be neglected up to densities of $\sim 10^{17}$ cm ^{3} [8].

The dominant process, limiting the gas density, is the surface recombination. In order to keep the gas density constant the losses have to be compensated by constantly feeding the cell with atomic hydrogen. Our calculations, based on the theory of such cells [8], show, that a very moderate feed rate of $\Phi \sim 1 \cdot 10^{15}$ atoms/s would provide a gas density of $7 \cdot 10^{15}$ cm^{-3}.

This can be compared with the measurement [12] of $3 \cdot 10^{15}$ cm^{-3}. The average lifetime of a "high field seeking" atom in the cell is $\sim$1 h.

2.5 Unpolarized contamination

The most important sources of unpolarized contamination in the target gas in absence of beam have been identified:

1) hydrogen molecules: $\sim 10^{-5}$;
2) high energy atomic states $|c\rangle$ and $|d\rangle$: $\sim 10^{-5}$;
3) excited atomic states $< 10^{-10}$;
4) other gasses, like helium and the residual gas in the cell: $\sim 10^{-3}$

The contributions 1)-3) are present when the cell is filled with hydrogen. They are difficult to measure directly and we have to rely on calculations. Nevertheless, the behavior of such storage cells has been extensively studied and is well understood [8]. The general parameters, like the gas lifetime, or the gas density are predicted with an accuracy better than a factor of 3. The estimates 1)-3) are about 100 times below the level of contamination of about 0.1% which may become important for polarimetry. In contrast, the contribution 4) can be easily measured with beam by taking an empty target measurement. Atomic hydrogen can be completely removed from the cell by heating a small bolometer inside the cell, which would remove the helium coating on this element, and catalyze a fast recombination of hydrogen on its surface. However, it is important to keep this contamination below several percent in order to reduce the systematic error associated with the background subtraction.

[4] This time was estimated using simulation, taking into account the gas density distribution along z and the repelling force in the magnetic field gradient.

3 Beam impact on storage cell

We have considered various impacts the $\mathcal{I}_b = 100$ μA CEBAF beam can inflict on the storage cell. The beam consists of short bunches with $\tau = \sigma_T \approx 0.5$ ps at a $\mathcal{F} = 499$ MHz repetition rate. The beam spot has a size of about $\sigma_X \approx \sigma_Y \sim 0.1$ mm. The most important depolarization effects we found are:

A) gas depolarization by the RF electromagnetic radiation of the beam: $\sim 3 \cdot 10^{-5}$;
B) contamination from free electrons and ions: $\sim 10^{-5}$;
C) gas excitation and depolarization by the ionization losses: $\sim 10^{-5}$;
D) gas heating by ionization losses: $\sim 10^{-10}$ depolarization and a $\sim$30% density reduction.

The effects A) and B) are described below.

3.1 Beam RF generated depolarization

The electromagnetic field of the beam has a circular magnetic field component, which couples to the $|a\rangle \to |d\rangle$ and $|b\rangle \to |c\rangle$ transitions. The transition frequency depends on the value of the local magnetic field in the solenoid and for the bulk of the gas ranges from 215 to 225 GHz. The spectral density function of the magnetic field can be presented in the form of Fourier series with the characteristic frequency of $\omega_\circ = 2\pi\mathcal{F}$. The Fourier coefficients are basically the Fourier transforms of the magnetic field created by a single bunch. The bunch length is short in comparison with the typical transition frequency ($\omega_{trans}\tau \sim 0.1$). The resonance lines of the spectrum (a reflection of the 499 MHz repetition rate) populate densely the transition range (see Fig. 2). The induced transition rate depends on the gas density at a given transition frequency. This rate was calculated taking into account the beam parameters and the field map of a realistic solenoid. Provided that the field of the solenoid is fine tuned to avoid the transition resonances for the bulk of the gas in the cell (see Fig. 2), the depolarization described has the following features:

- the transition rate is proportional to $\mathcal{I}_b^2$;
- the average rate of each of the two transitions is about $0.5 \cdot 10^{-4}$ of the target density per second;
- at the center around the beam the full transition rate is about 6% of the density per second.

In order to estimate the average contamination we take into account that each resonance line presented in Fig. 2 corresponds to a certain value of the solenoid field and, therefore, affects the gas at a certain z. Using a realistic field map of the solenoid we obtained that the average depolarization in the beam area will be reduced to about $\sim 0.3 \cdot 10^{-4}$ by the lateral gas diffusion and by the escape of the "low field seeking" atoms from the storage cell.

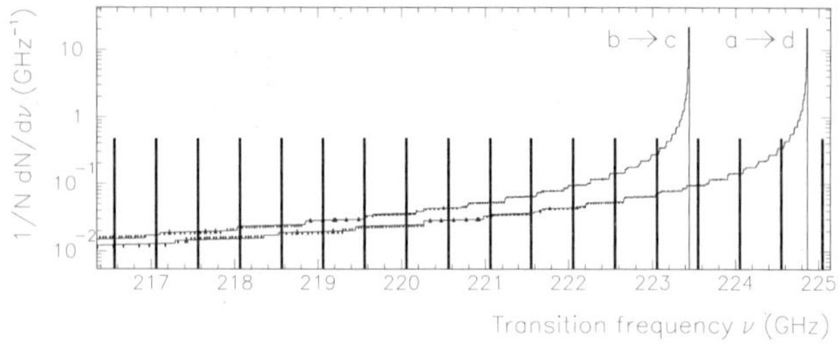

Fig. 2. Simulated spectra of the transitions on the axis of the hydrogen trap with the maximum field of 8.0 T. The density of atoms depends on the field as $exp(-\mu_e B/kT)$. The two *curves* show $\frac{1}{N}dN/d\nu_{ad}$ and $\frac{1}{N}dN/d\nu_{bc}$ - the relative number of atoms which can undergo $|a\rangle \rightarrow |d\rangle$ and $|b\rangle \rightarrow |c\rangle$ transitions at the given frequency, per one GHz. The resonant structure of the spectral function of the beam-induced electromagnetic field is shown as a set of *vertical bars*, 499 MHz apart

In order to study experimentally the depolarization effect discussed, one can tune the solenoid magnetic field to overlap a resonance line with the transition frequency of the gas at the cell center. This would increase the transition rate by a factor of ~70.

3.2 Contamination by free electrons and ions

The beam would ionize per second about 20% of the atoms in the cylinder around the beam spot . The charged particles would not escape the beam area due to diffusion, as the neutral atoms would do, but will follow the magnetic field lines, parallel to the beam. An elegant way to remove them is to apply a relatively weak ~1 V/cm electric field perpendicular to the beam. The charged particles will drift at a speed of $v = \mathbf{E} \times \mathbf{B}/B^2 \sim 12$ m/s perpendicular to the beam and leave the beam area in about 20 μs. This will reduce the average contamination to a 10^{-5} level.

4 Application of the atomic target to Møller polarimetry

This feasibility study was done for the possible application of the target discussed to the existing Møller polarimeter in Hall A at JLab [5].The results are, however, more generic and are largely applicable to other facilities with "continuous" electron beams.

The beam polarization at JLab is normally about 80%, at beam currents below 100 μA. Scaling the results of the existing polarimeter to to the hydrogen target discussed we estimated that at 30 μA a 1% statistical accuracy will be achieved in about 30 min. This is an acceptable time, in particular if the measurements are done in parallel with the main experiment.

There is no obvious way to measure directly the polarization of the hydrogen atoms in the beam area. The contamination from the residual gas is measurable. The rest relies on calculations. All calculations show that the polarization is nearly 100%, with a possible contamination of <0.01%, coming from several contributions. The impact of the most important of these contributions can be studied, at least their upper limits, by deliberately increasing the effect. For example, the beam RF induced transitions can be increased by a factor of ~70, by fine tuning of the solenoid magnetic field. The contribution from the charged particles in the beam area can be varied by a factor up to $\sim 10^4$, by changing the cleaning electric field.

The systematic errors, associated with the present Hall A polarimeter, when added in quadrature give a total systematic error of about 3% [5]. Scaling these errors to the hydrogen target option reduces the total error to about 0.3%.

5 Conclusion

The considerations above show that a stored, longitudinally electron-spin-polarized atomic hydrogen can be used as a pure, 100% electron polarized gas target. A thickness of at least $6 \cdot 10^{16}$ electrons/cm^2 can be reached with a target diameter of 4 cm and a length of 20 cm along the beam. The polarized hydrogen gas should be stable in the presence of a 100 μA CEBAF beam. A Møller polarimeter, equipped with such a target would provide a superb systematic accuracy of about 0.5%, while providing a 1% statistical accuracy in about 30 min of running at a beam current of 30 μA .

References

1. C.J. Horowitz, S.J. Pollock, P.A. Souder, R. Michaels: Phys. Rev. C **63**, 025501-1-18 (2001), [arXiv:nucl-th/9912038]
2. M. Baylac et al.: Phys. Lett. B **539**, 8–12 (2002), [arXiv:hep-ex/0203012]
3. M. Woods [SLD Collaboration]: arXiv:hep-ex/9611005
4. P.S. Cooper et al.: Phys. Rev. Lett. **34**, 1589–1592 (1975)
5. A.V. Glamazdin et al.: Fizika B **8**, 91–95 (1999), [arXiv:hep-ex/9912063]
6. M. Hauger et al.: Nucl. Instrum. Methods A **462**, 382–392 (2001), [arXiv:nucl-ex/9910013]
7. L.G. Levchuk: Nucl. Instrum. Methods A **345**, 496–499 (1994)
8. I.F. Silvera, J.T.M. Walraven: "Spin Polarized Atomic Hydrogen," in *Progress in Low Temperature Physics*, vol. X (Amsterdam: Elsevier Science Publisher B.V., 1986) 139–370
9. I.F. Silvera: Phys. Rev. B **29**, 3899–3904 (1984)
10. I.F. Silvera, J.T.M. Walraven: Phys. Rev. Lett. **44**, 164–168 (1980)
11. T. Roser et al.: Nucl. Instrum. Methods A **301**, 42–46 (1991)
12. M. Mertig, V.G. Luppov, T. Roser, B. Vuaridel: Rev. Sci. Instrum. **62**, 251–252 (1991)
13. M.D. Miller, L.H. Nosanow: Phys. Rev. B **15**, 4376–4385 (1977)

Eur Phys J A (2005) **24**, s2, 127–128
DOI: 10.1140/epjad/s2005-04-029-7

EPJ A direct
electronic only

Progress report on the A4 Compton backscattering polarimeter

Yoshio Imai[a], for the A4 Collaboration

Inst. für Kernphysik, Universität Mainz, J.-J.-Becher-Weg 45, D-55128 Mainz, Germany

Received: 1 November 2004 / Published Online: 8 February 2005

Abstract. The A4 collaboration at the Dept. of Nuclear Physics, University of Mainz, is conducting experiments on parity violation in the elastic electron-nucleon-scattering which require the use of polarized beams. In order to measure the absolute beam polarization, we have installed a Compton backscattering polarimeter in front of the target, using for the first time the internal cavity concept. A maximum intracavity intensity of 90 W has been measured, and in August 2003, first backscattered photons have been detected. This article describes the new design concept and the current status and results.

PACS. 29.27.Hj Polarized beams – 41.85.Qg Beam analyzers, beam monitors and Faraday cups – 42.60.By Design of specific laser systems – 42.60.Da Resonators, cavities, amplifiers, arrays and rings

1 Introduction

The A4 experiment at the Mainz Microtron (MAMI) is designed to determine the strange quark contribution to the nucleon properties by measuring the parity-violating cross-section asymmetry in the elastic electron-nucleon scattering with polarized beams. The measured asymmetry is related to the physics asymmetry via

$$A_{exp} = P_e A_{phys} \qquad (1)$$

where P_e is the (longitudinal) beam polarization. For an absolute measurement of P_e, a Compton backscattering polarimeter using a new design concept has been installed in the A4 beamline.

2 Compton polarimetry

The Compton cross-section for polarized light on polarized electrons can be written as follows [1]:

$$\frac{d\sigma}{d\Omega} = \frac{d\sigma_0}{d\Omega} + Q\frac{d\sigma_1}{d\Omega} - VP^e_{long}\frac{d\sigma_{long}}{d\Omega} - VP^e_{tr}\cos\varphi\frac{d\sigma_{tr}}{d\Omega} \qquad (2)$$

P_{long}, P_{tr} denote the longitudinal and transverse electron polarizations, Q, V the Stokes parameters describing the light polarization and φ the azimuthal scattering angle. When using purely circular light ($Q = 0$), switching the light helicity ($V = \pm 1$) leads to an asymmetry in the spatial and energetic distributions of the backscattered photons, from which P_{long} and P_{tr} can be extracted.

The measurement time depends on the cross-section, the asymmetry, and the luminosity $\mathcal{L}$ via [2]

$$t \propto \frac{1}{\mathcal{L}\langle\sigma\rangle\langle A^2\rangle} \qquad (3)$$

[a] comprises part of PhD thesis

Table 1. Luminosity requirements for green light, P_e=0.8

$\Delta P_e/P_e$	$\mathcal{L}(E_e = 855 MeV)$	$\mathcal{L}(E_e = 570 MeV)$
10 %	1.15 kHz/barn	2.51 kHz/barn
5 %	4.59 kHz/barn	10.05 kHz/barn
3 %	12.76 kHz/barn	27.91 kHz/barn
1 %	114.86 kHz/barn	251.16 kHz/barn

where $\langle\sigma\rangle$ is the detector efficiency-weighted average of the unpolarized cross-section over the energy range, and $\langle A^2\rangle$ the cross-section- and efficiency-weighted mean-squared asymmetry. Asymmetry and cross-section are mostly fixed by kinematics and available devices, so only the luminosity can be optimized. Table 1 shows the luminosity required to achieve various accuracies within 15 min in absence of background. When calculating the expected luminosity for reasonable setup parameters (green laser light, 10 W of output power), the maximum value is about 4 kHz/barn even for optimum light focusing. It strongly depends on the crossing angle and decreases by a factor of 20 within 20 mrad. It is therefore desireable to use a collinear geometry and necessary to increase the laser intensity.

3 Polarimeter layout

One possibility to increase the available intensity is to feed the laser beam into a Fabry-Pérot resonant cavity. This concept has been reported to work successfully [3] but is difficult to build because the small bandwidth makes a frequency stabilization of the laser necessary. The A4 polarimeter implements for the first time the internal cavity concept [4]: lasers already consist of a F-P cavity with

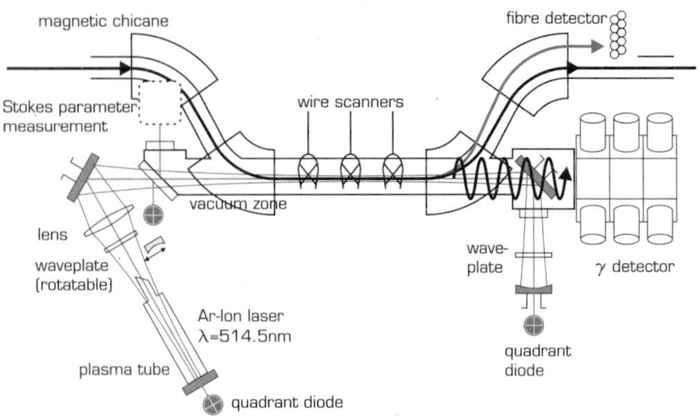

Fig. 1. Schematic view of the laser resonator. It is installed in a magnetic chicane in front of the A4 target

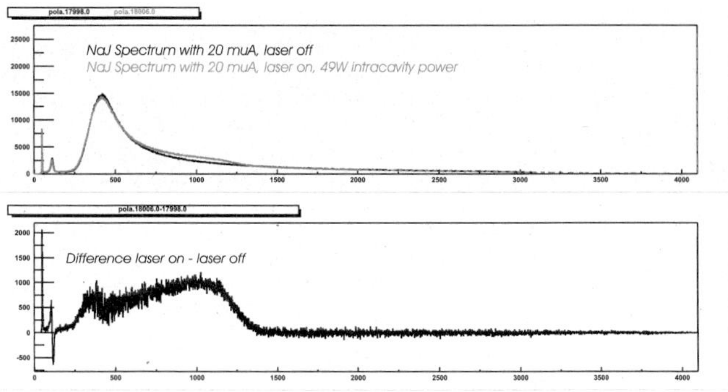

Fig. 2. ADC spectra in the NaI calorimeter with and without laser beam. The intracavity power was 49W

a lasing medium, and the output light is only a fraction of the internally circulating light. Our method is to extend the cavity length, use high-reflecting mirrors on both ends and guide the electron beam through the laser resonator where it interacts with the high intra-cavity power. Since the laser medium will adapt to cavity length fluctuations, no frequency stabilization is necessary; however, the achievable maximum power is lower than with an external cavity.

Figure 1 shows an overview drawing. The laser is an Ar-ion laser delivering 10 W at 514.5 nm in factory configuration. The lens is used to preserve the original beam profile in the medium while optimizing it in the interaction region. The cavity is now 7.8 m long and therefore vulnerable to vibrations of the optical elements. Since the influence of vibrations to the beam axis depends on the optics spacing and the vulnerability of the luminosity to beam axis fluctuations depends on the focusing, Monte-Carlo simulations of the effective luminosity as a function of vibration amplitude have been performed for various focusings. The final value is a focusing of $z_R = 2.5$ m with a maximum luminosity of 2.1 kHz/barn per 10 W of power. Also, a stabilization system for the laser beam position has been designed. The position is measured with quadrant diodes and stabilized using piezo-actuated mirrors [5]. In the interaction region, three wire scanners measure the positions of both beams simultaneously to establish overlap. The backscattered photons are detected in a NaI calorimeter. The electrons involved in the scattering lose energy and are displaced with respect to the main beam. A scintillating fibre array behind the chicane is used to detect them in coincidence with the photons to improve the data quality.

The circular polarization of the light is created using two quarter waveplates, one being rotatable to select the helicity. Two waveplates are necessary because the polarization optics is installed inside a resonator. One of the vacuum windows is used as a beam splitter to measure the polarization state. The extracted light (0.6%) is transmitted through a rotating quarter waveplate and a Glan-Laser prism; the intensity is thereby modulated with modulation amplitudes proportional to the beam's Stokes parameters.

4 Status and results

The magnetic chicane has been set up in December 2002. It does not affect the beam quality on the target. The laser system has also been installed and works reproducibly. After installation of the polarization optics, intra-cavity intensities of up to 90 W have been measured in single-line (514.5 nm) configuration. Procedures to bring electron and laser beem to overlap have been established, and backscattered photon spectra have been recorded with the NaI calorimeter. A calibration procedure for the detector has been established which uses muons from cosmic radiation with trigger-defined track lengths inside the detector. The fibre array detector has been commissioned, and the data quality was improved by imposing a coincidence condition between electrons and photons. The signal-to-noise ratio was increased from 1:7.1 to 1:2.1. The laser beam stabilization system has been installed, and tests have shown a significant reduction of beam axis fluctuations [5]. The measurement device for the laser Stokes parameters has been installed and tested. The next steps will comprise the improvement of the vacuum and an analysis of stress birefringence in the vacuum windows. The NaI calorimeter is φ-averaging and the polarimeter therefore only sensitive to longitudinal polarization. It is planned to install a position-sensitive detector to measure also the transverse polarization.

With this setup, a statistical accuracy of 2.5% without and about 5% with background seems to be achievable in 15 min. Systematic uncertainties can arise from detector response and the Stokes parameter measurement. We are currently working to control and minimize them.

This work has been supported by the Deutsche Forschungsgemeinschaft and the U.S. DoE.

References

1. F. Lipps, H. Tolhoek: Physica **XX**, (1954) 85–99, 395–405
2. G. Bardin et al.: *Conceptual Design Report of a Compton Polarimeter for Cebaf Hall A* (JLab, 1996)
3. N. Falletto et al.: N.I.M. A **459**, 412–425 (2001)
4. M. Düren: HERMES internal report 00–005, (2000)
5. J. Diefenbach: this volume

Eur Phys J A (2005) **24**, s2, 129–130
DOI: 10.1140/epjad/s2005-04-030-2

EPJ A direct
electronic only

The transmission Compton polarimeter of the A4 experiment

A simple simultaneous monitor for the longitudinal electron beam polarisation

Christoph Weinrich[a], for the A4 Collaboration

Johannes-Gutenberg-Universtität, Institut für Kernphysik, J.-J.-Becher-Weg 45, 55099 Mainz, Germany

Received: 1 November 2004 / Published Online: 8 February 2005

Abstract. A transmission Compton polarimeter as a simultaneous, relative monitor for the longitudinal polarisation of a stabilised, polarised electron beam (polarisation degree ~ 80 %) has been constructed in a new design. It is located in the vacuum between the target and the beam dump of the A4 parity violation experiment at MAMI. The analysing power is ~ 80 ppm at 854 MeV and ~ 115 ppm at 570 MeV. The measurement precision for the polarimeter asymmetry (which is proportional to the longitudinal beam polarisation degree) is ~ 3 ppm within 5 min. at 854 MeV and ~ 7 ppm within 5 min. at 570 MeV and the systematical error is about 0.5 ppm + 1 - 2 %. This simple polarimeter consists mainly of two graphite scatterers, a water-cooled Samarium-Cobalt permanent magnet, an aluminum secondary electron emission monitor, an understructure and electronics.

PACS. 29.27.Hj Polarized beams – 29.27.Fh Beam characteristics

1 Introduction

The A4 collaboration at MAMI measures (parity violating) beam spin asymmetries in elastic electron-nucleon scattering. The transmission Compton polarimeter serves as a relative monitor for the longitudinal polarisation between the absolute Møller polarimeter measurements and for spin angle measurements.

2 Functionality and setup

In a transmission Compton polarimeter the polarised electron beam (with flipping polarisation direction) produces proportionally polarised bremsstrahlung in some target. The polarisation of the bremsstrahlung beam is then analysed by measuring the asymmetry in its transmission through a magnet, arising from the polarisation dependence of Compton scattering. The beam electrons superposing the bremsstrahlung beam have to be diminished in front of the magnet because they dilute the measured asymmetry. The quantitatively unknown background from electromagnetic shower production by the photons (and electrons) in the magnet make the polarisation measurement only relative.

The A4 transmission Compton polarimeter is located in the 60 cm wide exit beam pipe between the target and the beam dump. Its compact design allows the beam to pass by and reach the beam dump without beeing dissipated too much. Electron background diminution is achieved by scattering, i.e. spreading the electron beam to a limited extent. For space saving and background minimisation, a permanent magnet is used as analysing magnet and secondary electron emission for the transmission measurement. Dimensions and positions of the main polarimeter parts were optimised on the basis of calculations of electron scattering, bremsstrahlung and pair production and Compton scattering. Figure 1 shows the design of the polarimeter. Polarised bremsstrahlung is produced in the liquid hydrogen target and in two graphite scatterers. These also spread the electron beam and strongly reduce the number of electrons impinging the magnet. The analysing magnet is an axially magnetised Sm_2Co_{17} (Vacomax 225 HR^{TM}) permanent magnet. It is highly remanent and coercive and heat-proof. The estimated electron polarisation in the magnet is about 3.2 % [1]. The length of the magnet has been chosen in order to nearly minimise the relative error for the measured asymmetry. The magnet is cooled by water of the beam dump cooling water circuit. A copper/CuFeP body is shrink-fitted to the magnet and conducts the heat to the cooling pipes. The aluminum converter is used to measure the photons transmitted through the magnet. Therein the photons produce electron-positron pairs, which generate a secondary electron emission signal. The signal of the 1st scatterer is used as the reference (normalising) signal for the calculation of the transmission (which is the ratio of the photon flux in front of and behind the magnet). To allow the measurement of secondary electron emission currents, the scatterers, the magnet and the converter are electrically isolated by an aluminum shielded ceramic isolator. The

[a] comprises part of PhD thesis

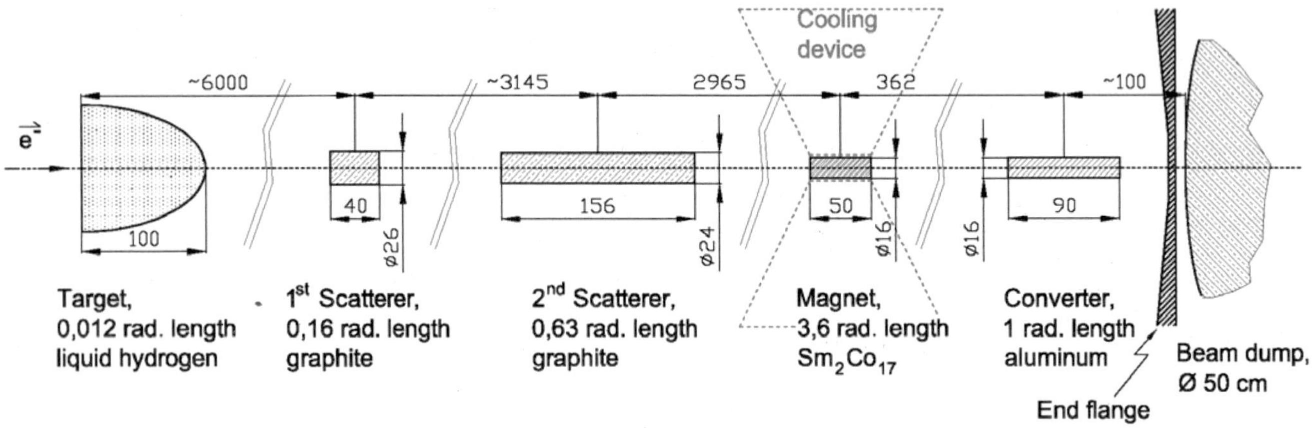

Fig. 1. The design of the A4 transmission Compton polarimeter (all parts are symmetric around the axis)

2nd scatterer, the magnet and the converter are mounted on one adjustable (MayTek™) aluminum carriage which can be rolled into the exit beam pipe and is kept in position by pins. The 1st scatterer is mounted on a proper frame. The rods holding the scatterers are made of titanium for heat resistance. The signal cables that are connected to the polarimeter pieces to conduct the secondary electron emission currents are radiation resistant in the rear ("hot") region. The polarimeter electronics consists of MAMI and A4 standard parts: amplifying current-voltage converters close to the signal source with differential output and integration-ADC- and histogramming-modules as designed for the A4 beam monitors (histogramming modules are used in timing mode). The signals are integrated over the 20 ms measurement gates between which the polarisation is flipped in the pattern (+P,-P,-P,+P), the sign of P beeing chosen pseudo-randomly before repetition. The transmission asymmetry $A^{Pola} = (T^+ - T^-)/(T^+ + T^-)$ with $T = S_C/S_{S1}$ ($S_{C/S1}$: converter/1. scatterer signal; +/-: polarisation state) is calculated offline using the timings of the (pedestal corrected) signals. A challenge was the pedestal correction of the converter signal. This pedestal is strongly drifting probably caused by activation of the material. We solved this problem by measuring the signal ratio of the converter and 1st scatterer signals $c = \Delta S_C/\Delta S_{S1}$ in switching off the beam. The converter pedestal is then calculated as $S_C^0 = \bar{S}_C - c \cdot \bar{S}_{S1}$ ($\bar{S}_{C/S1}$: signal averages over polarisation 4-tuple). Since the signal ratio is drifting slowly as well it has to be measured regularly in order to keep the systematical error small.

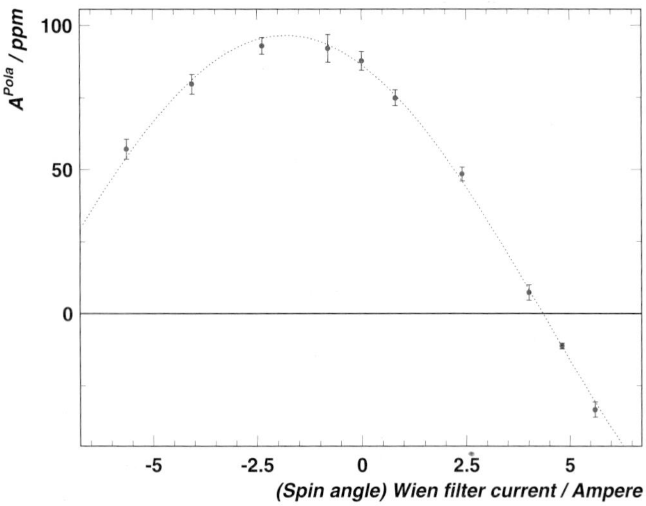

Fig. 2. Result of the spin rotation measurement at 570 MeV with a cosine-fit. The ordinate is the current of the Wien filter dipole which is proportional to the spin angle. The abscissa is the polarimeter asymmetry

3 Results

The measured analysing power of the polarimeter is about 80 ppm at 854 MeV and $\sim$ 115 ppm at 570 MeV beam energy. The variance (precision) of the measured polarimeter asymmetry (for 5 min. runs) is $\sim$ 3 ppm at 854 MeV and $\sim$ 7 ppm at 570 MeV. Sensitivity to helicity correlated beam parameters was found to be $\lesssim$ 30 ppm/μm for position differences (at target), $\lesssim$ 0.015 ppm/ppm for current asymmetries and negligible for energy (preliminary, probably overestimated values at 570 MeV). Averaged over a data sample ($\sim$ 500 runs) typical position differences are $\sim$ 50 nm, typical current asymmetries are less than 1 ppm. The signal ratio measurement error (used for pedestal correction) contributes as a systematical error of 1-2 % of the measured asymmetry. Runwise decorrelation of the measured polarimeter asymmetry of the beam parameters (based on the polarisation 4-tuples) had negligible effect under normal beam conditions. Figure 2 shows the result of a spin rotation measurement using a Wien filter. Spin rotation measurements were also used to determine the spin angle at the experiment.

References

1. M. Huth: personal communication

Eur Phys J A (2005) **24**, s2, 131–131
DOI: 10.1140/epjad/s2005-04-031-1

EPJ A direct
electronic only

Stabilization system of the laser system of the A4 Compton backscattering polarimeter

Jürgen Diefenbach[a]

Institut für Kernphysik, Universität Mainz, J. J. Becherweg 45, D-55099 Mainz, Germany

Received: 1 November 2004 / Published Online: 8 February 2005

Abstract. The A4 Compton backscattering polarimeter is used to measure the degree of spin polarization of the 855 MeV electron beam provided by the MAMI accelerator facility. Therefore special care must be taken to optimize and stabilize the overlap of electron and laser beam in order to get the highest luminosity for shortest measurement times. For this purpose an active stabilization system for the laser beam position of our intra-cavity polarimeter's optical resonator has been developed and commissioned.

PACS. 07.50.Qx Signal processing electronics – 29.27.Hj Polarized beams – 42.60.By Design of specific laser systems – 42.60.Lh Efficiency, stability, gain, and other operational parameters

1 Introduction

The laser resonator of the A4 backscattering polarimeter has a total length of 7.8 m. Therefore the proper alignment of the laser mirrors is sensitive to any mechanical instabilities caused for example by seismic vibrations, the flow of cooling water through laser tube and dipole magnets, air turbulences etc. Furthermore there are slow drifts of the laser pointing stability due to small changes in the temperature of the laser's plasma tube, e.g. from changes in the tube current. All this affects the laser beam position and leads to variations of the overlap between electron and laser beam and of the laser power, both causing changes in the luminosity of the polarimeter.

2 Hardware

These changes in the laser beam position and intra-cavity power have been measured using quadrant photodiodes installed along the laser cavity.

To suppress the changes in the beam position, three of the four laser mirrors are mounted on piezo driven platforms. These platforms can be used to optimize the laser resonator alignment so that the position of the laser beam and consequently its power can be stabilized.

From the measured beam positions, tilt angles for the piezo mounted mirrors are calculated in the decoupling circuits. An automatic shutdown circuit protects the system from resonant oscillations that could cause damage to the piezo platforms. A passive single-pole lowpass filter at 35 Hz provides the control loops with stability. The mechanical resonance of each piezo platform depends on the geometry and the momentum of inertia of the platform and the attached laser mirror, and has been measured as a transfer function. Then an active pole-zero cancellation circuit was tuned carefully to extinct electronically the resonance structure of the piezo platform measured before. The -6 dB bandwidth of the disturbance rejection of the stabilization system in a test setup has been increased from 210 Hz to 800 Hz by this method.

The system is fully remotely programmable using MAMI standard electronics. The elements of the decoupling matrices and the control loop amplifications are respectively given by eighteen and six 12-bit multiplying DACs.

3 First results and outlook

The stabilization system has been operated so far for test purposes with four out of its six control loops running.

When reconceived in a naive picture, where all beam position noise is projected onto tilting vibrations of the (so far) two stabilizable piezo-mounted laser mirrors, one can compute the mean tilt amplitudes of the mirrors from the obtained laser beam position data and decoupling matrices. Assuming this, the effective tilt amplitude was found to be reduced from about 90 μrad without stabilization to about 10 μrad with the stabilization system running and the pointing stability on the diodes from 8 to 12 μrad without to 0.8 μrad with stabilization. Long-term drifts of the beam went down from up to 21 μm/h to 0.3 μm/h.

Furthermore it should be possible to optimize the performance of the stabilization system by retuning the pole-zero cancellation circuits and then increasing the loop amplifications. Also a DAC module will be designed to add offsets to the position signals and thereby shifting the laser beam to provide fine-tuning of the overlap of both beams.

[a] comprises part of PhD thesis

Eur Phys J A (2005) **24**, s2, 133–133
DOI: 10.1140/epjad/s2005-04-032-0

EPJ A direct
electronic only

Electron beam line design of A4 Compton backscattering polarimeter

Jeong Han Lee[a]

Johannes Gutenberg Universität Mainz, Institut für Kernphysik, J.J. Becherweg 45, 55299 Mainz, Germany

Received: 1 November 2004 / Published Online: 8 February 2005

Abstract. The new beam line has been built to measure the polarization of the electron. The transverse position and emittance of the MAMI beam are determined from measurements of the beam size on three wire scanners. The position measurement has been done and the emittance measurement is still progressing.

PACS. 29.27.Eg Beam handling; beam transport – 29.27.Fh Beam characteristics

1 Introduction

The A4 collaboration measures the parity-violating asymmetry for elastic electron scattering off an unpolarized proton. To extract the physics asymmetry, it will be crucial to determine the polarization of the electron beam to high accuracy. For this reason, Compton polarimeter has been installed. The one of advantages of Compton polarimeter is that it is not a destructive method. That means the continuous on-line monitoring of the electron polarization would be possible with the same electron beam conditions.

2 Beam conditions in the chicane

A beam line, which is called a *chicane*, was designed to measure the polarization of the electron. The chicane has the same four dipole magnets, which separate the scattered photons from the electrons, and contributes to the dispersion functions due to the magnets. The dispersion must be eliminated to preserve the same beam properties by introducing two quadrupoles. TRANSPORT [1], which uses a matrix formalism to design a beam transport system, was used to design the chicane. The chicane has been built and is fully operational.

The motion of particles beam can be represented by the ellipse (e.g., Twiss) parameters, β, α, and γ. For transport lines, there is no periodic boundary condition about the parameters, which depend on the initial beam profiles. However, although the chicane is the transport line, two requirements must be fulfilled to achieve the good luminosity and to keep the same beam properties at the beginning and at the end of the chicane; the first one is that the electron beam has a waist at the middle of the chicane($\alpha_w = 0$ and $\beta_w = 1/\gamma_w$) and the second one is the Twiss parameters are the same at the beginning and end of the chicane($\beta_f = \beta_i$, $\alpha_f = \alpha_i$, and $\gamma_f = \gamma_i$). Moreover, it is necessary for the good luminosity to determine the Rayleigh length of the electron beam. The Rayleigh length can be determined by using the equation $x = \sqrt{\epsilon(\beta - 2\alpha s + \gamma s^2)}$, where ϵ stands for the emittance, s denotes the location of the beam, and x is the beam size at s. The design value of the Rayleigh length is 9.617 meter.

3 Position and emittance measurement

To make the scattering between the electron and the photon possible, it might be necessary to determine a transverse position of both of the beams. Moreover, to get the actual values of the Twiss parameters and the emittance of the electron beam [2], it would be needed to measure the size of the beam. The achievement of the two purposes will be done by three wire scanners that are the most commonly tool used to diagnose a beam profile in the middle of the chicane. However, the wire scanner, which A4 uses, is not the common type but the four-bar linkage type. The equation of motion of the wire scanner is therefore needed. The analysis of the wire scanner and the beam position measurement were done.

The horizontal and vertical emittance of the MAMI electron beam are $\epsilon_h = 7.760\pi \times 10^{-3}$mm · mrad and $\epsilon_v = 1.017\pi \times 10^{-3}$mm · mrad. The actual values of Twiss parameters and the emittance can be deduced from the measured electron beam profiles. It is a challenge to make them coincide with their design values.

References

1. D.C. Carey, K.L. Brown, F. Rothacker: SLAC-R-0530
2. M.G. Minty, F. Zimmermann: *Measurement and Control of Charged Particle Beams* (Springer, 2003), 99

[a] comprises part of PhD thesis

IV Experimental techniques in PV electron scattering

IV-3 Detection

Eur Phys J A (2005) **24**, s2, 137–140
DOI: 10.1140/epjad/s2005-04-033-y

EPJ A direct
electronic only

Background substraction in parity violation experiments

J. Van De Wiele and M. Morlet

Institut de Physique Nucléaire/IN2P3/CNRS, F-91406 Orsay Cedex, France

Received: 15 October 2004 / Published Online: 8 February 2005

Abstract. The importance of the knowledge of the background in parity violating (PV) experiments is shown. Some improvements in Monte Carlo simulations are presented and discussed.

PACS. 01.30.Cc Conference proceedings – 25.30.Bf Elastic electron scattering – 13.60.Le Meson production – 13.60.-r Photon and charged lepton interactions with hadrons

It is well known that the asymmetry in P.V. experiments due to the exchange of the Z_0 boson is small ($\approx 10^{-6} - 10^{-5}$). Much care has to be taken in the measurement of such a small quantity. Since a few years, impressive improvements in technical aspects have been achieved and some of them have been presented in this workshop. Without such improvements, the extraction of the physical quantity would be obtained with too large a systematic error and so would be meaningless.

In any experiment, simulation of all the processes which populate the "good" events as well as some "background" events, is a good tool to be sure that the experiment and in particular the experimental set-up is under control.

Simulation of very small effects is not an easy task for many reasons:

- The accuracy of the simulation depends strongly on the statistics and standard methods, which are time consuming, therefore may become inefficient.
- Some of the physical effects, which are usually considered as small and therefore are neglected, may contribute.
- It is necessary to improve the description of some processes which are usually treated only in an approximate way.
- Accurate models and data needed do not exist.

The basic formula is the following:

$$A_{\mathrm{Phys}} = A_{\mathrm{Meas}} \frac{1 + \sum (A_i/A_0 \quad \sigma_i/\sigma_0)}{1 + \sum \sigma_i/\sigma_0}$$

In this expression the index "0" stands for the elastic events and the index "i≠0" stands for any background event. If the background asymmetry vanishes, the denominator acts as a dilution factor to the physical asymmetry. As can be seen in Fig. 1, any measured spectrum of a physical quantity (energy or time of flight) shows up a the physical signal above some background events. In Table 1 A_{Phys}/A_{Meas} are given as a function of A_i/A_0 if the ratio of bad events to good events is equal to 10 %. In Fig. 2, this effect is plotted for different contamination rates. The change in the sign of A_i/A_0 could be dramatic.

In principle, it is possible to experimentally study the background but the statistical precision could be poor as compared to the elastic peak. In some cases, it is possible only to extrapolate the the background and the experimental study becomes less efficient (see Fig. 3)

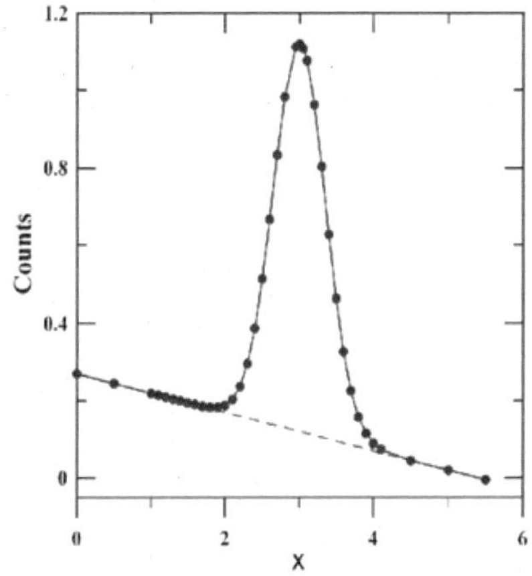

Fig. 1. Typical measured spectrum. X is a measured quantity -energy or time of flight-

Table 1. Ratio A_{Phys}/A_{Meas} as a function of A_i/A_0 for σ_i/σ_0 =10 %

A_i/A_0	A_{Phys}/A_{Meas}
-3.	0.63
-2.	0.73
-1.	0.82
0.	0.91
1.	1.00
2.	1.09
3.	1.18

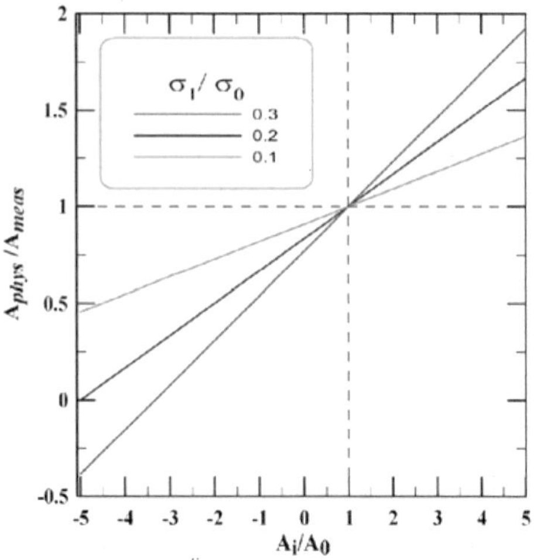

Fig. 2. Ratio A_{Phys}/A_{Meas} as a function of A_i/A_0 for several values of σ_i/σ_0

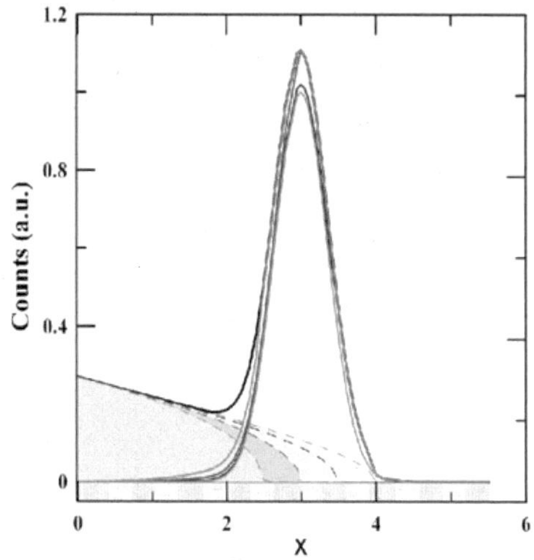

Fig. 3. Different extrapolations of the background

Monte Carlo (M.C.) simulations are supposed to be a powerful tool to understand both the elastic observables and the background features. They give complementary information to the measurement. This method is powerful if the physical laws under consideration are well known. Monte Carlo results are accurate if we are able to generate a large number of events.

M.C. is based on a one-to-one correspondence between uniform random number $\eta \in]0,1[$ and a physical law. For example, in one dimension

$$\eta = \frac{\mathcal{N}}{\mathcal{D}}, \quad \mathcal{N} = \int_{x_{min}}^{x} f(x')\, dx', \quad \mathcal{D} = \int_{x_{min}}^{x_{max}} f(x')\, dx',$$

and in two dimensions:

$$\eta_1 = \frac{\mathcal{N}_1}{\mathcal{D}_1}, \quad \mathcal{N}_1 = \int_{x_{1min}}^{x_1} f_1(x'_1)\, dx'_1, \quad \mathcal{D}_1 = \int_{x_{1min}}^{x_{1max}} f_1(x'_1)\, dx'_1,$$

$$\text{with } f_1(x_1) = \int_{x_{2min}(x_1)}^{x_2} f(x_1, x'_2)\, dx'_2, \quad \text{and } \eta_2 = \frac{\mathcal{N}_2}{\mathcal{D}_2},$$

with

$$\mathcal{N}_2 = \int_{x_{2min}(x_1)}^{x_2} f(x_1, x'_2)\, dx'_2, \quad \mathcal{D}_2 = \int_{x_{2min}(x_1)}^{x_{2max}(x_1)} f(x_1, x'_2)\, dx'_2.$$

Using the definition of M.C. method is time consuming because we need to invert the expression in the numerator to get the physical quantity. To increase the efficiency of the M.C. method, it is possible to introduce some weights W to take into account of the cross-sections.

Example 1: $a + b \longrightarrow 1 + 2$

$$\eta_\theta = \frac{\theta_1 - \theta_{1min}}{[\Delta\theta_1]}, \quad \eta_\phi = \frac{\phi_1 - \phi_{1min}}{[\Delta\phi_1]},$$

$$W = \frac{\mathcal{L}}{N_T}\, [\Delta\theta_1]\, [\Delta\phi_1]\, \frac{d^2\sigma}{d\Omega_1}\, \sin\theta_1.$$

where $\mathcal{L}$, N_T are respectively the luminosity and the number of random events. $[\Delta\theta_1] = \theta_{1max} - \theta_{1min}$ is the angular range for the polar angle and $[\Delta\phi_1] = \phi_{1max}(\theta_1) - \phi_{1min}(\theta_1)$ is the angular range for the azimuthal angle.

Example 2: $a + b \longrightarrow 1 + 2 + 3$

$$\eta_\theta = \frac{\theta_1 - \theta_{1min}}{[\Delta\theta_1]}, \quad \eta_\phi = \frac{\phi_1 - \phi_{1min}}{[\Delta\phi_1]},$$

$$\eta_E = \frac{E_1 - E_{1min}(\theta_1)}{[\Delta E_1]},$$

$$W = \frac{\mathcal{L}}{N_T}\, [\Delta\theta_1]\, [\Delta\phi_1]\, [\Delta E_1]\, \frac{d^3\sigma}{d\Omega_1\, dE_1}\, \sin\theta_1.$$

$[\Delta E_1] = E_{1max}(\theta_1) - E_{1min}(\theta_1)$ is the energy range. The two methods are equivalent when these weights are correctly introduced.

Another method to improve the efficiency of the simulation is to calculate the cross-sections of the needed processes. In the following part, we will concentrate on the A4- and the G^0 experiment and will give some specific examples.

In the A4 experiment [4], the rate of inelastic electron produced by one pion electro-production $e + p \longrightarrow e' + p' + \pi^0$ and $e + p \longrightarrow e' + n' + \pi^+$ reactions has been measured and calculations with effective lagrangians for E_e lower than 1 GeV are accurate enough. It is more difficult to calculate the number of photons coming from the π^0 decay. With an electromagnetic shower calorimeter, it is impossible to disentangle electrons and photons. A part of these photons have the same energy as the elastic electrons and thus contribute to the background under the elastic peak. The standard method – calculation of the π^0 electro-production followed by π^0 decay after Lorentz boost - is not appropriate to calculate the rate of photons if we want to take into account energy loss and external radiative corrections. Furthermore, it is impossible

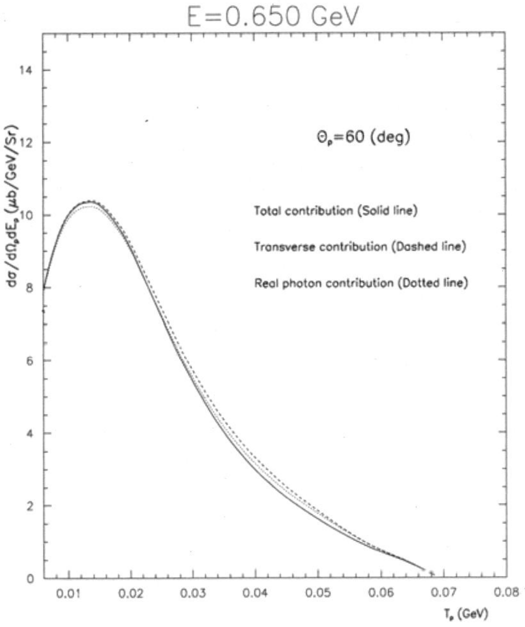

Fig. 4. Inclusive proton differential cross-section at 650 MeV

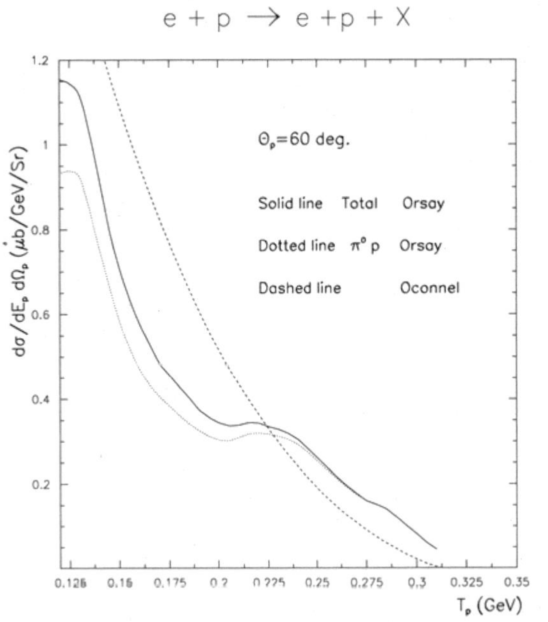

Fig. 5. Comparison between our calculation and the EPC code developed by LightBody and O'Connel

to estimate the false asymmetry carried by the photons. Nevertheless, it is possible to include all these effects if we calculate directly the photon production cross-section (the lagrangian for the $\pi^0 \longrightarrow 2\gamma$ is known):

$$\frac{d^3\sigma}{d\Omega_\gamma\; dE_\gamma} = \int \frac{d^3\sigma(h_e)}{d\Omega_{\pi^0}\; dE_{\pi^0}}\; \mathcal{K}\; \delta(\; \widetilde{E}_\gamma - E_\gamma\;)\; d\Omega_{\pi^0}\; dE_{\pi^0},$$

$$\mathcal{K} = \frac{m_\pi^2}{4\pi}\frac{1}{E_{\pi^0} - p_{\pi^0}\mathbf{u}_{\pi^0}\cdot\mathbf{u}_\gamma}, \qquad \widetilde{E}_\gamma = \frac{m_\pi^2}{E_{\pi^0} - p_{\pi^0}\mathbf{u}_{\pi^0}\cdot\mathbf{u}_\gamma}.$$

The π^0 differential cross-section, which depends on the helicity of the incident electron beam, is calculated through the models described above. From such accurate calculations, we conclude that the photon contamination does not depend on the helicity and acts only as a dilution factor. We note here that the inclusive pion or proton cross-sections has to be calculated by replacing the usual flux factor Γ which is divergent when $\theta_{e'} \to 0$ and $m_e = 0$ by $\overline{\Gamma}$:

$$\Gamma \;=\; \frac{\alpha}{2\pi^2}\frac{E'_e}{E_e}\frac{E_\gamma}{Q^2}\frac{1}{1-\varepsilon} \;\longrightarrow \overline{\Gamma} = \frac{\alpha}{8\pi^2}\frac{|\mathbf{p'}_e|}{|\mathbf{p}_e|}E_\gamma F_{virt}.$$

The expression of F_{virt} is given in [1].

In G^0 experiment phase I, with an electron beam at 3 GeV, we are interested to know the inclusive pion and proton cross-section. Unfortunately, at such an energy, the cross-section is not well known and the calculations performed with effective lagrangians on one hand, accurate at energies below 1 GeV, and with Regge models on the other hand, well suited at energies greater than 5-6 GeV, are not very reliable. LightBody and O'Connel have developed a code (EPC) to compute such cross-sections but it is based on high energy data and the validity of the extrapolation to our energy is questionable. We have developed a code with an other approach. The inclusive spectrum is obtained by integrating the five times differential cross-section over the electron angles. Because of the exchange of a virtual photon in the electro-production reaction, the main contribution in the integration is expected to come from the terms with small values of q^2. We then assume that the square of the matrix element may be extracted from photo-production measurements which exist between 200 MeV and 3 GeV [2]. We have checked this approximation at an energy of 650 MeV where an exact calculation with an effective lagrangian can be used. The results are displayed in Fig. 4. The agreement between electro-production and photo-production is better than 5%. An event generator based on this model has been written for the G^0 collaboration. The angular distribution of one pion photo-production data have been included. For two pion (or more) photo-production reactions, there are only few angular distribution data but several measurements of the total cross-section exist. All the channels up to 3 pions have been included. A comparison with the EPC code is shown in Fig. 5. The Time of Flight (TOF) proton spectrum includes, in addition to the electro-production, some contribution from photo-production reactions. This photo-production is due to the competition, in any material, between electro-production of the incoming electron and the real bremsstrahlung photons. The rate of inelastic protons is proportional to the number of bremsstrahlung photons, more precisely to $I_\gamma(E_0, E_\gamma, t)$ which is the number of photons in the energy bin E_γ, $E_\gamma + dE_\gamma$ after an electron, initially with an an energy E_0, has passed through a target of thickness t measured in unit of the radiation

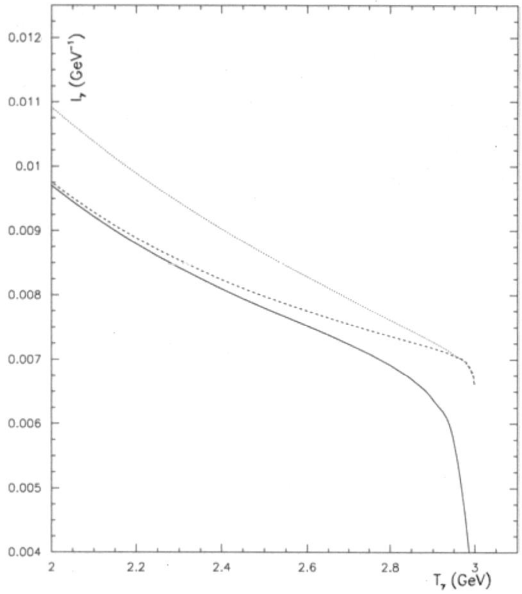

Fig. 6. Value of $I_\gamma(E_0, E_\gamma, t)$ for a LH_2 target (t=20 cm, E_0= 3 GeV) as a function of E_γ: Approximate formula (*dotted line*), complete screening approximation (*dashed line*) and exact calculation (*solid line*)

length of the material [3]. This number of photons is also proportional to the bremsstrahlung cross-section:

$$\frac{d\sigma_b}{dE_\gamma}(E, E_\gamma) = \frac{A}{NX_0}\frac{1}{E_\gamma}\left(y^2 - \frac{4}{3}y + \frac{4}{3}\right) g(y), \qquad y = \frac{E_\gamma}{E}$$

where X_0 is the radiation length of the material. In the standard calculations, we assume a complete screening approximation, with $g(y) = 1$. Within this approximation, Tsai and Van Whitis have derived an analytical expression:

$$[I_\gamma(E_0, E_\gamma, t)]_{approx} = \frac{1}{E_\gamma}\,\frac{(1 - \frac{E_\gamma}{E_0})^{\frac{4}{3}t} - e^{-\frac{7}{9}t}}{\frac{7}{9} + \frac{4}{3}\ln(1 - \frac{E_\gamma}{E_0})}$$

As was stated by Y. Tsai, this approximate expression may be a poor if the detailed shape at the high-energy tip of the Bremsstrahlung is needed. This is our case because high-energy inelastic protons, produced by high-energy photons, will have the same TOF compared to elastic protons. We have performed some exact calculations of $I_\gamma(E_0, E_\gamma, t)$ and new expressions have been derived for hydrogen and aluminium. For this material, we have used the Thomas-Fermi Molière Model [3]. Comparison between the approximate expression, the complete screening case and the exact calculation for liquid hydrogen is shown in Fig. 6. The approximate formula or the complete screening calculations overestimate the high-energy number of photons.

References

1. S. Ong, J. Van de Wiele: Phys. Rev. C **63**, 024614 (2001)
2. P. Corvisiero et al.: Nucl. Instr. Meth. A **346**, 433 (1994)
3. Y.-S. Tsai: Rev. Mod. Phys. **46**, 815 (1974); Y.S. Tsai, Van Whitis, Phys. Rev. **149**, 1248 (1966)
4. F.E. Maas et al.: Phys. Rev. Lett **93**, 022002 (2004)

Eur Phys J A (2005) **24**, s2, 141–141
DOI: 10.1140/epjad/s2005-04-034-x

EPJ A direct
electronic only

Redesign of the A4 calorimeter for the measurement at backward angles

Boris Gläser[a], for the A4 Collaboration

Institut für Kernphysik, Johannes Gutenberg-Universität Mainz, J.-J.-Becher-Weg 45, 55099 Mainz, Germany

Received: 1 November 2004 / Published Online: 8 February 2005

Abstract. Backward angle measurements of the parity violating asymmetry in elastic electron proton scattering are planned with the A4 calorimeter. At present the experiment measures at forward angles. For the measurement at backward angles the support of the PbF_2-detector has to be redesigned. For this purpose it will be positioned on a rotatable platform, so that measurements at forward and backward angles will be possible with little effort. We will highlight the new experimental setup and its special features.

PACS. 07.05.Fb Design of experiments – 13.40.Gp Electromagnetic form factors

1 Introduction

Backward kinematics measurements in parity violating electron proton scattering on hydrogen and deuterium are planned with the A4 calorimeter. For this purpose the support for the PbF_2-calorimeter will be positioned on a rotatable platform, so that measurements at forward and backward angles will be possible.

2 The calorimeter

The A4 Calorimeter is at present mounted on a linear moveable platform. The support is used to adjust the focus of the crystal calorimeter on the target. A special requirement for a possible backward angle setup comes through the fact, that not only the detector but also the cryogenic hydrogen target with its scattering chamber and the luminosity monitors have to be adjusted. A new support for the calorimeter and the scattering chamber is needed.

3 The scattering chamber and the luminosity monitors

Figure 1 shows the solid angle of the crystals of the calorimeter and of the luminosity monitors. An elongation chamber for the main scattering chamber will be installed, so that the luminosity monitors have in forward and backward scattering position the same solid angle with respect to the target centre (Fig. 1).

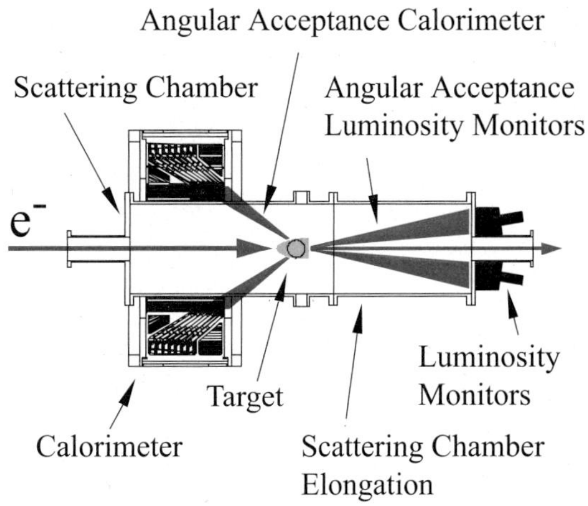

Fig. 1. New Scattering Chamber. The electron beam comes from the left

4 The new experimental setup

We have designed a new, common support structure so that the calorimeter and the scattering chamber are supported on a rotatable platform.
To ensure a vibration free, shock free and easy positioning of the platform, it is supported on three hydraulic oil sliding feet. By its small stroke (1/10 mm) contrary to air cushions (stroke some mm), an easy adjustment without impacts during set off is ensured. Vibrations during rotating, that can damage the fragile crystals, are minimal too.

The support platform, the scattering chamber elongation and the hydraulic oil sliding feet are already delivered and will be installed in December 2004. The new setup will be in operation by end of February 2005.

[a] Comprises part of diploma thesis

Eur Phys J A (2005) **24**, s2, 143–143
DOI: 10.1140/epjad/s2005-04-035-9

EPJ A direct
electronic only

Performance of the G^0 superconducting magnet system

Steven E. Williamson, for the G^0 Collaboration

University of Illinois, with support from the U.S. National Science Foundation under grant PHY94-10768, USA

Received: 15 October 2004 / Published Online: 8 February 2005

Abstract. At the heart of the G^0 Spectrometer is the toroidal superconducting magnet system (SMS). The SMS has been in use at Jefferson Lab since the fall of 2002. Experience with the operation and reliability of the magnet over that period is reported. Some measured performance parameters are compared with the magnet specification.

PACS. 29.30.Aj Charged-particle spectrometers: electric and magnetic

The G^0 superconducting magnet system (SMS) is an iron-free toroid with zero magnification optics. Its field, peaking at 3.5 T (3 T in conductor) is generated by eight coils, each with 144 turns (4X36 windings), in a common cryostat. The stored energy is 6.6 MJ at the normal operating current of 5 kA.

Coil locations were measured at room temperature after installation at Jefferson Lab using photogrammetry to locate 128 targets (16 on each coil). Design and measured target locations were compared, while adjusting the overall position and orientation of the magnet for a best fit. The average deviation of measurements from the ideal was found to be 1.6 mm, less than the 2.0 mm specification. The locations of the coils, when cooled, were deduced from known coefficients of thermal expansion.

A measurement of the Q^2 associated with each focal plane detector was extracted [1] from the difference between the time-of-flight of elastic protons and of π^+ particles. This difference is sensitive to the particle trajectory through the magnet and thus to the magnetic field configuration. Measurements were compared to a simulation based on the design magnetic field and found to agree to a precision of 100 ps, which implies an uncertainty on Q^2 within the 1% requirement of the experiment.

The SMS cooldown, specified to take 7 days, actually required about 21 days. This rate was limited by the requirement that ΔT between inlet and coil average be < 75 K. Heat load to LHe was specified to be < 40 W, but boil-off studies indicate a load of about 107 W. The steady-state LHe requirement of the magnet at full power was found to be about 8 g/s, consistent with the measured heat load with some additional load from the supply lines.

During a fast dump of magnet stored energy, the current decays with a 10.4 s time constant into the 0.05 Ω dump resistor. This implies an inductance of 0.52 H which matches the design inductance of 0.53 H. Redundant quench protection systems, a "digital" system (DQP), which relied on the operation of the control system programmable logic controller (PLC), and an independent "analog" system (AQP), were used to trigger a fast dump when a quench was detected. The DQP initially suffered from the failure of series "safety" resistors on voltage taps due to thermal cycling. Circuitry was added to detect broken resistors. For each coil, a battery provided an isolated current, which circulated through the coil and adjacent voltage tap safety resistors. Diodes were used to ensure that the isolated current was only seen by the corresponding input stage to the DQP. Offsets voltages produced by the battery current were measured and subtracted by the PLC software. The absence of the offset voltage was the signature for a broken resistor. After the first commissioning run (October 2002 to January 2003), the safety resistors were re-located outside of the cryostat.

About 160 of the 3270 hours of available data collection time during commissioning and production running were lost due to magnet problems. This is about 48% of all lost data collection time. Most (70.3%) of the magnet problems were caused by radiation damage to control system components. A typical failure began with a halt of PLC program execution due to a radiation-related memory error, which caused the "heart-beat" interlock to open. This shut down the power supply. A transient at the start of the shut-down caused the AQP to erroneously detect a "quench" and initiate a fast-dump. Eddy-current heating then evaporated LHe in the coils and reservoir requiring a minimum 2.5-hour recovery time. LHe supply and return problems were the second largest cause (18.7%) of magnet related lost time.

References

1. G.Batigne: Proceedings of the Fourth International Conference on Perspectives in Hadronic Physics, Trieste, May 12–16, 2003

Eur Phys J A (2005) **24**, s2, 145–145
DOI: 10.1140/epjad/s2005-04-036-8

EPJ A direct
electronic only

Cherenkov counter for the G^0 backward angle measurements

Benoît Guillon, for the G^0 collaboration

Laboratoire de Physique Subatomique et de Cosmologie, 53 avenue des Martyrs, 38000 Grenoble, France

Received: 15 October 2004 / Published Online: 8 February 2005

Abstract. The G^0 program consists of a set of parity violation experiments performed at Jefferson Lab (Va, USA) dedicated to the determination of the strange quark contribution to the charge and magnetization distributions of the nucleon [1]. This paper describes the final design of the Cherenkov counter used to reject the significant charged background of the G^0 backward angle measurements.

PACS. 29.40.Ka Cherenkov detectors

1 Introduction

The main focus of the G^0 experiment is to measure the neutral weak form factors G_E^Z and G_M^Z of the nucleon, over a large momentum transfer range. This will allow us to determine the strange quark contributions to the charge and magnetization densities of the nucleon. For this purpose, parity-violating asymmetries in elastic e-p scattering have already been measured at forward angles over the Q^2 range 0.1-1 $(GeV/c)^2$, and will be measured at backward angles for Q^2 values of 0.3, 0.5 and 0.8 $(GeV/c)^2$. At backward angles we also measure quasi-elastic scattering from a LD_2 target to extract precisely the axial form factor [1]. However, negatively charged pions, as well as their decay products (μ^-), will produce a significant background to the elastic and quasi-elastic rates detected by the G^0 spectrometer. Aerogel Cherenkov counters afford the best π/e discrimination at these energies and allow implementation in the current G^0 setup. LPSC Grenoble has designed and constructed half of the 8 Cherenkov counters needed.

2 The aerogel Cerenkov counter

The aerogel refractive index ($n = 1.03$) was fixed by the momentum distributions of the background. It should be less than $1/\beta$, but as large as possible to maximize light yield. The aerogel was supplied by Matsuhita Electric Works in the format of tiles (113x113x10 mm^3). The geometry of the counter and the necessary number of photomultipliers (PMT) was studied by using simulations [2] (Geant, Litrani) and validated with measurements made on a prototype. To allow the mounting of the aerogel in France and a safe transportation to Jlab, the Cerenkov box (see Fig. 1) is composed of two parts. The first part is the aerogel radiator which is 5 cm thick, and the second one is the lightbox itself. To ensure the best photon collection using 4 PMTs, the inner parts of these two boxes are covered by three layers of diffuse reflecting paper (Millipore). The XP4572b 5 inch tubes (Photonis) are very sensitive to magnetic fields. As they will be located in the fringe field of the G^0 magnet (4.4 mT in the axial direction and 11 mT in the transverse one), efficient magnetic shielding is needed. The design finally retained after tests at the LCMI Laboratory in Grenoble is made of three layers of 2 mm soft iron separated by 2 mm of air, and one layer of 0.8 mm of μ-metal. During these tests, one had to find the best compromise (light yield and field) for the positioning of the tubes inside the shielding. It was found that a back-off of 15 cm was sufficient. The 4 Cerenkov counters have been also successfully tested with cosmic rays and typical total numbers of 8 photoelectrons were measured with a collection time lower than 25 ns.

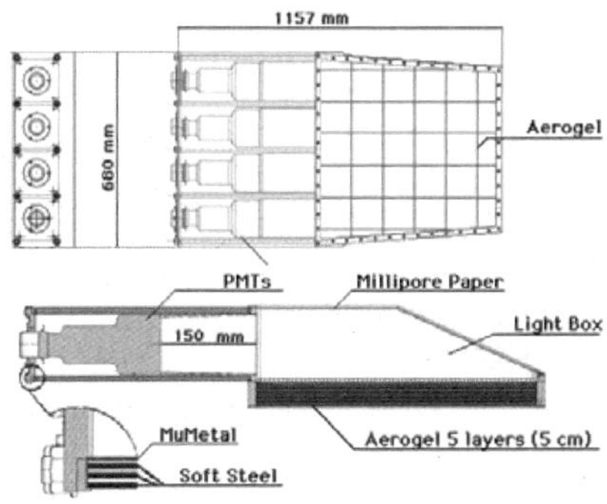

Fig. 1. Cherenkov design and magnetic shielding

References

1. D.H Beck, the G^0 collaboration: Jefferson Lab proposal E04-115 (originally E91-017)
2. G. Quéméner, S. Kox: G^0 Report G0-00-52 (2000)

Eur Phys J A (2005) **24**, s2, 147–147
DOI: 10.1140/epjad/s2005-04-037-7

EPJ A direct
electronic only

A bin-per-bin dead-time control technique for time-of-flight measurements in the G^0 experiment: The differential buddy

L. Bimbot, for the G^0 Collaboration

Institut de Physique Nucléaire d'Orsay, BP n° 1, F-91406 Orsay-Cedex, France

Received: 15 October 2004 / Published Online: 8 February 2005

Abstract. The general principle is presented. The application to the G^0 experiment [1] was enabled by the specificity of the time encoding ASIC component. On the poster, some encountered difficulties are exposed, together with a possible software remedy. An internal report is in preparation [2].

PACS. 01.30.Cc Conference proceedings – 25.30.Bf Elastic electron scattering – 07.05.Dz Control system – 07.05.Hd Data acquisition – 07.05.Kf Data analysis

Since PAVI02 the G^0 experiment successfully passed its commissioning and the first phase of data taking at forward angles [3]. This poster contribution aims to depict an experimental approach to control the dead-time in the time-of-flight histogramming of events.
Every part of the electronics - Constant Fraction Discriminator (CFD), Mean Timer (MT) and Time to Digital Converter (TDC) - has its own dead-time contributing. Different types of events must be considered: single CFD (only one of the left or right CFDs sees an event), single MT (only one of the Front or Back MTs sees an event) and good events (4 CFD signals leading to TDC encoding). Only the last type is subject to the *differential buddy* treatment. The correction for dead-time from other types of events is done with the help of a slow, event by event, acquisition made in parallel with a fastbus setup, and giving the probability of each kind of event [4] . The standard procedure of linear regression [5] for helicity correlated beam properties will help to remove residual dead-time effects. The idea, here, is to measure the loss of good events resulting from acquisition blocking. The direct counting of these events is impossible because they occur when the electronics is busy, analyzing a previous event, but an image can be obtained by checking, for each event if an associated detector (buddy), supposed to count identically, is busy or not. This is experimentally possible because it involves two different channels. However, as one needs time to process the signals, the comparison is made after a delay corresponding exactly to one pulse of beam duration. There are two basic assumptions: that counting is identical despite a spatial rotation (180°) and a time translation (32 nsec).

After solving tuning and analyzing problems, it is possible to use the data from the *buddy histograms* to estimate the dead-time of each detector and its stability (see figures below). The dead-time is the ratio, within proton peak limits, of the corrected number of events having encountered the buddy busy, to the total number of events.

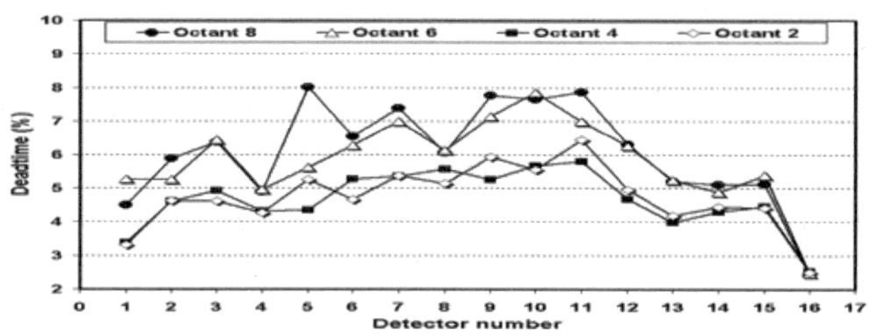

Fig. 1. Measured dead-time according to detector number

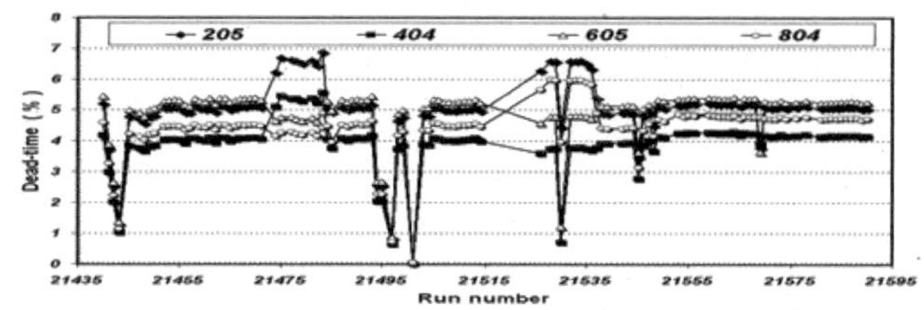

Fig. 2. Evolution of measured dead-time for specific channels

Special thanks to J.C. Cuzon, A. Gauvin, H. Guler, G. Quemener, J. Lenoble and R. Sellem for discussions and technical assistance.

References

1. see http://www.npl.uiuc.edu/exp/G0/G0Main.html
2. http://www.npl.uiuc.edu/exp/G0/docs/docs.html, Internal report G0-04-009
3. P.G. Roos: talk at this workshop
4. G. Batigne: Thesis, LPSC 03-41, 2003, p. 147–172
5. http://g0web.jlab.org/manual/Analysis_manuals.html, A. Biselli et al.: G^0 Analysis replay engine, p. 29

V Hadronic structure. . . and more

V-1 Test of the SM at low energy

Eur Phys J A (2005) **24**, s2, 151–154
DOI: 10.1140/epjad/s2005-04-038-6

EPJ A direct
electronic only

A precise measurement of $sin^2\theta_W$ at low Q^2 in Møller scattering

Antonin Vacheret and David Lhuillier, representing the E-158 Collaboration

CEA-Saclay, Dapnia/SPhN, 91191 Gif-sur-Yvette, France

Received: 15 December 2004 / Published Online: 8 February 2005

Abstract. The E-158 experiment has performed the first measurement of the parity-violating asymmetry in electron-electron (Møller) scattering using a 50 GeV polarized electron beam and a fixed unpolarized liquid hydrogen target in End Station A at SLAC. Our preliminary results is: $A_{PV} = -128\pm14(stat)\pm12(syst) \times 10^{-9}$. From this quantity we extract $\sin^2\theta_W(Q^2 = 0.026GeV/c^2)_{\overline{MS}} = 0.2403\pm0.0010(stat)\pm0.0009(syst)$, consistent with the Standard Model prediction.

PACS. 11.30.Er Charge conjugation, parity, time reversal, and other discrete symmetries – 12.15.Lk Electroweak radiative corrections – 12.15.Mm Neutral currents – 13.66.Lm Processes in other lepton-lepton interactions – 13.88.+e Polarization in interactions and scattering – 14.60.Cd Electrons

1 Introduction and physics motivation

In the scattering of longitudinally polarized electrons from unpolarized targets, the reversal of the helicity of incoming electrons is equivalent to applying the parity symmetry. Hence the quantity $A_{PV} = (\sigma_R - \sigma_L)(\sigma_R + \sigma_L)$, where $\sigma_{R,L}$ is the cross section for incident right(left)-handed electrons, is a pseudo-scalar arising from the parity violating part of the interaction in the scattering process. To first order this corresponds to the interference of the neutral weak and electromagnetic amplitudes [1]. When considering the Møller process, A_{PV} is proportional to the electron's weak charge, written at tree level as $Q^e_W = 1 - 4\sin^2\theta_W$ where θ_W is the weak mixing angle. The motivation of the E158 experiment is to measure $\sin^2\theta_W$ at low energy [2], far away from the Z-pole ($Q^2 \ll M_Z^2$) where precise measurements of this parameter have already been performed by the SLD and LEP collaborations [3]. The high accuracy we are seeking ($< 1\%$ relative error on $\sin^2\theta_W$) requires to include radiative corrections in the expression of A_{PV}:

$$A_{PV} = \frac{G_F Q^2}{\sqrt{2}\pi\alpha} \frac{1-y}{1+y^4+(1-y)} \mathcal{F}_b\, Q^e_W(Q^2) \qquad (1)$$

where G_F and α are the Fermi and fine structure constants, Q^2 is the square of the four-vector momentum transfer and $y = Q^2/s$, where s is the square of the center-of-mass energy. The $\mathcal{F}$ factor accounts for photon radiation effects in initial and final states. Most of the loop and vertex electroweak corrections are absorbed into the definition of an effective weak mixing angle $\sin^2\theta_W(Q^2)$ which acquires a Q^2 dependence. Taking advantage of the large cross section and little theoretical uncertainty of the purely leptonic Møller reaction, a precise measurement of these electroweak radiative corrections probes physics beyond the Standard Model (SM) at the TeV scale.

The figure of merit of the measurement peaks at high energy and to less extent at center-of-mass scattering angles around 90°. With a 50 GeV beam and $Q^2 = 0.026$ GeV/c^2, A_{PV} is predicted to be $\simeq$320 parts per billion (ppb) at tree level. Electroweak radiative corrections [4] reduce $\sin^2\theta_W$ by 3%. However this quantity remains numerically very close to 1/4 resulting in an experimental asymmetry reduced by more than 40%. Hence the difficulty of measuring an extremely small asymmetry is compensated by a great sensitivity to $\sin^2\theta_W$.

2 Beam

The development of the intense and highly polarized 50 GeV beam in SLAC End Station A (ESA) is a key element in the measurement of A_{PV}. Longitudinally polarized electrons are produced by optical pumping of a strained GaAs photo-cathode [5] by circularly polarized laser light. The sign of the laser circular polarization state determines the electron beam helicity. The beam is pulsed at 120 Hz with an intensity of 5.10^{11} electrons in a 300 ns pulse. The time structure consists of quadruplets of two consecutive pulses with pseudo-randomly chosen helicities, followed by their complements, yielding two independent left-right pulse pairs every 33 ms. This sequence is phase-locked to the 60 Hz of the power line in order to reduce electronic noise.One of the main experimental challenges when measuring a physics asymmetry at the few 100 ppb level is to keep the beam parameters (intensity, energy, position, angle) with negligible left-right asymmetries. The strategy used to reduce these asymmetries is three-fold [6]: a passive minimization of helicity correlated intensity and position differences that result from imperfections in the laser light and the photo-cathode response is first performed by a careful optimization of all the optical elements in the laser path. Then helicity-dependent corrections are applied to the laser beam in a feedback loop using periodic average measurement of beam asymmetries at upstream and downstream ends of the accelerator. These measurements are performed by cavity monitors implemented by pairs on the beam line [7]. This redundancy is critical to access the intrinsic resolution of the monitors and be able to read the beam parameters with a precision far better than their pulse to pulse jitter. Finally a set of slow helicity reversals, performed on a 1 or 2 days basis, further suppress effects of the remaining asymmetries. A half-wave plate can be inserted across the laser beam of the source to flip its polarization. Also the main linac can be operated at 45.6 or 48.7 GeV resulting in a 180° extra spin precession in the arc leading to ESA. In both cases the electron beam helicity is flipped, hence the sign of A_{PV}, while certain classes of beam asymmetries remain unaffected yielding to partial cancellations. Thanks to these active minimizations, cumulative beam asymmetries are reduced to < 200 ppb in intensity, < 10 ppb in energy, and < 5 nm in position, orders of magnitude below the "natural" beam asymmetries one could expect from the accelerator.

3 Instrumentation in ESA

The apparatus in ESA is illustrated in Fig. 1. The high luminosity and good background rejection required for the measurement of the tiny Møller asymmetry are achieved by the use of an extended target, a spectrometer/collimator system and integrating detectors.

The target cell [8] is a 1.57 m long cylinder filled with liquid hydrogen circulating at $\simeq$5 m/s. Turbulences are enhanced by Aluminum meshes in the fluid path allowing the absorption of $\simeq$500 W deposited by the beam while keeping density fluctuations below 40 ppm per pulse pair. These fluctuations are monitored by "luminosity monitors" consisting of eight ionization chambers arranged around the beam pipe downstream the detectors and intercepting particles scattered at $\simeq$1 mr. The huge rate provides great sensitivity to target boiling as well as a powerful check of the null asymmetry expected at such forward angles. The choice of hydrogen as an electron target results from the best Z/A ratio.

The low Q^2 and high energy kinematics lead to very forward laboratory scattering angles and both the primary beam and the scattered particles propagate through the spectrometer. A horizontal 3-dipole chicane defines the energy acceptance and shields the detectors from direct line-of-sight to the target. At the exit of the chicane the primary beam is put back on its axis and the main acceptance collimator, 3QC1B, selects particles between 4.4 and 7.5 mr. Then a series of 4 quadrupoles separates spatially the Møller-scattered electrons from the Mott (ep scattering) background at the detector plane 60 meters downstream the target. Selected Møller electrons form a ring

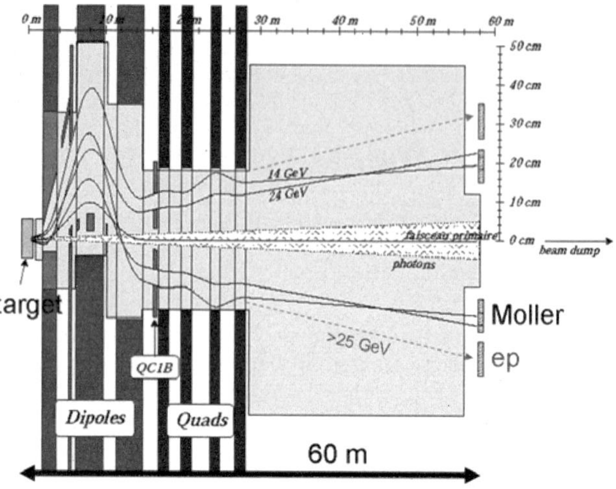

Fig. 1. Overview of the experimental setup in ESA. The horizontal axis is shrunk by a factor $\simeq$50 in order to enhance the horizontal deviation of the dipole chicane and the focusing of the Møller events toward the inner ring of detectors

approximately symmetric about the beam axis with an energy range of 13-24 GeV and containing $\simeq 2.10^7$ electrons per pulse. For the asymmetry measurement, the particles are intercepted by the Møller and ep detectors, consisting of 25 cm thick concentric cylinders with a 15-25 cm and 25-35 cm radius coverage respectively. They are assembled by layering planes of flexible fused-silica fibers between elliptical copper plates so as to withstand a 100 Mrad radiation dose. The fibers are oriented at the Cherenkov angle and gathered in bunches at the back of the detector. They direct the Cherenkov light into shielded PMTs via air light-guides. A total of 60 bunches provides radial and azimuthal segmentation (Fig. 2). The asymmetry is simply measured by extracting the fractional difference in the integrated calorimeter response for incident right- and left-handed beam pulses. This technique allows reaching very high counting rate with no dead time. The drawback is a "blind" detector integrating all events in the acceptance, including background. Therefore a complete set of auxiliary detectors is implemented to monitor the flux and the asymmetry of the different backgrounds.

A profile detector is located just upstream of the main calorimeter providing a complete radial and azimuthal map of the charged particles flux. Numerous profile measurements allow a very accurate calibration of the simulations (Fig. 2). The dominant background is the ep flux with 8% contamination of the Møller flux. The small contribution of photons and neutrons to the calorimeter response is measured in calibration runs. The pion flux and asymmetry are measured behind the Møller detector and a heavy lead shielding using a set of 10 Cherenkov detectors.

Finally the beam polarization is measured once every 2 days using a polarized supermendur foil target just upstream the LH2 target. Thanks to a dedicated retractable collimator, doubly-polarized Møller events are isolated inside a small element of the acceptance of the same spec-

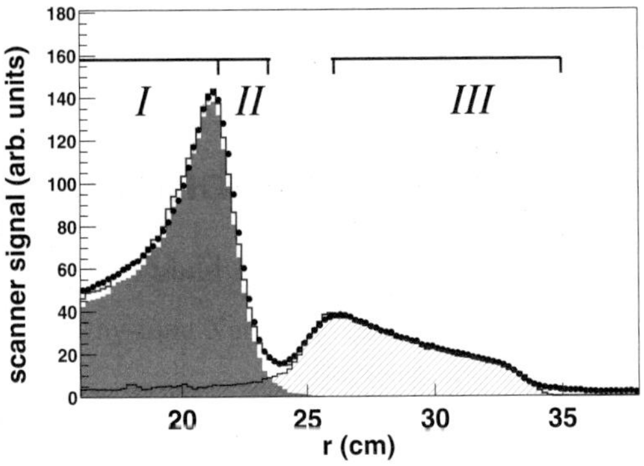

Fig. 2. Radial profile of charged particles at the calorimeter. The *points* are data, the histograms are simulations of the Møller (*shaded*) and *ep* (*hatched*) flux. For the detection of Møller electrons, regions I+II are segmented in three rings of 10, 20 and 20 PMTs respectively

trometer as for the main experiment. They are detected by a small tungsten/quartz stack fixed at the end of an air light-guide with PMT readout. This setup is mounted on the profile detector frame.

4 Analysis and preliminary results

Every 33 ms the data from all beam monitors and detector channels are collected for the 4 beam pulses of a quadruplet. The experimental detector asymmetries (A_{exp}) and beam parameters differences (Δx_i, $i = E, x, y, \theta_x, \theta_y, Q$) for the corresponding two pairs are extracted. The analysis is performed with a blinding offset added to the detector asymmetries. This offset is picked-up by a random algorithm in a range comparable to the expected physics asymmetry and kept secret until the very last stage of the analysis, preventing any psychological influence on the final results. The cuts applied to the data (16% of rejected events) select periods of stable beam and operational equipement in an helicity-independent way. In total $\approx 4 \times 10^8$ pulse pairs satisfy all criteria for the three runs taken in 2002-3. Final results have already been published for the first run [9], we present here preliminary results for the complete statistics.
The detector signals are corrected for differences in the right-left beam properties as measured by the beam monitors according to

$$A_{raw} = A_{exp} - \sum_i \alpha_i \Delta x_i \qquad (2)$$

where the coefficients α_i are the sensitivities of the detector to each beam parameter. They are determined by two independent methods. The first one applies an unbinned least squares linear regression to the pulses used for physics, taking advantage of the beam parameters jitter. The second method, so-called "dithering", uses calibration subset of the pulses (4% duty cycle), where each parameter is modulated periodically around its average value

Table 1. Corrections ΔA and dilutions f to A_{raw} for run III

Source	ΔA (ppb)	f
Beam (first order)	1 ± 1	-
Beam (higher order)	0 ± 4	-
Transverse polarization	-4 ± 2	-
$e^- \, p \to e^- \, p(+\gamma)$	-8 ± 2	0.058 ± 0.007
$e^- \, (\gamma) \, p \to e^- \, p(+\gamma)$	-22 ± 6	0.009 ± 0.003
High energy photons	3 ± 3	0.004 ± 0.002
Synchrotron photons	0 ± 2	0.0015 ± 0.0005
Neutrons	-1 ± 1	0.0006 ± 0.0002
Pions	1 ± 1	0.001 ± 0.001

by an amount large compared to beam jitter. Because of its very forward kinematics E-158 is quite sensitive to beam fluctuations therefore the extra lever arm given by the large amplitudes of the dithering method is not critical to extract accurate α coefficients. Final analysis relies on the first method, statistically more powerful, and uses the dithering as a cross-check of systematic errors. These corrections remove both the beam-induced random and systematic effects. The RMS of the A_{raw} distribution is reduced to $\approx$200 ppm, very close to the expected pure statistical width and a factor 2 smaller than the RMS of the A_{exp} distribution. The cumulative asymmetry correction is listed in Table 1 for run III. The impressively small systematic error of 1 ppb follows from the excellent agreement between the two correction methods when applied on the same data sample.

After correcting the effects of all the beam parameters non-statistical fluctuations of A_{raw} around its mean value are observed in the, most sensitive, outer ring of the Møller detector, pointing to extra systematic effects at the sub-pulse scale. Such effects couldn't be treated with the above-mentioned methods which only deal with mean values per pulse. To study and correct the induced false asymmetries, the signals of the beam monitors are oversampled with 4 time-slices inside the duration of a beam pulse. The final correction removes all the non-statistical fluctuations with no significant effect on the mean value and a total systematic error of 4 ppb.
After a complete regression a remaining azimuthal modulation of A_{raw} is observed due to a small non-zero component of the transverse beam polarization leading to a correction of -4 ± 2 ppb. A separate study limited the bias due to beam spot-size fluctuations on A_{raw} to 1 ppb, using data from a retractable wire array.
The physics asymmetry is finally formed from A_{raw} by correcting for background contributions, detector linearity ϵ and beam polarization P_e:

$$A_{phys} = \frac{1}{P_e \epsilon} \frac{A_{raw} - \sum_i \Delta A_i}{1 - \sum_i f_i} \qquad (3)$$

where ΔA_i and f_i are the asymmetry corrections and dilutions for various background sources listed in Table 1.

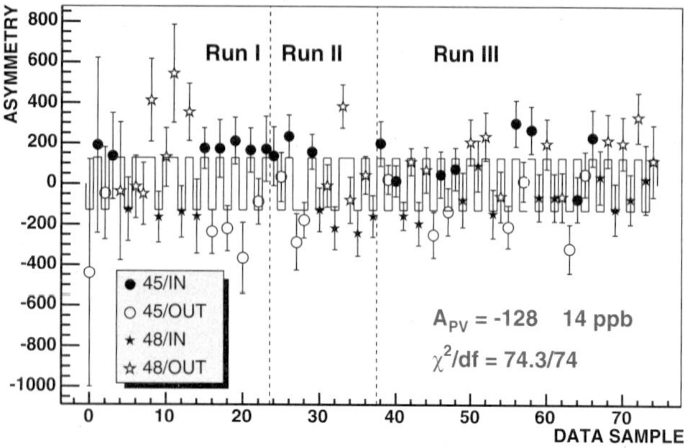

Fig. 3. Møller physics asymmetry. The *solid line* indicates the expectation for the asymmetry for all combinations of beam energy and half-wave plate state

The largest correction is due to electrons from inelastic electron- and photon-proton interactions. The associated asymmetry was measured during the first run in the *ep* detector and reasonable assumptions were used for the kinematic extrapolation to the Møller region. The flux contamination is determined from a simulation validated with the numerous profile detector data.
The averaged beam polarization is 88±5% with dominant errors arising from the knowledge of the magnetization of the foil target and the background. The linearity of the calorimeter response is determined to be $\epsilon = 0.99 \pm 0.01$.
As a final check the mean A_{phys} is plotted for each of the energy and half-wave plate configurations. Figure 3 show that data are compatible with a perfect sign reversal, giving confidence in the removal of all spurious asymmetries. After removing the blinding offset the grand average of the sign corrected asymmetries gives the final physics result:

$$A_{PV} = -128 \pm 14\,(stat.) \pm 12\,(syst.)\,\text{ppb} \tag{4}$$

establishing parity violation in Møller scattering with the most precise measurement of any asymmetry in electron scattering. Using (1) we can extract the weak charge of the electron and then $\sin^2\theta_W(Q^2)$. The average kinematics as well as photon radiation effect are determined from a Monte Carlo simulation: $Q^2 = 0.026$ (GeV/c)^2, $y \simeq 0.6$ and $\mathcal{F}_b = 1.01 \pm 0.01$. We find

$$\sin^2\theta_W(Q^2 = 0.026\,\text{GeV/c}^2)_{\overline{MS}} = 0.2403 \pm 0.0010\,(stat) \pm 0.0009\,(syst) \tag{5}$$

consistent at the 1.2 σ level with the SM expectation [4] $\sin^2\theta_W(0)_{\overline{MS}} = 0.2385 \pm 0.0006\,(theory)$. This result provides significant new limits on physics beyond the SM. For example the scale Λ_{LL} of a new left-handed contact interaction [10] is set to $\Lambda^+_{LL} \geq 6.4$ TeV and $\Lambda^-_{LL} \geq 13.9$ TeV for potential positive and negative deviations respectively, and the lower limit for the mass of a new Z_χ boson [4] is set to $M_{Z_\chi} \geq 860$ TeV at 95% C.L.
For a coherent comparison between all precise measurements of $\sin^2\theta_W$ Fig. 4 shows all results evolved to the Z-pole. At low energy the E-158 accuracy is comparable

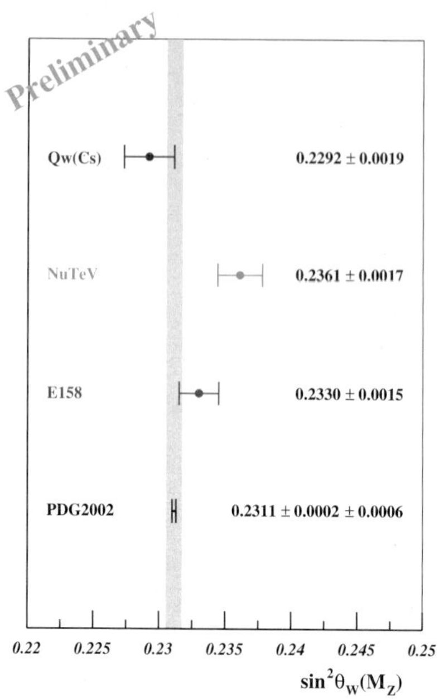

Fig. 4. Summary of $\sin^2\theta_W(\overline{MS})$ measurements evolved to the Z-pole. $Q_W(Cs)$: parity violation in Cesium atoms, NuTeV: ν-nuclei scattering, PDG2002: world average, see [14]

to measurements in the Cesium atoms [11,12] and in ν-nucleon scattering [13]. The significant deviation of the NuTeV result can't be attributed to new physics yet as detailed analysis of nuclear effects is still going on. The sensitivity of the new E-158 result to new physics is complementary to the existing data and the measurement of an asymmetry with a 15 ppb error is a benchmark for future high accuracy experiments [15].

References

1. Y.B. Zel'dovich: Sov. Phys. JETP **94**, 262 (1959)
2. K.S. Kumar, R.S. Holmes, E.W. Hughes, P.A. Souder: Mod. Phys. Lett. A **10**, 2979 (1995)
3. LEP and SLD collaborations: hep-ex/0312023 (2003)
4. A. Czarnecki, W.J. Marciano: Int. J. Mod. Phys. A **15**, 2365 (2000); J. Erler, A. Kurylov, M.J. Ramsey-Musolf: Phys. Rev. D **68**, 016006 (2003); F.J. Petriello: Phys. Rev. D **68**, 033006 (2003); A. Ferroglia, G. Ossola, A. Sirlin: eprint hep-ph/0307200 (2003)
5. T. Maruyama et al.: Nucl. Inst. Meth. A **492**, 199 (2002)
6. T.B. Humenski et al.: Nucl. Inst. Meth. A **521**, 261 (2004)
7. H. Whittum, Y. Kolomensky: Rev. Sci. Instrum. **70**, 2300–2313 (1999)
8. J. Gao et al.: Nucl. Instrum. Meth. A **498**, 90 (2003)
9. P.L. Anthony et al.: Phys. Rev. Lett. **92**, 181602 (2004)
10. M.J. Ramsey-Musolf: Phys. Rev. C **60**, 015501 (1999)
11. C.S.Wood et al.: Science **275**, 1759 (1997)
12. V.A. Dzuba, V.V. Flambaum, J.S.M. Ginges: Phys. Rev. D **66**, 076013 (2002); A. Derevianko, B. Ravaine, W.R. Johnson: physics/0401043, (2004)
13. K.S. McFarland: these proceedings and references therein
14. *Review of Particle Properties*, K. Hagiwara et al.: Phys. Rev. D **66**, 010001 (2002)
15. K.S. Kumar: these proceedings and references therein

Eur Phys J A (2005) **24**, s2, 155–158
DOI: 10.1140/epjad/s2005-04-039-5

EPJ A direct
electronic only

Qweak: A precision measurement of the proton's weak charge

Gregory R. Smith, for the Qweak Collaboration

Jefferson Lab, Newport News, VA 23606, USA

Received: 15 October 2004 / Published Online: 8 February 2005

Abstract. The $Q_{\rm weak}$ experiment at Jefferson Lab will measure the parity-violating asymmetry in e–p elastic scattering at very low Q^2 using a longitudinally polarized electron beam and a liquid hydrogen target. The experiment will provide the first measure of the weak charge of the proton, Q_w, to an accuracy of 4%. Q_w is simply related to the weak mixing angle θ_w, providing a precision test of the Standard Model. Since the value of $\sin^2\theta_w$ is approximately 1/4, the weak charge of the proton $Q_w^p = 1 - 4\sin^2\theta_w$ is suppressed in the Standard Model, making it especially sensitive to the value of the mixing angle and also to possible new physics. The experiment employs an 85% polarized, 180 μA, 1.2 GeV electron beam, a 35 cm liquid hydrogen target; and a toroidal magnet to focus electrons scattered at $8° \pm 2°$, corresponding to $Q^2 \sim 0.03$ (GeV/c)^2. With these kinematics the systematic uncertainties from hadronic processes are strongly suppressed. To obtain the necessary statistics this 2200 hours experiment must run at an event rate of over 6 GHz. This requires current (integrating) mode detection of the scattered electrons, which will be achieved using synthetic quartz Cherenkov detectors. A tracking system will be used in a low-rate counting mode to determine the average Q^2 and the dilution factor of background events. The theoretical context of the experiment and the status of its design are discussed.

PACS. 24.80.+y Nuclear tests of fundamental interactions and symmetries – 25.30.Bf Elastic electron scattering

1 Introduction

We describe a new experiment at Jefferson Lab to make the world's first measurement of the weak charge of the proton. $Q_{\rm weak}$ is a well defined experimental observable with a definite prediction in the Standard Model (SM). The experiment would constitute the first SM test at Jefferson Lab. To lowest order, the weak charge can be expressed as $Q_w^p = 1 - 4\sin^2\theta_{\rm w}$, where $\sin^2\theta_w \approx 0.23$ is the weak mixing angle. The goal of the $Q_{\rm weak}$ experiment [1] is a 4% measurement of Q_w^p (combined statistical and systematic errors), which corresponds to a 0.3% measurement of $\sin^2\theta_w$. A measurement of this precision is possible because hadronic corrections are small at low Q^2 [2], and measured by many different experiments aimed at electromagnetic and weak hadronic form factors.

2 Physics motivation

The SM makes a firm prediction for $Q_{\rm weak}$ based on the running [3,4] of $\sin^2\theta_w$ from the Z^0 pole. As shown in

[1] The Qweak Collaboration:
D. Armstrong[1], T. Averett[1], J. Birchall[2], T. Botto[3], J.D. Bowman[4], A. Bruell[5], R. Carlini[5], S. Chattopadhyay[5], C. Davis[6], J. Doornbos[6], K. Dow[3], J. Dunne[7], R. Ent[5], J. Erler[8], W. Falk[2], M. Farkhondeh[3], J.M. Finn[1], T. Forest[9], W. Franklin[3], D. Gaskell[5], K. Grimm[1], F.W. Hersman[10], M. Holtrop[10], K. Johnston[9], R. Jones[11], K. Joo[11], S. Kôwalski[3], C. Keppel[12], M. Kohl[3], E. Korkmaz[13], L. Lee[6], Y. Liang[14], A. Lung[5], D. Mack[5], S. Majewski[5], J. Mammei[15], R. Mammei[15], D. Meekins[5], G. Mitchell[4], H. Mkrtchyan[16], N. Morgan[15], A. Opper[14], S. Page[2], S. Pentillo[4], M. Pitt[15], M. Poelker[5], T. Porcelli[13], W.D. Ramsay[2], M. Ramsey-Musolf[17], J. Roche[5], N. Simicevic[9], G.R. Smith[5], T. Smith[18], R. Suleiman[15], S. Taylor[3], E. Tsentalovich[3], W.T.H. vanOers[2], S. Wells[9], W.S. Wilburn[4], S. Wood[5], H. Zhu[10], and C. Zorn[5].
[1]College of William & Mary
[2]University of Manitoba
[3]Massachusetts Institute of Technology
[4]Los Alamos National Laboratory
[5]Thomas Jefferson National Accelerator Facility
[6]TRIUMF
[7]Mississippi State University
[8]Universidad Nacional Autonoma de Mexico
[9]Louisiana Tech University
[10]University of New Hampshire
[11]University of Connecticut
[12]Hampton University
[13]University of Northern British Columbia
[14]Ohio University
[15]Virginia Polytechnic Institute
[16]Yerevan Physics Institute
[17]California Institute of Technology
[18]Dartmouth College

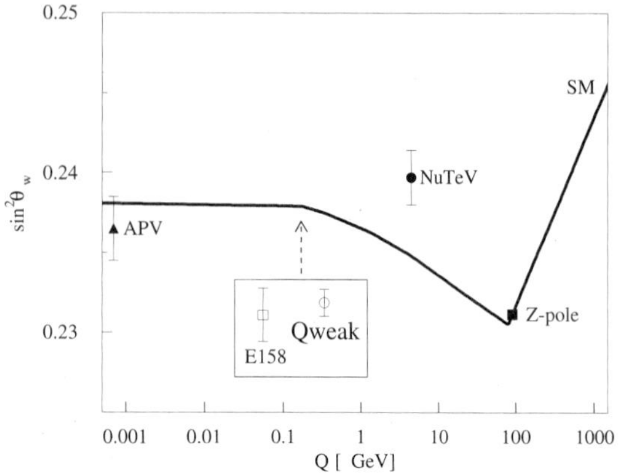

Fig. 1. Running of the weak mixing angle [3] in the Standard Model, calculated in the $\overline{\text{MS}}$ scheme. Shown are results from atomic parity violation (APV) [8–10], NuTeV [7], and the Z pole [5]. The *error bar* shown for the E158 [6] Møller experiment is that anticipated for their final result

Fig. 1 the value of the weak mixing angle is predicted to change (in the $\overline{\text{MS}}$ renormalization scheme) by $\sim 3\%$ from the energy scale of the Z pole (where a decade of precision measurements have been made at high energy colliders like SLAC and LEP [5]) to the low energy scale of Q_{weak}. With the proposed 0.3% measurement of the weak mixing angle, the Q_{weak} experiment will make a $\sim 10\sigma$ verification of this effect. A deviation from the SM prediction would imply new physics beyond the SM. In the case of the Q_{weak} experiment, there is even sensitivity to which SM extension is responsible for the deviation, especially when taken in conjunction with the results of other experiments. Should the result of this experiment agree with the prediction of the SM, then the result would dramatically constrain possible extensions to the SM.

The Q_{weak} experiment [1] at Jefferson Lab (JLab) will make such a precision measurement of the asymmetry between cross-sections for positive and negative helicity electrons in polarized elastic electron-proton scattering. The asymmetry violates parity, and arises from the interference of electromagnetic and weak amplitudes (photon and Z^0 boson exchange). At this low energy scale, the asymmetry is a measure of the weak charge of the proton, Q_w^p, which is the strength of the weak vector coupling of the Z^0 boson to the proton. At tree level the value of Q_{weak} is expected to be ~ 0.072. In the limit of small scattering angle and small momentum transfer ($Q^2 \to 0$), the asymmetry is given by [2]:

$$\frac{\sigma_+ - \sigma_-}{\sigma_+ + \sigma_-} = [\frac{-G_F}{4\pi\alpha\sqrt{2}}][Q^2 Q_w^p + Q^4 B(Q^2)] \quad (1)$$
$$\approx -0.3 \text{ ppm at } Q^2 = 0.03 \text{ GeV}^2$$

where $B(Q^2)$ is a contribution from electromagnetic and weak form factors. It is sensitive to $G_{E,M}^{\gamma}$ and $G_{E,M}^{Z}$, the target of many recent and ongoing experiments at Jefferson Lab, Bates, and MAMI. As 1 makes clear, the B term is suppressed as Q^2 goes to zero by an extra factor of Q^2 relative to Q_{weak}. On the other hand, the experimental asymmetry grows with Q^2. At Jefferson Lab a Q^2 of ~ 0.03 GeV^2 has been chosen as the most ideal compromise between these two competing considerations.

The expected measured asymmetry A_{meas} of -0.29 ppm consists of three main terms. The biggest is the term sought in this experiment, namely A_{Qw}, the first term in 1. It has an expected value of -0.19 ppm, or about 2/3 of the expected A_{meas}. The second term comes from the hadronic form factors discussed above, and has a magnitude expected to be about 31% of A_{meas}, or about -0.09 ppm. It will be well constrained by the precise measurements provided by HAPPEX, PVA4, and G0. The third term is almost negligible, and amounts to only about 3% of the expected A_{meas}. It accounts for the axial contribution, contains G_A^e, and includes substantial electroweak radiative corrections. It will be constrained by SAMPLE and G0. When added in quadrature, the two background terms A_{hadronic} and A_{axial} contribute only about 2% to the expected final error on Q_{weak}. We re-emphasize that the bulk of this so-called background contribution can be accounted for solely with the results of precision experiments aimed at hadronic form factors, and does not rely on theoretical calculations or models. In addition, the Q_{weak} collaboration is considering augmenting the existing experimental information on $B(Q^2)$ by performing a short independent measurement at a Q^2 slightly higher than 0.03 GeV^2. This should not be necessary if the ongoing form factor experiments achieve their proposed goals.

The Q_{weak} experiment will provide tight constraints on SM parameters associated with potential new physics. For example, the up and down quark couplings in the SM electron-quark Lagrangian:

$$L_{SM}^{PV} = -\frac{G_F}{\sqrt{2}}\bar{e}\gamma_\mu\gamma_5 e \sum C_{1q}\bar{q}\gamma^\mu q$$

are constrained by Q_W^p according to

$$Q_W^p(SM) = -2(2C_{1u} + C_{1d}) \sim 0.0721 \, .$$

This constraint is complementary to other existing constraints, and when combined with those other measurements, dramatically reduces the allowed phase space for new physics.

The physics reach of this experiment can be expressed in terms of its precision as:

$$\frac{\lambda}{g} \sim \frac{1}{\sqrt{\sqrt{2}G_F|\Delta Q_W^p|}} \approx 4.6 \text{ TeV}.$$

The discovery potential of weak charge measurements will be unmatched until the LHC turns on. If new physics is uncovered at the LHC, such as an extra Z boson, then precision experiments like Q_{weak} undertaken at low Q^2 will be essential to determine the charges, coupling constants, etc.

There are many reasons to expect that the SM is a low-energy effective theory of some more fundamental description of nature. Neutral current experiments like Q_{weak} provide important consistency checks of the SM, are complementary to direct searches for new physics, and can help distinguish and characterize the new physics once it is found [3,4]. One of the most plausible extensions to the SM which can be tested for by the Q_{weak} experiment is the presence of extra Z's. The present lower bound for Z's provided by Tevatron experiments is only 0.6 TeV, which leaves plenty of discovery potential for the Q_{weak} experiment. Z's would resolve the $\sim 2\sigma$ discrepancy from the SM in the generation number ($N_\nu = 2.986 \pm 0.008$) derived from the measured Z^0 lineshape. Furthermore, Z's would improve the agreement of both the APV and NuTeV experiments with the SM. Another SM extension the Q_{weak} experiment is particularly sensitive to is R parity violation, one of the most interesting varieties of SUSY. Finally, the Q_{weak} experiment will provide a sensitive test for leptoquarks.

Weak charge and mixing angle results can be obtained in other types of experiments. The SLAC E158 [6] experiment is a purely leptonic Møller scattering experiment recently completed at a similar Q^2. Their preliminary results are consistent with the SM. The Fermilab NuTeV experiment [7] explored νA scattering at ~ 10 GeV/c^2, and found an intriguing 3σ discrepancy with the SM. The interpretation of this discrepancy remains somewhat controversial for now. Finally, there have been atomic parity violation (APV) measurements [8–10] made of the Cs atom (Q^2=0). Those results have been plagued by extraordinarily difficult and changing theoretical corrections (see, for example [11,12]), however, at present their result is consistent with the SM. To provide some perspective, a 13% measurement of Q^p_{weak} in our e–p experiment would provide the same sensitivity to new physics as a 1% measure of $Q_{weak}(N, Z)$ using APV, but without the theoretical uncertainties. On the other hand, our ability to test for extensions to the SM will be greatest when the results of the Q_{weak} experiment are combined with those from the complementary triad of experiments referred to above, especially E158.

3 The qweak experiment

The conceptual design of the Q_{weak} experiment is illustrated in Fig. 2. The experiment consists of scattering a longitudinally polarized 1.2 GeV electron beam by a 35 cm liquid hydrogen target. Elastically scattered electrons at $8° \pm 2°$ are selected by a collimation system, and then focused by a large toroidal resistive magnet onto a set of eight synthetic quartz Čerenkov detectors. At the average experimental momentum transfer of $Q^2 = 0.03$ GeV^2 the expected Q_{weak} asymmetry is small, -0.3 parts per million (ppm). The expected event rate for scattered electrons of ~ 6 GHz precludes counting the individual events. Instead, the experiment will use current mode detection and low noise front end electronics.

In brief, the various experimental systems are as follows. The liquid hydrogen target will be patterned after the successful design of the G0 and SAMPLE targets except that a centrifugal pump will deliver several times the flow velocity achieved in those targets. To meet the more challenging requirements of the Q_{weak} experiment, the beam will be rastered into a uniform 4mm x 4mm pattern. The target cell will be 35cm in length. With the proposed beam current of 180μA, this target will require almost 2.5kW of cooling power. The ability of the target to provide the necessary small boiling contribution to the asymmetry width will be backed up by an array of high rate, small angle quartz luminosity monitors near the end of the Jefferson Lab Hall C beam line.

The toroidal magnet consists of eight resistive coils each approximately 3.7m long and 1.5m tall composed of a double pancake of Cu conductor 3.8cm x 5.8cm in cross section, with a 2cm diameter central cooling channel. The magnet will provide an $\int \mathbf{B} \cdot d\mathbf{l} \sim 0.7$ T-m with a 9500 A, 1.2 MW DC power supply. The coils will be supported in an aluminum stand, and will permit a 1 foot diameter, Pb-shielded beam pipe to pass through the central axis. Simulations show that this magnet will cleanly separate photons and inelastic events from the elastic events of interest at the focal plane.

The detectors themselves will be relatively insensitive to γ's, n's, and π's, be capable of withstanding $\geq$ 300 kRad, and operate at counting statistics. Eight fused silica Čerenkov detectors will be used, each approximately 2m x 16cm x 2.54cm. These bars are characterized by n=1.47, $\theta_{Cerenkov} = 47°$, and a total internal reflection angle of 43°. The bars will be read out by S20 photocathodes at each end, and should provide about 100 photoelectrons per event.

A tracking system will be used with the beam current reduced by four orders of magnitude, allowing individual events to be observed. This will enable both a measurement of the dilution of the Čerenkov detector signal by background and a precise determination of the average Q^2. The tracking system components will rotate to cover all octants, and will include the following sets of detectors. A gas electron multiplier closest to the target will serve as a vertex detector. Horizontal drift chambers near the magnet entrance will measure the scattering angle. These will be augmented by a mini-torus to sweep away the otherwise dominant Møller electrons. At the focal plane, vertical drift chambers will map the analog response of the Čerenkov system. Finally, large scintillation counters will provide a charged particle trigger.

Precision beam polarimetry is required in order to have a polarization contribution to the systematic error of less than 1.5%. To achieve this a two-pronged approach is being pursued. First, efforts are already underway to increase the operating current of the existing $\leq 10\mu$A, $\leq 1\%$ Møller polarimeter to 100 μA beam currents. Methods being investigated include kicking the beam across wire Møller targets, and use of rotating foil targets. In addition, a Compton polarimeter is being developed for JLab Hall C which would provide continuous relative measurement of the beam polarization, normalized by the Møller polarimeter.

Other requirements on beam properties have been determined by extensive simulations. The Q_{weak} beam property specifications have for the most part already been achieved at Jefferson Lab during the G0 and HAPPEX

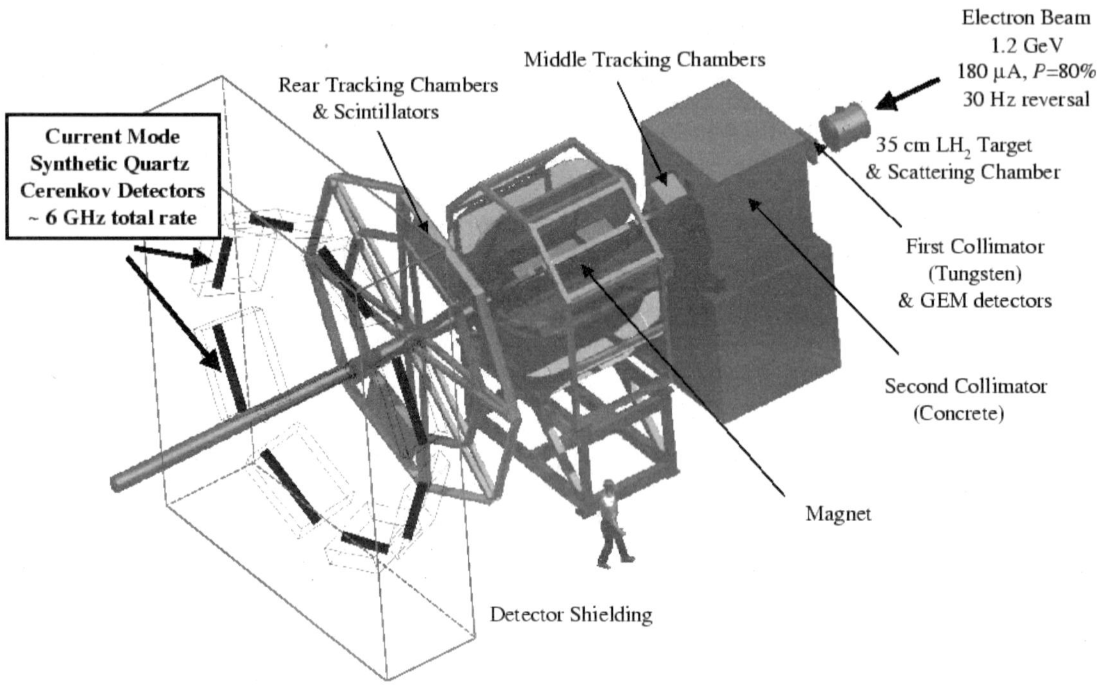

Fig. 2. Conceptual design for the Q_{weak} experimental setup, in Hall C at Jefferson Lab. The eight quartz detectors, each 2 m×16 cm×2.5 cm, are shown inside their re-entrant shielding enclosure. The spectrometer provides clean separation of elastic and inelastic electrons at its focal plane

experiments. The Q_{weak} requirements are very similar to those for HAPPEX-Pb, which runs with standard equipment and therefore should be scheduled well before Q_{weak}. That experiment, however, does not enjoy the common mode rejection inherent in a toroidal spectrometer like Q_{weak}. In addition, intensity and position feedback will be separated in the Q_{weak} experiment by making use of position information just upstream of the target. In spite of the fact that the beam properties at Jefferson Lab seem to already be very close to what is required for the Q_{weak} experiment, the collaboration is working with the JLab source group to explore the possibility of implementing helicity reversal faster than the canonical 30 Hz, which shall considerably ease the demands this experiment places on many beam properties as well as target density fluctuations, among others.

The Q_{weak} physics proposal was approved in January, 2002 by the JLab PAC, and has since become an important new thrust of the JLab scientific program. The collaboration presented a successful technical design review in January 2003. A management plan is in place for the project, and all the required funding has been secured. The project is being supported by the U.S. Department of Energy, the National Science Foundation, and the Natural Sciences and Engineering Research Council of Canada. The Q_{weak} experiment will proceed in two stages. First, a statistics limited run with a low power target will aim for a $\leq$8% result on the asymmetry in 2007. This will be followed by runs totalling about 2200 hours at 180 μA in order to achieve a 4% result. The collaboration is investigating whether the absolute limits of the technique can be pushed any further, recognizing that the physics impact of an even more precise experiment would be enormous.

References

1. Qweak Collaboration, R. Carlini et al.: Jefferson Lab Proposal E-02-020.
   Proposal and collaboration member list available at `http://www.jlab.org/qweak/`.
2. M.J. Musolf et al.: Phys. Rep. **239**, 1 (1994)
3. J. Erler, A. Kurylov, M.J. Ramsey-Musolf: Phys. Rev D **68**, 016006 (2003)
4. A. Kurylov, M.J. Ramsey-Musolf, S. Su: `hep-ph/0303026`, to appear in Phys. Rev. D
5. Particle Data Group: K. Hagiwara et al.: Phys. Rev. D **66**, 010001 (2002); In particular, the section *Electroweak Model and Constraints on New Physics*
6. P.L. Anthony et al.: Phys. Rev. Lett. **92**, 181602 (2004)
7. NuTeV Collaboration, G.P. Zeller et al.: Phys. Rev. Lett. **88**, 091802 (2002)
8. N.H Edwards et al.: Phys. Rev. Lett. **74**, 2654 (1995)
9. P.A. Vetter et al.: Phys. Rev. Lett. **74**, 2658 (1995)
10. C.S. Wood et al.: Science **275**, 1759 (1997)
11. M.Y. Kuchiev, V.V. Flambaum: Phys. Rev. Lett. **89**, 283002 (2002)
12. A.I. Milstein, O.P. Sushkov, I.S. Terekhov: Phys. Rev. Lett. **89**, 283003 (2002)

Eur Phys J A (2005) **24**, s2, 159–160
DOI: 10.1140/epjad/s2005-04-040-0

EPJ A direct
electronic only

The Qweak tracking system

Measurement of the average Q^2 and the dilution by background events

Klaus H. Grimm, for the Qweak Collaboration

The College of William & Mary, Department of Physics, Williamsburg, VA 23187-8795, USA

Received: 01 November 2004 / Published Online: 8 February 2005

Abstract. The Q^p_{weak} experiment will measure the parity-violating elastic e-p scattering asymmetry to extract the weak charge of the proton. The experiment employs a toroidal magnet to focus electrons scattered at $8^\circ \pm 2^\circ$, corresponding to $Q^2 \sim 0.03$ $(\mathrm{GeV/c})^2$, on eight Čerenkov detectors located in the focal plane of the spectrometer. Since the asymmetry is proportional to Q^2, it is crucial to obtain an accurate measure of the acceptance-averaged value of Q^2. A tracking system will be used in a low-rate counting mode, allowing individual events to be observed. This will enable a determination of the average Q^2 by measuring the scattering angle and interaction vertex, for mapping the response across the surface of the Čerenkov detectors, and for the dilution of the Čerenkov detector signal by background.

PACS. 25.30.Bf Elastic electron scattering – 29.40.Gx Tracking and position-sensitive detectors

1 Introduction

In elastic electron-proton scattering the parity-violating asymmetry is proportional to Q^2. In the Q^p_{weak} experiment [1,2] a collimator system, optimized by GEANT simulations, defines the average value of the accepted Q^2 by setting kinematical cuts on the electron scattering angles. For Q^p_{weak} it is critical to determine the acceptance averaged value of Q^2 for the electrons from the e-p elastic scattering events of interest with $\simeq$ 1% accuracy. This translates into a 0.6 mrad precision in the measure of the scattering angle. Since Q^p_{weak} is an integrating experiment, the Q^2 dependence of the *ep* cross-section will bias the average detected Q^2. This and any position-dependent detector bias must be taken into account since Q^2 may be correlated with position at the focal plane. Although in principle these matters can be simulated, it is essential to check the Monte Carlo using an ancillary calibration measurement: to measure the scattering angle, interaction vertex, shape of the focal plane distributions, and the position-dependent detector bias. This information will be extracted from ancillary measurements at low beam current in which the Čerenkov detectors will be read out in pulse mode and individual particles tracked through the toroidal spectrometer with a tracking system.

2 Tracking system design concept

The tracking system consists of a set of drift chambers at three locations along the track. It will be capable of mapping two opposing octants simultaneously where

[1] The Qweak Collaboration:
D. Armstrong[1], T. Averett[1], J. Birchall[2], T. Botto[3], J.D. Bowman[4], A. Bruell[5], R. Carlini[5], S. Chattopadhyay[5], C. Davis[6], J. Doornbos[6], K. Dow[3], J. Dunne[7], R. Ent[5], J. Erler[8], W. Falk[2], M. Farkhondeh[3], J.M. Finn[1], T. Forest[9], W. Franklin[3], D. Gaskell[5], K. Grimm[1], F.W. Hersman[10], M. Holtrop[10], K. Johnston[9], R. Jones[11], K. Joo[11], S. Kowalski[3], C. Keppel[12], M. Kohl[3], E. Korkmaz[13], L. Lee[6], Y. Liang[14], A. Lung[5], D. Mack[5], S. Majewski[5], J. Mammei[15], D. Meekins[5], G. Mitchell[4], H. Mkrtchyan[16], N. Morgan[15], A. Opper[14], S. Page[2], S. Pentillo[4], M. Pitt[15], M. Poelker[5], T. Porcelli[13], W.D. Ramsay[2], M. Ramsey-Musolf[17], N. Simicevic[9], G.R. Smith[5], T. Smith[18], R. Suleiman[15], S. Taylor[3], E. Tsentalovich[3], W.T.H. vanOers[2], S. Wells[9], W.S. Wilburn[4], S. Wood[5], H. Zhu[10], and C. Zorn[5].
[1]College of William & Mary
[2]University of Manitoba
[3]Massachusetts Institute of Technology
[4]Los Alamos National Laboratory
[5]Thomas Jefferson National Accelerator Facility
[6]TRIUMF
[7]Mississippi State University
[8]Universidad Nacional Autonoma de Mexico
[9]Louisiana Tech University
[10]University of New Hampshire
[11]University of Connecticut
[12]Hampton University
[13]University of Northern British Columbia
[14]Ohio University
[15]Virginia Polytechnic Institute
[16]Yerevan Physics Institute
[17]California Institute of Technology
[18]Dartmouth College

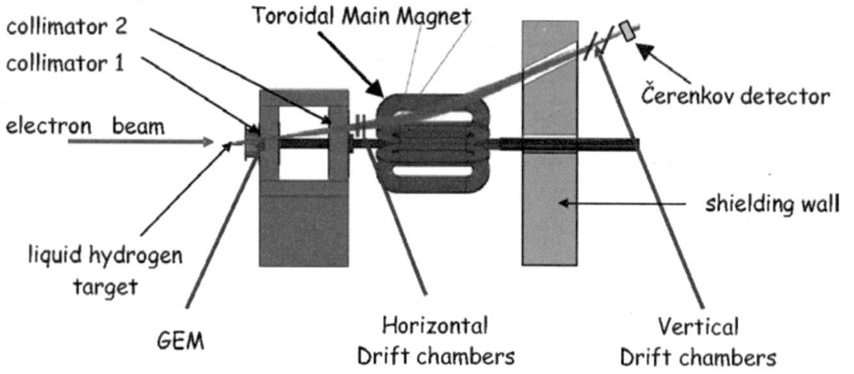

Fig. 1. Conceptual design of the Q^p_{weak} tracking system. The toroidal magnetic field spatially separates inelastic events from elastic events which will be focus on a fused silica (quartz) Čerenkov detector per octant located in the focal plane of the spectrometer. The tracking for single particle will be performed at low beam current using a set of drift chambers at three locations in two opposite octants at a time

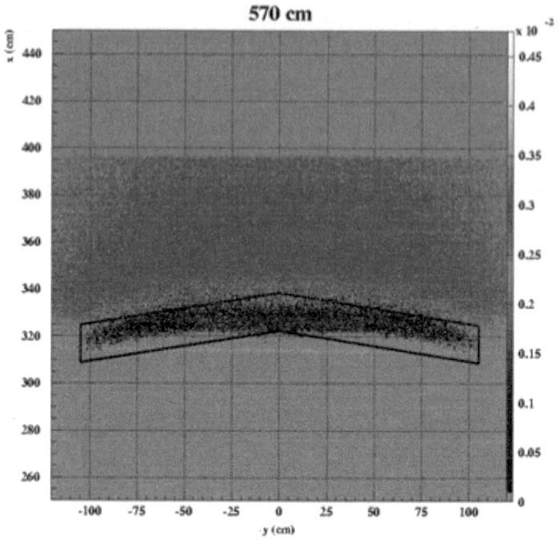

Fig. 2. GEANT simulation of the spatial separation of the elastic and inelastic events in the focal plane. The Λ-shaped Čerenkov detector made from joint quartz bars will cover/detect the elastic events with a $\simeq$0.02% inelastic rate contribution.

the tracking chambers will be mounted on a rotating wheel assembly. The Čerenkov detectors are operated in a low beam current counting mode (I$\sim$10 nA,) equivalent to an e-p elastic rate of $\sim$50 kHz/octant) for background/acceptance studies using the tracking system, and in a classic integrating mode (I$\sim$180 μA, equivalent to an e-p elastic rate of $\sim$800 MHz/octant) during high beam current production running where the tracking system will be retracted.

A Triple GEM (Gas Electron Multiplication) chamber is situated between the primary and secondary collimator. A GEM was chosen because of its high rate capability, fast time response and good position resolution, but it does not provide any track angle information. Each chamber has an active area of 16 cm$\times$16 cm, consists of 3200 channels supporting a rate of 10 MHz and an expected position resolution of $\simeq$200 μm.

A pair of horizontal drift chambers (HDC) separated by 50 cm is situated between the secondary collimator and the main magnet. Each chamber has an active area of 53 cm$\times$49 cm, and consists of 374 signal wires. The expected position resolution will be $\simeq$200 μm. The HDCs in combination with the GEM will measure the scattering angle, interaction vertex, and establish the trajectory upstream of the main magnet. The expected vertex resolution along the target will be $\simeq$1 mm ; the reconstruction of the scattering angle will be archived with an angular resolution of $\simeq$0.6 mrad. The HDCs will see both elastic and inelastic events and are in direct line of sight to the target. A low field toroidal sweeping magnet will be installed between the GEMs and the HDCs to prevent Møller events ($\sim$30-70 MeV, $\sim$10 MHz) from reaching the HDCs. The sweeping magnet will remain on during production run to maintain the same conditions between production running and Q^2 calibration runs.

A pair of vertical drift chambers (VDC) separated by 30 cm will be situated in front of the Čerenkov detectors, after the shielding wall. Each chamber has an active area of 200 cm$\times$ 50 cm, and consists of 800 signal wires. The expected position resolution will be $\sim$100 μm. The VDCs will map out the position-dependent Čerenkov detector response across the surface of each quartz bar. With a known field map of the main and sweep magnets and with the local track information of GEMs, HDCs, and VDCs it is possible to reconstruct the particle momentum. This allows the separation between elastic and inelastic track events needed for the mapping and background measurements. The global tracking in a toroidal magnetic fields is not a trivial issue since with a non-homogeneous field there is no simple analytic expression to describe the track. The Q^p_{weak} main magnet has a moderate focusing for electrons regarding the polar scattering angle but it has a strong defocusing in the azimuthal angle, so each track will have an unique trajectory parametrization. Therefore the final track fitting method will depend upon a iterative simulation and comparison method similar to existing algorithms used by BLAST and CLAS. In addition to the tracking technique for the background determination long calibration time-of-flight spectra out to 500 ns will be taken in order to verify that there are no long-lived backgrounds or delayed lights coming out of the quartz detectors that would require a correction to the central value of the measured asymmetry.

References

1. Qweak Collaboration, R. Carlini et al.: Jefferson Lab Proposal E-02-020, available at: `http://www.jlab.org/qweak/`
2. See G.R. Smith, in these proceedings, *Qweak: A Precision Measurement of the Proton's Weak Charge*

Eur Phys J A (2005) **24**, s2, 161–164
DOI: 10.1140/epjad/s2005-04-041-y

EPJ A direct
electronic only

Neutral currents and strangeness of the nucleon from the NuTeV experiment

Kevin S. McFarland[a]

University of Rochester, Rochester, NY 14610, USA

Received: 15 December 2004 / Published Online: 8 February 2005

Abstract. The NuTeV neutrino experiment ran in 1997-1998 at Fermilab and accumulated the world's highest statistics samples of high energy (20-300 GeV) separated neutrino and anti-neutrino interactions. The NuTeV collaboration has used this data to extract the electroweak parameter, $\sin^2\theta_W$, from the measurement of the ratios of neutral current to charged current neutrino and antineutrino deep inelastic scattering cross sections. This result, though consistent with previous neutrino electroweak measurements, is not consistent with predictions. One interpretation involves the possibility that the strange quark sea carries significantly more momentum in the nucleon than the anti-strange sea. We report on the direct study of this possibility from measurements of charged-current interactions on strange quarks in our neutrino and anti-neutrino beams.

PACS. 12.15.Mn Neutral Currents – 13.60.Hb Cross-sections in inelastic lepton-hadron scattering

1 Introduction and motivation

Neutrino scattering played a key role in establishing the structure of the Standard Model of electroweak unification, and it continues to be one of the most precise probes of the weak neutral current available experimentally today. With the availability of copious data from the production and decay of on-shell Z and W bosons for comparison, contemporary neutrino scattering measurements serve to validate the theory over many orders of magnitude in momentum transfer and provide one of the most precise tests of the weak couplings of neutrinos. In addition, precise measurements of weak interactions far from the boson poles are inherently sensitive to processes beyond our current knowledge, including possible contributions from leptoquark and Z' exchange [1] and new properties of neutrinos themselves [2].

The ratio of neutral current to charged current cross-sections for either ν or $\overline{\nu}$ scattering from isoscalar targets of u and d quarks can be written as [3]

$$R^{\nu(\overline{\nu})} \equiv \frac{\sigma(\overset{(-)}{\nu} N \to \overset{(-)}{\nu} X)}{\sigma(\overset{(-)}{\nu} N \to \ell^{-(+)} X)} = (g_L^2 + r^{(-1)} g_R^2), \quad (1)$$

where

$$r \equiv \frac{\sigma(\overline{\nu} N \to \ell^+ X)}{\sigma(\nu N \to \ell^- X)} \sim \frac{1}{2}, \quad (2)$$

and $g_{L,R}^2 \equiv (\epsilon_{L,R}^u)^2 + (\epsilon_{L,R}^d)^2$, isoscalar combinations of quark chiral couplings to the Z. Many corrections to 1 are required in a real target [4], but those most uncertain result from the suppression of the production of charm quarks in the target, which is the CKM-favored final state for charged-current scattering from the strange sea. This uncertainty has limited the precision of previous measurements of electroweak parameters in neutrino-nucleon scattering [5,6,7]. One way to reduce the uncertainty on electroweak parameters is to measure the observable

$$\begin{aligned} R^- &\equiv \frac{\sigma(\nu_\mu N \to \nu_\mu X) - \sigma(\overline{\nu}_\mu N \to \overline{\nu}_\mu X)}{\sigma(\nu_\mu N \to \mu^- X) - \sigma(\overline{\nu}_\mu N \to \mu^+ X)} \\ &= \frac{R^\nu - r R^{\overline{\nu}}}{1-r} = (g_L^2 - g_R^2), \end{aligned} \quad (3)$$

first suggested by Paschos and Wolfenstein [8] and valid under the assumption of equal momentum carried by the u and d valence quarks in the target. Since $\sigma^{\nu q} = \sigma^{\overline{\nu}\,\overline{q}}$ and $\sigma^{\overline{\nu} q} = \sigma^{\nu \overline{q}}$, the effect of scattering from sea quarks, which are symmetric under the exchange $q \leftrightarrow \overline{q}$, cancels in the difference of neutrino and anti-neutrino cross-sections. Therefore, the suppressed scattering from the strange sea does not cause large uncertainties in R^-. R^- is more difficult to measure than R^ν, primarily because the neutral current scatterings of ν and $\overline{\nu}$ yield identical observed final states which can only be distinguished through *a priori* knowledge of the initial state neutrino.

The experimental details and theoretical treatment of cross-sections in the NuTeV electroweak measurement are described in detail elsewhere [4]. In brief, we measure the experimental ratio of neutral current to charged current candidates in both a neutrino and anti-neutrino beam. A Monte Carlo simulation is used to express these exper-

[a] presenting on behalf of the NuTeV Collaboration

imental ratios in terms of fundamental electroweak parameters. This procedure implicitly corrects for details of the neutrino cross-sections and experimental backgrounds. For the measurement of $\sin^2\theta_W$, the sensitivity arises in the ν beam, and the measurement in the $\overline{\nu}$ beam is the control sample for systematic uncertainties, as suggested in the Paschos-Wolfenstein R^- of 3.

1.1 QCD corrections

Equations 1 and 3 assume targets symmetric under the exchange of u and d quarks, and that quark seas consist of quarks and anti-quarks with identical momentum distributions. The NuTeV analysis corrects for the significant asymmetry of d and u quarks that arises because the NuTeV target, which is primarily composed of iron, has a $5.74 \pm 0.02\%$ fractional excess of neutrons over protons. However, this correction is made under the assumption of isospin symmetry,

$$\overset{(-)}{u}_p(x) = \overset{(-)}{d}_n(x),\ \overset{(-)}{d}_p(x) = \overset{(-)}{u}_n(x)$$

This assumption, if significantly incorrect, could produce a sizable effect in the NuTeV extraction of $\sin^2\theta_W$ [9,10,11, 12]. Similarly, the cancellation of charm production from the strange quarks (3) assumes that the momentum distributions of the strange and anti-strange seas are identical, i.e., $s(x) = \overline{s}(x)$. NuTeV's analysis is done to leading order in pQCD, but perhaps surprisingly, the NLO corrections to R^- are very small [13,14,15].

Dropping the assumptions of symmetric heavy quark seas and isospin symmetry, but assuming small deviations in all cases, the effect of these deviations on R^- is [18]:

$$\delta R^- \approx \frac{(U_p - \overline{U}_p - D_n + \overline{D}_n) - (D_p - \overline{D}_p - U_n + \overline{U}_n)}{2(U_p - \overline{U}_p + D_p - \overline{D}_p)} \times (3\Delta_u^2 + \Delta_d^2) + \frac{S_p - \overline{S}_p}{U_p - \overline{U}_p + D_p - \overline{D}_p}(2\Delta_d^2 - 3(\Delta_d^2 + \Delta_u^2)\epsilon_c), \quad (4)$$

where $\Delta_{u,d}^2 = (\epsilon_L^{u,d})^2 - (\epsilon_R^{u,d})^2$, Q_N is the total momentum carried by quark type Q in nucleon N, and ϵ_c denotes the ratio of the scattering cross section from the strange sea including kinematic suppression of heavy charm production to that without kinematic suppression.

NuTeV does not exactly measure R^-, in part because it is not possible experimentally to measure neutral current reactions down to zero recoil energy. To parameterize the exact effect of the symmetry violations above, we define the functional $F[\sin^2\theta_W, \delta; x]$ such that

$$\Delta\sin^2\theta_W = \int_0^1 F[\sin^2\theta_W, \delta; x]\, \delta(x)\, dx, \quad (5)$$

for any symmetry violation $\delta(x)$ in PDFs. All of the details of the NuTeV analysis are included in the numerical evaluation of the functionals shown in Fig. 1. For this analysis,

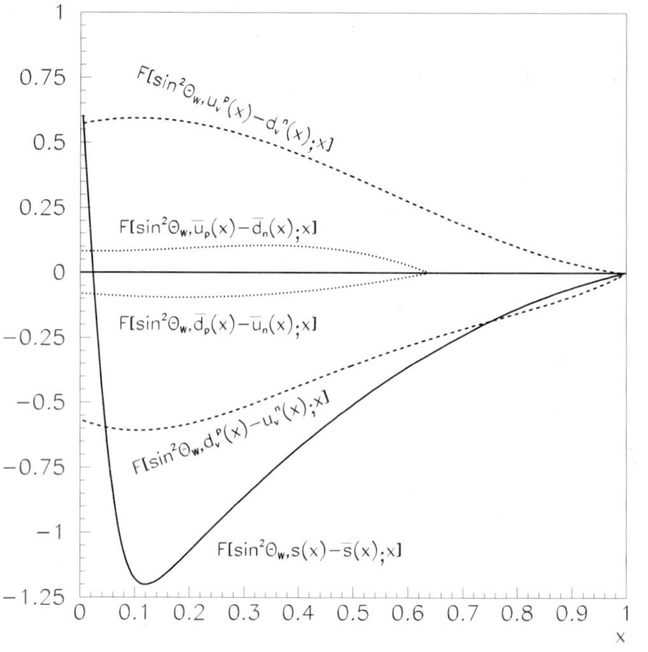

Fig. 1. The functionals describing the shift in the NuTeV $\sin^2\theta_W$ caused by not correcting the NuTeV analysis for isospin violating u and d valence and sea distributions or for $\langle s(x)\rangle \neq \langle \overline{s}(x)\rangle$. The shift in $\sin^2\theta_W$ is determined by convolving the asymmetric momentum distribution with the plotted functional

it can be seen that the level of isospin violation required to shift the $\sin^2\theta_W$ measured by NuTeV to its standard model expectation would be, e.g., $D_p - U_n \sim 0.01$ (about 5% of $D_p + U_n$), and that the level of asymmetry in the strange sea required would be $S - \overline{S} \sim +0.007$ (about 30% of $S + \overline{S}$).

2 Electroweak results

As a test of the electroweak predictions for neutrino nucleon scattering, NuTeV performs a single-parameter fit to $\sin^2\theta_W$ with all other parameters assumed to have their standard values, e.g., standard electroweak radiative corrections with $\rho_0 = 1$. This fit determines

$$\sin^2\theta_W^{(\text{on-shell})} = 0.22773 \pm 0.00135(\text{stat.}) \pm 0.00093(\text{syst.}) - 0.00022 \times (\frac{M_{\text{top}}^2 - (175\ \text{GeV})^2}{(50\ \text{GeV})^2}) + 0.00032 \times \ln(\frac{M_{\text{Higgs}}}{150\ \text{GeV}}). \quad (6)$$

A fit to the precision electroweak data, excluding neutrino measurements, predicts a value of 0.2227 ± 0.00037 [16, 17], approximately 3σ from the NuTeV measurement. In the on-shell scheme, $\sin^2\theta_W \equiv 1 - M_W^2/M_Z^2$, where M_W and M_Z are the physical gauge boson masses; therefore, this result implies $M_W = 80.14 \pm 0.08$ GeV Although this deviation is statistically significant, it is not immediately apparent what the cause of this discrepancy might be.

We discuss, in turn, possibilities of new physics, nuclear effects, large isospin violation and an asymmetric strange sea.

2.1 New physics

The primary motivation for embarking on the NuTeV measurement was the possibility of observing hints of new physics in a precise measurement of neutrino-nucleon scattering. NuTeV is well suited as a probe of non-standard physics for two reasons: first, the precision of the measurement is a significant improvement, most noticeably in systematic uncertainties, over previous measurements [5,6,7], and second NuTeV's measurement has unique sensitivity to new processes when compared to other precision data. In particular, NuTeV probes weak processes far off-shell, and thus is sensitive to other tree level processes involving exchanges of heavy particles. Also, the initial state particle is a neutrino, and neutrino couplings are the most poorly constrained by the Z^0 pole data, since they are primarily accessed *via* the measurement of the Z invisible width. An often useful low-energy parameterization of new physics is to consider a unit-coupling "contact interaction" in analogy with the Fermi effectively theory of low-energy weak interactions. Assuming a contact interaction described by a Lagrangian of the form

$$-\mathcal{L} = \sum_{H_q \in \{L,R\}} \frac{\pm 4\pi}{\left(\Lambda^{\pm}_{LH_q}\right)^2} \times \{\overline{l_L}\gamma^\mu l_L \overline{q_{H_q}}\gamma_\mu q_{H_q} + l_L\gamma^\mu \overline{l_L}\,\overline{q_{H_q}}\gamma_\mu q_{H_q} + \text{C.C.}) \quad (7)$$

the NuTeV result can be explained by an interaction with mass scale $\Lambda^{+}_{LL} \approx 4 \pm 0.8$ TeV. However, *post hoc* it appears that well motivated and complete models for such an interaction seem to be difficult to find [12].

An extra $U(1)$ gauge group giving rise to a heavy Z' boson is a possibility [12,19], but the $U(1)$ gauge group suggested by NuTeV would not necessarily be one motivated by models of unification of known forces [1]. There are few other precision measurements of neutrino neutral current interactions. Measurements of neutrino-electron scattering from the CHARM II experiment [20] and the direct measurement of $\Gamma(Z \to \nu\overline{\nu})$ from the observation of $Z \to \nu\overline{\nu}\gamma$ at the Z^0 pole [16] provide measurements of a few percent precision. The two most precise measurements come from the inferred Z invisible width [16] and the NuTeV result. Both of the precise rate measurements are significantly below the expectation. Again, models how such a deviation of neutrino couplings might fit against other constraints from the data are difficult to find, and the most complete attempts come from models which mix heavy and light neutrinos to form the eigenstates of the neutral weak interaction [21].

2.1.1 Nuclear Effects

It has been suggested by several authors, correctly, that differences in nuclear effects between charged and neutral current neutrino scattering could affect the NuTeV $\sin^2\theta_W$ analysis. There are constraints on process dependent nuclear effects, notably the agreement between F_2 from charged-lepton and neutrino charged-current scattering [22]. However, effects such Vector Meson Dominance in shadowing [23] or models of anti-shadowing [24] could still affect the NuTeV result. The former model predicts large increases in R^ν and $R^{\overline{\nu}}$ as measured by NuTeV of 0.6% and 1.2%, respectively. This effect, however, largely cancels in R^-, and furthermore the NuTeV $\sin^2\theta_W$ data itself disfavors this model through its separate measurements of R^ν and $R^{\overline{\nu}}$, which are both below predictions, while this model *increases* those very predictions. The latter model of anti-shadowing would primarily effect $R^{\overline{\nu}}$ and increase it modestly. This effect would be consistent with, though not favored by, the NuTeV data, and although it could reduce slightly the measured $\sin^2\theta_W$ from R^-, it would not improve the overall agreement of the NuTeV R^ν and $R^{\overline{\nu}}$ with the data.

2.2 Isospin violation

Several recent classes of models predict isospin violation in the nucleon [9,10,11]. The earliest estimation in the literature, a bag model calculation [9], predicts large valence asymmetries of opposite sign in $u_p - d_n$ and $d_p - u_n$ at all x, which would produce a shift in the NuTeV $\sin^2\theta_W$ of -0.0020. A more complete calculation done by Thomas *et al.* [10] concludes that asymmetries at very high x are larger, but the asymmetries at moderate x are smaller and of opposite sign at low x. This calculation is sensitive to the amount of smearing allowed in the energy of the remaining diquark at the bag scale after scattering, agreeing qualitatively with the Sather result with no smearing, but reducing the effect to a negligible -0.0001 when assuming smearing of order Λ_{QCD}. The effect is also evaluated in the Meson Cloud model [11], and there the asymmetries are much smaller at all x, resulting in a modest shift in the NuTeV $\sin^2\theta_W$ of $+0.0002$. Finally, Thorne *et al.* [25] have proposed that isospin violation may arise from QED corrections to PDFs, and have estimated the possible size of the effect in R^- to be ~ -0.002. However, this calculation is very sensitive to assumptions about the quark mass, and the value above assumes quark masses of a few MeV are appropriate. The assumption of constituent quark masses would drastically reduce the size but would retain the sign of the effect.

Models for isospin violation aside, the NuTeV data itself cannot provide a significant independent constraint on this form of isospin violation. A recent global analysis has also attempted to constrain this possibility, but found no sufficiently significant constraint. It allows isospin violation large enough to move NuTeV into agreement with prediction or large enough to double the discrepancy [26]. We conclude that NuTeV may have indeed found strong evidence for large (compared to even generous predictions) isospin violation in PDFs in a direction favored by most models, but that there is no independent evidence to support this hypothesis.

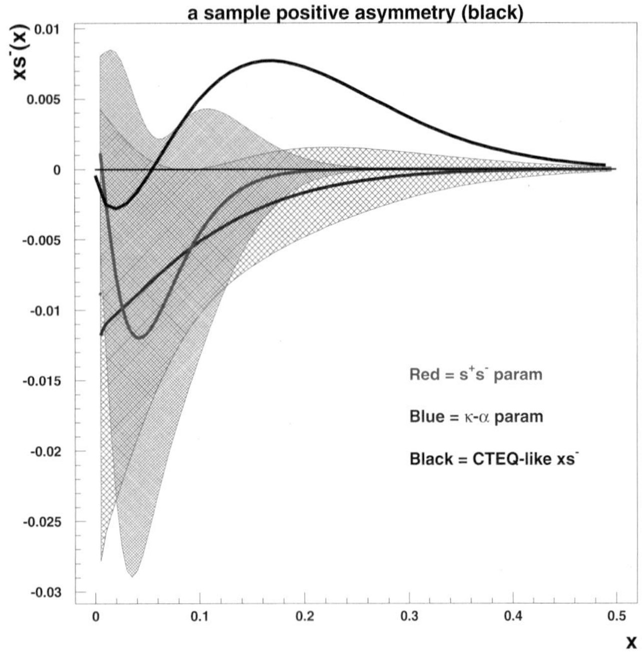

Fig. 2. NuTeV's results for the strange quark momentum asymmetry, $s(x) - \overline{s}(x)$, using different parameterizations. The "$\kappa - \alpha$" parameterization, allowing differences in magnitude and power in (1-x) is shown in *blue*; a parameterization suggested by CTEQ for s^+,s^- is shown in *red*; and a sample positive asymmetry using that parameterization with the central value of [31] is shown in *black*

2.2.1 Strange Sea Asymmetry

If the strange sea is generated by purely perturbative QCD processes, then neglecting electromagnetic effects, one expects $\langle s(x) \rangle = \langle \overline{s}(x) \rangle$. However, it has been noted that non-perturbative QCD effects can generate a significant momentum asymmetry between the strange and anti-strange seas [27,28,29,30]. Outside the context of the NuTeV electroweak data, measurements of this momentum asymmetry constrain the properties of the intrinsic strange sea of the nucleon and helps to discriminate among the models suggested above.

By measuring the processes $\nu_N, \overline{\nu}_N \to \mu^+\mu^- X$, the CCFR and NuTeV experiments constrain the difference between the momentum distributions of the strange and anti-strange seas. A recent analysis from the CTEQ collaboration [31] has claimed that this data favors a positive $S - \overline{S}$, perhaps large enough to explain one sigma of the NuTeV discrepancy. However, the NuTeV fully NLO QCD analysis [32], does not confirm these results and instead weakly prefers negative $S - \overline{S}$, as illustrated in Fig. 2. The constraint represented by this analysis, with an uncertainty that translates to less than one NuTeV standard deviation in $\sin^2\theta_W$, makes it unlikely that a positive $S - \overline{S}$ is responsible for what we observe.

Acknowledgements. We gratefully acknowledge support for this work from the U.S. Department of Energy, the National Science Foundation and the Alfred P. Sloan Foundation. The NuTeV experiment benefited greatly from significant contributions from the Fermilab Particle Physics, Computing, Technical and Beams Divisions.

References

1. P. Langacker et al.: Rev. Mod. Phys. **64**, 87 (1991)
2. K.S. McFarland, D. Naples et al.: Phys. Rev. Lett. **75**, 3993 (1995)
3. C.H. Llewellyn Smith: Nucl. Phys. B **228**, 205 (1983)
4. G.P. Zeller et al.: Phys. Rev. Lett. **88**, 091802 (2002)
5. K.S. McFarland et al. Eur. Phys. Jour. **C1**, 509 (1998)
6. A. Blondel et al.: Zeit. Phys. C **45**, 361 (1990)
7. J. Allaby et al. Zeit. Phys. C **36**, 611 (1985)
8. E.A. Paschos, L. Wolfenstein: Phys. Rev. D **7**, 91 (1973)
9. E. Sather: Phys. Lett. B **274**, 433 (1992)
10. E.N. Rodionov, A.W. Thomas, J.T. Londergan: Mod. Phys. Lett. A **9**, 1799 (1994)
11. F. Cao, A.I. Signal: Phys. Rev. C **62**, 015203 (2000)
12. S. Davidson, S. Forte, P. Gambino, N. Rius, A. Strumia: hep-ph/0112302
13. K. McFarland, S.-O. Moch: Proceedings of Mini-Workshop on Electroweak Precision Data and the Higgs Mass, hep-ph/0306052
14. S. Kretzer: Procedings of the Rencontres de Moriond on QCD and High-Energy Hadronic Interactions, hep-ph/0405221
15. B. Dobrescu, R.K. Ellis: Phys. Rev. D **69**, 114014 (2004)
16. "A Combination of Preliminary Electroweak Measurements and Constraints on the Standard Model", CERN-EP/2001-98, hep-ex/0112021
17. M. Grünewald: private communication, for the fit of [16] without neutrino-nucleon scattering data included
18. G.P. Zeller et al.: "On the effect of asymmetric strange seas and isospin-violating parton distribution functions on $\sin^2\theta_W$ measured in the NuTeV experiment," hep-ex/0203004
19. E. Ma, D.P. Roy: Phys. Rev. D **65**, 075021 (2002)
20. P. Vilain et al: Phys. Lett. B **335**, 248 (1994)
21. W. Loinaz, N. Okamura, T. Takeuchi, L.C.R. Wijewardhana: Phys. Rev. D **67**, 073012 (2003)
22. U.K. Yang et al., [CCFR/NuTeV Collaboration]: Phys. Rev. Lett. **86**, 2742 (2001)
23. G.A. Miller, A.W. Thomas: hep-ex/0204007
24. S. Brodsky, I. Schmidt, J.-Y. Yang: SLAC-PUB-9677, USM-TH-136, hep-ph/0409279
25. A. Martin, R. Roberts, W. Stirling, R. Thorne: IPPP-04-62, DCPT-04-124, CAVENDISH-HEP-2004-28, hep-ph/0411040
26. R.S. Thorne (Cambridge U.): Int. J. Mod. Phys. A **19**, 1074 (2004)
27. A.I. Signal, A.W. Thomas: Phys. Lett. B **191**, 205 (1987)
28. M. Burkardt, B.J. Warr: Phys. Rev. D **45**, 958 (1992)
29. S. Brodsky, B. Ma: Phys. Lett. B **381**, 317 (1996)
30. W. Melnitchouk, M. Malheiro: Phys. Lett. B **451**, 224 (1999)
31. S. Kretzer et al.: Phys. Rev. Lett. **93**, 041802 (2004); F. Olness et al.: MSU-HEP-030701, BNL-NT-03-17, RBRC-329, hep-ph/0312323
32. D. Mason [NuTeV Collaboration]: Proceedings of 39th Rencontres de Moriond on QCD and High-Energy Hadronic Interactions, hep-ex/0405037

V Hadronic structure. . . and more

V-2 PV in nuclear systems

Eur Phys J A (2005) **24**, s2, 167–170
DOI: 10.1140/epjad/s2005-04-042-x

EPJ A direct
electronic only

Parity violation in astrophysics

C.J. Horowitz

Nuclear Theory Center and Department of Physics, Indiana University, Bloomington, IN 47405, USA

Received: 15 October 2004 / Published Online: 8 February 2005

Abstract. Core collapse supernovae are gigantic explosions of massive stars that radiate 99% of their energy in neutrinos. This provides a unique opportunity for large scale parity or charge conjugation violation. Parity violation in a strong magnetic field could lead to an asymmetry in the neutrino radiation and recoil of the newly formed neutron star. Charge conjugation violation in the neutrino-nucleon interaction reduces the ratio of neutrons to protons in the neutrino driven wind above the neutron star. This is a problem for r-process nucleosynthesis in this wind. On earth, parity violation is an excellent probe of neutrons because the weak charge of a neutron is much larger than that of a proton. The Parity Radius Experiment (PREX) at Jefferson Laboratory aims to precisely measure the neutron radius of ^{208}Pb with parity violating elastic electron scattering. This has many implications for astrophysics, including the structure of neutron stars, and for atomic parity nonconservation experiments.

PACS. 25.30.Bf Elastic electron scattering – 11.30.Er Charge conjugation, parity, time reversal, and other discrete symmetries – 26.60+c Nuclear matter aspects of neutron stars – 97.60.Bw Supernovae

1 Introduction

Can there be large scale parity violation in astrophysics? For example, the 1960s science fiction movie *Journey to the Far Side of the Sun* visits a parity double of earth. In the real world, core collapse supernovae provide a unique opportunity for macroscopic parity violation. These gigantic explosions of massive stars are dominated by neutrinos that transport 99% of the energy. This is because no other known particle can diffuse through the very dense matter of the collapsed star. In Sect. 2 we discuss how the weak interactions of these neutrinos may lead to large scale parity or charge conjugation violation.

On earth, parity violation can be used to probe neutrons because the weak charge of a neutron is much larger than that of a proton. In Sect. 3 we describe the Parity Radius Experiment (PREX) at Jefferson Laboratory that aims to use parity violating electron scattering to measure the neutron radius in ^{208}Pb. This measurement has many implications for astrophysics including properties of neutron stars, nuclear structure, and atomic parity nonconservation.

2 Parity violation in supernovae

In this section we explore large scale parity or charge conjugation violation in core collapse supernovae. When the center of a massive star has burned to ^{56}Fe, thermonuclear reactions cease and the core of the star collapses, in a fraction of a second, all the way to nuclear densities. This forms a proto-neutron star with half the mass of the sun and a radius of order 10 kilometers. The proto-neutron star is initially hot and lepton rich. Over a period of seconds, its very large gravitational binding energy of over 100 MeV/nucleon is radiated away in neutrinos. Note, that any electromagnetic or strongly interacting particles diffuse very slowly at these high densities. This immense neutrino burst involves 10^{58} neutrinos and carries about 3×10^{53} ergs. In an historic first, about 20 neutrinos were detected from SN1987A.

The proto-neutron star is so dense that even weakly interacting neutrinos diffuse. They interact thousands of times before leaving the star. Could there be a signature of all of these weak interactions? One possibility is parity violation in a strong magnetic field. This may lead to an asymmetry in the number of neutrinos emitted along the field compared to those emitted against the field, see Fig. 1. The neutrinos carry away so much momentum that a few % asymmetry in the neutrino radiation can lead to a recoil of the neutron star at hundreds of kilometers per second [1]. Indeed, many neutron stars are observed to have large galactic velocities that presumably came from kicks at birth of 500 km/s or larger. Unfortunately, explicit calculations of the neutrino asymmetry from parity violation have proved complicated and uncertain. A very strong magnetic field, of order 10^{15} gauss or more, may be required to produce the few percent asymmetry that can explain the observed velocities of neutron stars.

Charge conjugation (C) violation is closely related to parity (P) violation. In the standard model, the product CP is approximately conserved so the large P violation

Send offprint requests to: e-mail: horowit@indiana.edu

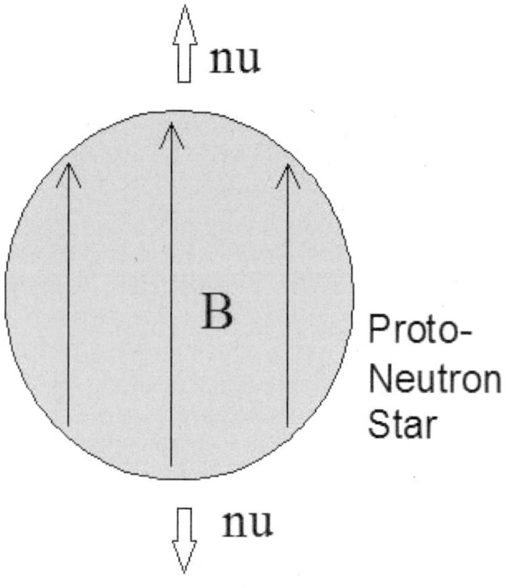

Fig. 1. Parity violation in a strong magnetic field could lead to an up/ down asymmetry in the neutrino flux

implies large C violation. As a result of C violation, neutrino nucleon cross sections are systematically larger than antineutrino-nucleon cross sections. This changes the composition of the neutrino driven wind during a supernova and may be important for nucleosynthesis. The intense neutrino flux blows some baryons off of the neutron star and into a high entropy wind. As this wind cools, the nucleons can condense to form heavy nuclei. This wind is the most promising site for r-process nucleosynthesis. In the r-process, seed nuclei rapidly capture several neutrons to form about half of the elements heavier than Fe.

The ratio of neutrons to protons in the wind is set by the rates of neutrino and antineutrino capture.

$$\nu_e + n \to p + e^- \tag{1}$$

$$\bar{\nu}_e + p \to n + e^+ \tag{2}$$

Because of C violation, the cross section for 2 is significantly smaller than the cross section for 1. This decreases the ratio of neutrons to protons in the wind by 20%. As a result, the wind is unlikely to be significantly neutron rich [2] and present simulations do not have enough neutrons to synthesize all of the r-process elements. Furthermore, it appears very difficult to change conditions to make the wind significantly neutron rich. This is an important problem for r-process nucleonsynthesis in the neutrino driven wind. Possible alternative sites, although these also have problems, include neutron star mergers and accretion disks around black holes. The site of the r-process remains an important open problem in nuclear astrophysics.

3 The parity radius experiment (PREX)

Parity violation is a uniquely clean probe of neutron densities. This is because the weak charge of a neutron is much larger than that of a proton. Determining neutron densities of heavy nuclei has important implications for the neutron rich matter that is present in many astrophysical objects. In this section we describe the Parity Radius Experiment (PREX) at Jefferson Laboratory to measure parity violation in elastic electron scattering from ^{208}Pb, and mention some of its implications.

The parity violating asymmetry A for elastic electron scattering provides a purely electroweak probe of neutron densities. In Born approximation, A is

$$A = \frac{G_F Q^2}{4\pi 2^{1/2}\alpha} F_W(Q^2)/F_{ch}(Q^2), \tag{3}$$

and involves the ratio of the weak form factor $F_W(Q^2)$ to the charge form factor $F_{ch}(Q^2)$. The charge form factor is just the Fourier transform of the charge density at the momentum transfer Q of the experiment and is known from electron scattering. Likewise the weak form factor is the Fourier transform of the weak charge density. This directly gives the neutron density since most of the weak charge is carried by neutrons. At low momentum transfers A is very sensitive to the difference in neutron and proton root mean square radii.

A heavy nucleus such as ^{208}Pb is expected to have a neutron rich skin because of the neutron excess and because of the Coulomb barrier that prevents protons from being at large distances. The thickness of this skin has many implications for the properties of neutron rich matter as we discuss below. However the skin thickness has never been cleanly measured because all hadronic probes of neutron densities have significant uncertainties from the strong interaction. Parity violation provides a very clean way to measure the neutron skin thickness.

The Parity Radius Experiment (PREX) is proposed at Jefferson Laboratory to measure A for elastic scattering of 850 MeV electrons off ^{208}Pb at a scattering angle of six degrees [3]. The goal is to determine A to $\pm 3\%$. This allows one to determine the neutron rms radius to 1% (± 0.05 fm) [4]. Because it is a purely electro-weak reaction, the theoretical interpretation of the results is very clean.

3.1 Coulomb distortions and transverse analyzing power

A heavy nucleus has a large charge Z that will significantly distort the electron waves. This invalidates the simple Born approximation formula in 3. However, Coulomb distortions can be included exactly by numerically solving the Dirac equation for an electron moving in the Coulomb potential plus the very small (axial-vector) weak potential [5]. This involves summing over a very large number of partial waves, because of the long range of the Coulomb potential. However, there are standard techniques for improving the convergence of this sum. We find that Coulomb distortions reduce A by about 30% [5]. Since the experimental goal is a 3% measurement it is very important that distortions are accurately included.

One way to check the distortions is to look at the parity allowed transverse analyzing power A_y. If the initial

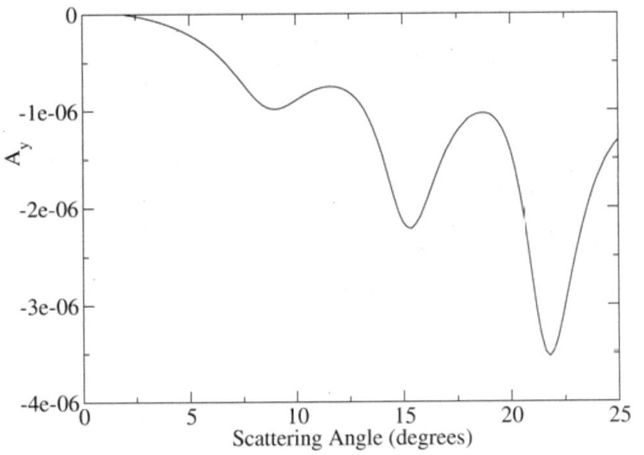

Fig. 2. Analyzing power A_y for elastic scattering of 850 MeV electrons from ^{208}Pb versus laboratory scattering angle

electron spin is transverse to the scattering plane, the analyzing power A_y gives the asymmetry for scattering to the left compared to scattering to the right. Time reversal invariance requires A_y to vanish in Born approximation. However, A_y is nonzero when Coulomb distortions are included. Therefore, A_y provides a direct test of the same Coulomb distortions. Furthermore, a nonzero A_y could be an important source of systematic errors. This is because the beam may have a small residual transverse polarization, and this coupled with A_y could lead to a false asymmetry that changes sign with the beam helicity. There is considerable interest in A_y for electron-proton scattering [6].

In Fig. 2 we show the transverse asymmetry A_y for the scattering of 850 MeV electrons off ^{208}Pb. This is based on an exact numerical solution of the Dirac equation that sums up any number of photon exchanges. The nucleus is assumed to remain in its ground state at all times. Intermediate states involving excited states of Pb are not included. However, the elastic scattering is coherent and leads to a cross section proportional to Z^2. Each intermediate state, involving the excitation of a given proton, is not coherent and contributes a cross section of order unity. Even when one sums over the Z protons this only gives a total contribution of order Z and this is small compared to the Z^2 coherent contribution. Therefore our elastic approximation should be good for a heavy nucleus like ^{208}Pb. We find that A_y is relatively large, comparable in magnitude to A and has significant structure in diffraction minima. It would be useful to measure A_y using a modest amount of running with transversely polarized beam. This can be done during PREX. Note, that A_y grows with Z because it is sensitive to Coulomb distortions. Therefore A_y is smaller for lighter nuclei. We will present more A_y results in a future publication [7].

3.2 Atomic parity nonconservation and PREX

Atomic parity nonconservation experiments provide important low energy tests of the standard model. Parity violation involves the overlap of atomic electrons with the weak charge density of the nucleus and this is primarily carried by the neutrons. Therefore, precision atomic experiments are sensitive to the neutron radius. In the future, the most accurate low energy parity test of the standard model may involve the combination of an atomic experiment and PREX to constrain the neutron density [4]. Note, that the Colorado Cs experiment is presently limited by atomic theory uncertainties in the electron density at the nucleus [8]. One way to reduce these electron wave function uncertainties is to measure the ratio of parity violation in two different isotopes. Such a ratio measurement can be very sensitive to the neutron radii of the isotopes. These radii can be constrained by PREX or future parity violating electron scattering experiments.

3.3 The equation of state of neutron rich matter, PREX, and neutron stars

Measuring the neutron radius in Pb constrains the Equation of State (EOS), pressure versus density, of neutron rich matter. The pressure forces neutrons in the skin of a heavy nucleus out against the surface tension. Therefore, the higher the pressure, the larger will be the neutron radius. Thus a measurement of the neutron radius in ^{208}Pb determines the pressure of neutron matter at just below normal nuclear density [9].

A neutron star is a gigantic nucleus, see for example [10] for an introduction. Its structure depends only on the equations of General Relativity and the EOS of neutron rich matter. The central density of a neutron star can be a few or more times the nuclear density. Therefore, measuring the radius of a neutron star, which is expected to be of order 10 km, will determine the EOS at high densities [11]. Astronomers are working hard to measure neutron star radii. One approach is to determine both the X-ray luminosity and surface temperature. Thermodynamics allows one to determine the surface area assuming black body radiation plus model corrections for non-black body effects in the atmosphere [12].

It is very interesting to compare the low density information on the EOS from the Pb radius with the high density information from a neutron star radius. There are many possible exotic phases of QCD at high densities. These include quark matter, strange matter, pion or kaon condensates, and color superconductivity. An exotic phase could show up as an abrupt softening of the EOS with increasing density. Note, if an exotic phase has a stiff EOS (high pressure) then it will not be thermodynamically favored.

The following scenario would provide an exciting indication of an exotic high density phase. PREX could measure a large Pb radius. This would show that the EOS is stiff at low density. If the radius of a neutron star then turns out to be small, this shows that the high density EOS is relatively soft. Although these measurements would not, by themselves, determine the nature of the high density phase, they would strongly suggest interesting behavior. It has proved very difficult to find other signatures of an exotic phase in the center of neutron stars. Finally, we note that PREX has many other implications for the solid crust of neutron stars [13] and for how neutron stars cool [14].

3.4 Neutron rich nuclei and PREX

PREX also has important implications for the structure of neutron rich nuclei that can be studied in radioactive beams. The Rare Isotope Accelerator (RIA) would produce such nuclei, and their properties are interesting for several reasons. For example as discussed in Sect. 2, r-process nucleosynthesis involves the capture of many neutrons to produce very neutron rich nuclei.

The radius of Pb determines the density dependence of the symmetry energy $S(\rho)$. The symmetry energy describes how the energy of nuclear matter rises as one moves away from equal numbers of neutrons and protons. It is very important for the structure of neutron rich nuclei. The energy of pure neutron matter at density ρ, $E_{neutron}(\rho)$, is approximately,

$$E_{neutron}(\rho) \approx E_{nuclear}(\rho) + S(\rho), \tag{4}$$

where $E_{nuclear}$ is the energy of symmetric nuclear matter ($N = Z$) and is largely known. Determining the pressure of neutron matter by measuring the neutron radius gives $dE_{neutron}/d\rho$ and this gives $dS/d\rho$. Thus PREX will determine the density dependence of the symmetry energy which is poorly constrained at present.

4 Conclusions: Parity violation and astrophysics

Core collapse supernovae are dominated by neutrinos. This provides a unique opportunity for large scale parity or charge conjugation violation. Parity violation in a strong magnetic field could lead to an asymmetry in the neutrino radiation and recoil of the newly formed neutron star. Charge conjugation violation in the neutrino driven wind above a neutron star reduces the ratio of neutrons to protons. This is a large problem for r-process nucleosynthesis in the wind.

On earth, parity violation probes neutrons because the weak charge of a neutron is much larger than that of a proton. The Parity Radius Experiment (PREX) at Jefferson Laboratory aims to measure the neutron radius of ^{208}Pb to 1% by using parity violating elastic electron scattering. Because it is an electroweak reaction, the theoretical interpretation of this measurement is very clean. The neutron radius of Pb has many implications for atomic parity nonconservation experiments, neutron stars, and the structure of neutron rich nuclei.

Acknowledgements. Some of this work was done in collaboration with Jorge Piekarewicz. We thank E. D. Cooper for help on the A_y calculations. This work was supported in part by DOE grant DE-FG02-87ER40365.

References

1. C.J. Horowitz, J. Piekarewicz: Nucl. Phys. A **640**, 281 (1998)
2. C.J. Horowitz: Phys. Rev. D **65**, 083005 (2002)
3. P.A. Souder, R. Michaels, G. Urciuoli, spokespersons: Jefferson Laboratory Experiment E-03-011, See also http://hallaweb.jlab.org/parity/prex
4. C.J. Horowitz, S.J. Pollock, P.A. Souder, R. Michaels: Phys. Rev. C **63**, 025501 (2001); See also http://cecelia.physics.indiana.edu/prex
5. C.J. Horowitz: Phys. Rev. C **57**, 3430 (1998)
6. F.E. Maas et al.: nucl-ex/0410013
7. E.D. Cooper, C.J. Horowitz: to be published
8. Wood CS, Bennett SC, Cho D, Masterson BP, Roberts JL, Tanner CE, Wieman CE, Science **5307**, 1759 (1997)
9. B.A. Brown: Phys. Rev. Lett. **85**, 5296 (2000)
10. J.M. Lattimer, M. Prakash: Science **304**, 536 (2004)
11. J. Carriere, C.J. Horowitz, J. Piekarewicz: Astrophys. J. **593**, 463 (2003)
12. Jose Pons et al.: Astrophys. J. **564**, 981 (2002)
13. C.J. Horowitz, J. Piekarewicz: Phys. Rev. Lett. **86**, 5647 (2001)
14. C.J. Horowitz, J. Piekarewicz: Phys. Rev. C **66**, 055803 (2002)

Eur Phys J A (2005) **24**, s2, 171–174
DOI: 10.1140/epjad/s2005-04-043-9

EPJ A direct
electronic only

Parity violation in nuclear systems

B. Desplanques

Laboratoire de Physique Subatomique et de Cosmologie (UMR CNRS/IN2P3–UJF–INPG),
F-38026 Grenoble Cedex, France

Received: 15 October 2004 / Published Online: 8 February 2005

Abstract. Parity violation in nuclear systems is reviewed. A few ingredients relevant to the description of the parity-violating nucleon-nucleon force in terms of meson exchanges are reminded. Effects in nuclear systems are then considered. They involve *pp* scattering, some complex nuclei and the deuteron system.

PACS. 24.80.+y Nuclear tests of fundamental interactions and symmetries

1 Introduction

A large number of parity-non-conserving (pnc) effects has been observed in various nuclear systems. While their expected size at low energy is of the order of 10^{-7} (for the amplitude), they can be strongly enhanced in some cases, due to the closeness of states with opposite parities or the suppression of the regular transition. Thus, effects of the order of 10^{-1} have been measured in neutron-nucleus scattering in the vicinity of low energy p-wave resonances (see [1] for some review). Qualitatively, such effects are understood. However, little quantitative information could be obtained on the pnc component of nucleon-nucleon (NN) forces expected to account for them. From the knowledge of this interaction, one can expect to learn about the pnc meson-nucleon coupling constants which they depend on and, thus, get information on the underlying hadronic physics. This one is complementary to the information that can be obtained from non-leptonic hyperon decays. It concerns in first place the πNN coupling that has been at the center of many theoretical and experimental works. This one can be most easily compared to non-leptonic hyperon decay amplitudes. Another less fundamental but important motivation for the study of pnc nuclear effects is the necessity to determine the effective strength of the various pnc NN amplitudes. These ones can indirectly contribute to other pnc effects, especially in electron scattering mainly discussed at this meeting. Though the effect is not large, its knowledge is required to determine the reliability of the information that is looked for in such high accuracy measurements. Some recent developments in the field are reviewed here.

The plan of the paper is as follows. In the second part, we briefly remind ingredients entering the pnc NN force, while emphasizing a few points of interest for the following part devoted to pnc effects. The third section is concerned with pnc *pp* scattering. This process is the only one that provides a calibration of the strength of pnc NN forces at the present time. A few nuclear pnc effects in complex nuclei, especially in ^{18}F and in ^{133}Cs, are discussed in the fourth section. The fifth section is devoted to pnc effects in the *np* system, including the deuteron. This particular field has been particularly active these last years. A conclusion and an outlook are presented in the sixth section.

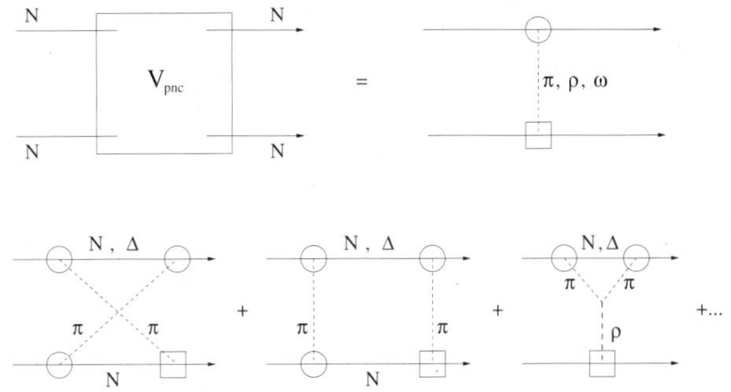

Fig. 1. Diagram representation of the pnc NN interaction

2 PNC NN potential: Ingredients

The pnc NN force is generally described as resulting from meson exchanges, π, ρ and ω. A diagrammatic representation is given in Fig. 1. One of the vertex, represented by a circle, corresponds to the strong interaction. The other one, represented by a box, corresponds to the weak, pnc, interaction. As isospin is not conserved, there are many couplings in some cases. They are:
- h^1_π, which governs the long range part of the force and necessarily involves the $\Delta T = 1$ component of the weak interaction
- $h^{0,1,2}_\rho$: $\Delta T = 0,\ 1,\ 2$
- $h^{0,1}_\omega$: $\Delta T = 0,\ 1$
- $h^{1'}_\rho$: $\Delta T = 1$, (different type ρNN coupling).

The vector-meson couplings determine the short-range part of the pnc NN force. As is well known, the contribution of this part is sensitive to short-range correlations in the strong NN interaction as well as to other correlations.

Many contributions going beyond the above ones have been considered in the literature. They involve for instance two-pion exchanges displayed in Fig. 1, either with the same coupling as for the one-pion exchange or with the same coupling as for the ρ exchange. In the last case, the ρ-exchange force acquires a longer range that could show up in the analysis of pnc effects in pp scattering. Some discussion and references could be found in [1].

At low energy, only gross features may be relevant. The NN interaction can then be parametrized by five $S \leftrightarrow P$ NN transition amplitudes [2]:
- $^1S_0 \leftrightarrow {}^3P_0$, $\Delta T = 0, 1, 2$, (pp, pn and nn forces)
- $^3S_1 \leftrightarrow {}^1P_1$, $\Delta T = 0$ (pn force)
- $^3S_1 \leftrightarrow {}^3P_1$, $\Delta T = 1$ (pn force).

It was shown that this description could be extended to higher energy by singularizing the pion-exchange contribution which, due to its long range, contributes sizeable $P \leftrightarrow D$ transition amplitudes [3]. Apart from the name, these works largely anticipated recent effective field-theory approaches [4], which also consider $P \leftrightarrow D$ transitions.

Many works have been devoted to the pnc meson-nucleon couplings, which enter NN interaction models. A large part of them, prior to the DDH work [5] or later, fit in this framework. Due to the lack of space, we again refer to [1] for references and detailed discussion. We only present here some estimates and make a few pertinent remarks. The sample of results given in Table 1 corresponds to the predictions of two significantly different models for the most relevant couplings, h^1_π, h^0_ρ and h^0_ω. They are based on a quark model (DDH), partly updated, and a chiral soliton model by Kaiser and Meissner (KM) [6] (see also [7]). Despite appearances, results turn out to be qualitatively similar. Discrepancies can be ascribed to the weight attributed to individual contributions in DDH. It is noticed that the dominant contribution to h^1_π is produced by strange quarks, of particular interest at this meeting while the consistency of this estimate with the QCD sum rules ones remains an open problem. It was proposed to use a chiral quark model to make a new estimate (Lee *et al.* [8], this conference). It is also noticed that DDH estimates, relying for a part on experimental data, should be less sensitive to "rescattering effects" evoked in the literature [9] whereas h^0_ω is likely to be negative.

Table 1. Meson-nucleon pnc coupling constants: a few estimates from different works. The question mark at the *last line* indicates that the original value could be actually close to 0

	DDH (range)	DDH("best")	KM
$10^7\ h^1_\pi$	$0 \leftrightarrow 11$	4.6	0.2
$10^7\ h^1_\pi$	$0 \leftrightarrow 2.5$ (update, $K = 3$)		0.8 - 1.3 (with $\bar{s}s$)
$10^7\ h^0_\rho$	$-31 \leftrightarrow 6$	-11	-4
$10^7\ h^0_\omega$	$-10 \leftrightarrow 6?$	-2	-6

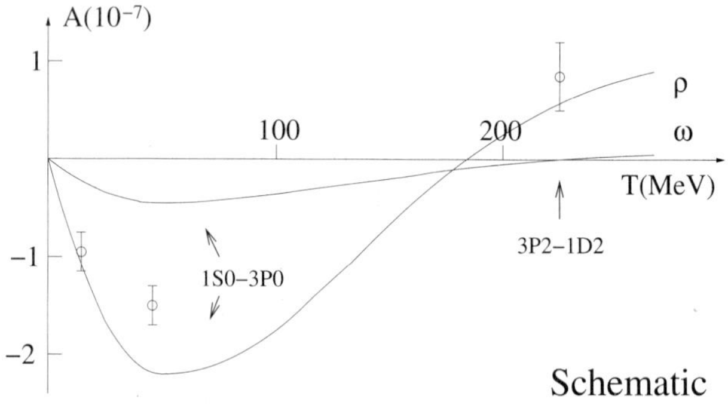

Fig. 2. Schematic representation of ρ- and ω-exchange contributions to the longitudinal asymmetry in pp scattering

3 Longitudinal asymmetry in pp scattering

The low-energy longitudinal asymmetry in pp scatering is the most important benchmark in the field at present. A complete theoretical analysis can be done. It shows that measurements at 13.6 and 45 MeV are in complete agreement with each other, thus fixing the strength of the $^1S_0 \leftrightarrow {}^3P_0$ pnc pp transition amplitude. For a given description of the strong interaction model, the strength of a combination of h^{pp}_ρ and h^{pp}_ω couplings or, in first approximation, the h^0_ρ and h^0_ω couplings, can be obtained. At higher energy, around 221 MeV, it was noticed that the contribution of the $^1S_0 \leftrightarrow {}^3P_0$ transition amplitude was vanishing, providing a window to determine the $^3P_2 \leftrightarrow {}^1D_2$ transition amplitude. By combining this measurement with the low-energy one, contributions due to ρ and ω exchanges can thus be disentangled.

A schematic representation of the two contributions with couplings close to the DDH "best guess" ones, is given in Fig. 2. It is seen that the ω-exchange one has a negligible contribution around 221 MeV while a ρ-exchange contribution alone is not doing badly. A better agreement is obtained by increasing the strength of this one and compensating for the overestimate at low energy by adding a ω-exchange contribution with a sign opposite to the DDH "best guess" or KM one. This provides a simple explanation for the couplings obtained by Carlson *et al.* [10]:

$$h^{pp}_\rho = -22.3\ 10^{-7},\ \ h^{pp}_\omega = +5.2\ 10^{-7}\ \ (\text{fit})$$
$$h^{pp}_\rho = -15.5\ 10^{-7},\ \ h^{pp}_\omega = -3.0\ 10^{-7},\ \ (\text{DDH}). \quad (1)$$

The fit evidences a striking feature as the value for the ω coupling has a sign opposite to expectations and, thus, can point to missing ingredients in predictions. We however notice that the significance of the result is not strong (the ρ alone is already giving a good account). A refined theoretical analysis and a more accurate measurement could be quite useful. It would be interesting to investigate for instance the role of a longer range ρ-exchange contribution mentioned in the second section.

4 Parity-non-conservation in complex nuclei

Many pnc effects in complex nuclei have been measured and analyzed. Considered individually, it is however difficult to draw some conclusion from them. Moreover, when they are considered together, it is not rare that the information obtained from one process is at the limit to contradict that one from another process. Two effects nevertheless deserve some attention: the circular polarization of photons emitted in the transition $0^-(1.08\,\mathrm{MeV}) \to 1^+(\mathrm{g.s.})$ in ^{18}F and the ^{133}Cs anapole moment. They are successively discussed in the following (references for both theory and experiment may be found in [1]).

The interest of the pnc effect in ^{18}F is that the calculation of the relevant pnc nuclear matrix element ($0^- \leftrightarrow 0^+$) can be checked by looking at the first forbiden β decay of the neighboring nucleus, ^{18}Ne. It implies the $\Delta T = 1$ part of the weak interaction and the measurement can thus provide information on the pnc πNN coupling, h^1_π. From the experimental limit of P_γ, one gets the following upper limit:

$$|h^1_\pi| \leq 1.3\,10^{-7}. \tag{2}$$

This result is supported by the absence of effect in two other processes in ^{21}Ne and in ^{93}Tc. In these cases, the contribution of the pion exchange is *a priori* large. To agree with the upper experimental limit, one has first to assume that the coupling h^1_π is not too large and, moreover, that the corresponding contribution be cancelled for a part by some isoscalar contribution (for ^{21}Ne). Contrary to ^{18}F, there is no available check on the relevant pnc nuclear matrix element. Accepting that this one be uncertain by up to a factor 3 would however give a limit on h^1_π similar to (2).

The ^{133}Cs anapole moment has been analyzed by different authors. To a large extent, this quantity involves a combination of the pnc NN force close to that one governing pnc effects in several odd-proton systems as different as $p\alpha$ scattering, ^{19}F, ^{41}K, ^{175}Lu, ^{181}Ta. At first sight, it appears that the above combination should be two times larger for the anapole moment than for the other processes. The discrepancy has the order of a typical uncertainty in the field but there was some belief that the estimate in the first case could be less uncertain than for other effects (for a part, it involves a long-range operator). On the other hand, the effects in the other odd-proton systems overdetermine the above combination of parameters. If these last processes are ignored, it appears that a large value of h^1_π, of the order of $1\,10^{-6}$, at the upper limit of the original DDH range, is needed. This is inconsistent both with the upper limit, (2), and the DDH updated range. We notice that the last calculation of the anapole moment [11] relies on an approximation that allows for an improved calculation in one respect but implies some contribution from orbitals below the Fermi level with a wrong sign in another respect. A correct account of these ones could enhance the theoretical estimate but will not reach a factor 2. The validity of a similar approximation, which omits 3-body terms, was discussed in [12].

5 Parity-non-conservation in the deuteron

Most recent pnc studies in nuclear systems have concentrated on the np system (deuteron and scattering state). This emphasis is largely motivated by both the feasibility of the corresponding experiments in a near future (see Stiliaris's talk at this conference) and a safer interpretability of possible effects. These ones include the photon-emission asymmetry in the thermal-energy radiative capture of polarized neutrons by protons, $\mathbf{n} + p \to d + \gamma$ [13,14,15,16], presently performed at LANSCE, the asymmetry in the deuteron photo-disintegration depending on the photon helicity [17,18,19,20], which could be performed at JLab, IASA, SPring-8, $\cdots$, the deuteron anapole moment [21,22, 23,24], the pnc deuteron electro-disintegration in relation with the SAMPLE experiment [25,16], and the longitudinal asymmetry and the neutron-spin rotation in np scattering [20]. Some earlier works could be quoted. The recent ones involve new methods (effective-field theories [13, 15]), improved NN interaction models (AV18+ $\cdots$ [15, 24,16,18,20]), more complete calculations (two-body currents [25,16]), and increased attention to gauge invariance [23,24,20]. A few remarks are made below about these different works.

The earlier pion-exchange contribution to the asymmetry in the thermal-neutron radiative capture on protons, $\mathbf{n} + p \to d + \gamma$,

$$A_\gamma = -0.11\, h^1_\pi, \tag{3}$$

is confirmed by recent estimates, indicating that the correction for a wrong $^1S_0$ np scattering length, was appropriately made. It is also found that the above estimate results from a strong cancellation when a calculation is performed without relying on the Siegert theorem [15,16]. Amazingly, this weak interaction problem provides information on the accuracy of effective-field theory methods employed for the strong interaction. The approach used in [13], for instance, overestimates (3) by 60% at leading order (almost a factor 2 for comparable ingredients).

On the basis of an estimate by Oka [26], it was thought that the study of the photon-helicity dependence of the deuteron photo-disintegration cross section could provide an alternative way to determine the coupling h^1_π. This motivated several works that disproved the above estimate and its main conclusion [17,18,19,20]. An account of the new results can be found in the Hyun's talk at this conference. For the inverse process near threshold ("Lobashov experiment"), it should be noticed that a circular polarization of photons as large as $1\ 10^{-7}$ is not excluded for some reasonable models of both the strong and the weak NN interaction [20].

The deuteron anapole moment is largely academic as there is not much hope it could be measured in a near future. It however provides a nice laboratory for studying the implications of gauge invariance, which is essential for getting a consistent estimate of this quantity. A contribution required by chiral symmetry [21], absent in [22], has thus been recovered in potential based approaches [23].

On the other hand, this last work confirms the conclusion obtained from the study of A_γ about the reliability of some effective-field theories. It is likely that an alternative approach [27], which is nothing but the one initiated by Danilov's work [2], extended later on to higher energy [3], should do better.

In pnc-electron experiments performed on the deuteron, aiming at determining the contribution of strange quarks to nucleon form factors, there was some concern about the role of a nuclear pnc effect. This one was studied in two different works [25,16] which showed that the effect, a few percents, would be negligible. Actually, the main role of pnc nuclear effects in this process (together with that one involving the proton) is an indirect one. They allow one to put limits on coupling constants that enter radiative corrections [28].

Parity-non-conservation in np scattering has been recently revisited [20]. The main feature evidenced by the new results is the dominance of the pion-exchange contribution, as far as the DDH "best guess" is used for the corresponding pnc coupling.

6 Discussion and conclusion

Many low-energy pnc nuclear effects, involving mainly protons, are within expectations. However, one has often to be satisfied with discrepancies up to a factor 2. This is not enough to constrain the different pnc meson-nucleon couplings if one refers to a potential approach or the low-energy NN scattering amplitudes if one rather relies on the less ambitious approach represented by effective field theories.

Most probably, the pnc πNN coupling, h^1_π, is small and within the DDH updated range. Some processes could require a significantly larger value but, in our opinion, they have not the weight of the other ones that point to a small value. Concerning the vector meson-nucleon couplings, there is a slight hint that the isoscalar ω one, h^0_ω, could have a sign opposite to expectations. This should motivate further studies to confirm the hint on the one hand, to see whether this opposite sign is conceivable.

An analysis in terms of couplings has some interest but, apart from the fact it assumes that multi-meson exchanges can be ignored, it does not necessarily provide a pertinent clue at which part of the pnc interaction is rather unconstrained. Looking at the various NN scattering amplitudes can thus represent a complementary view. Among the five amplitudes required for the description of pnc effects at low energy, only one (pp) is determined with a good accuracy. From the study of odd-proton systems, and after removing the contribution of the pp amplitude, a pn amplitude involving "unpolarized" neutrons can be obtained. Being derived indirectly, from complex systems moreover, the accuracy of this amplitude is not as good as for the pp one. For the three other amplitudes, which involve "polarized" neutrons (with "unpolarized" protons, with "unpolarized" neutrons and with "polarized" protons), only upper limits are known.

To determine this sector of the pnc NN interaction, appropriate experiments are heavily needed, preferentially with light systems where theoretical uncertainties are reduced. The np amplitude with "polarized" neutrons is better studied in the neutron-spin rotation. The nn amplitude could be obtained from the neutron-spin rotation in neutron-α scattering, after removing the previous contribution of the np amplitude. The best process for determining the last np amplitude, which involves both "polarized" neutrons and "polarized" protons, is the circular polarization of photons in the thermal neutron-proton radiative capture ("Lobashov experiment"). Evidently, the asymmetry A_γ, already mentioned, is part of the needed experiments. While it involves the difference in the two np amplitudes with a "polarized" neutron and an "unpolarized" proton on the one hand, the inverse configuration on the other hand, it also allows one to get information on the most debated pnc coupling, h^1_π.

Acknowledgements. We are very grateful to C. H. Hyun and C.-P. Liu for a stimulating and fruitful collaboration.

References

1. B. Desplanques: Phys. Rep. **297**, 1 (1998)
2. G.S. Danilov: Sov. J. Nucl. Phys. **14**, 443 (1972); Phys. Lett. **18**, 40 (1965); Phys. Lett. B **35**, 579 (1971)
3. B. Desplanques, J. Missimer: Nucl. Phys. A **300**, 286 (1978)
4. B. Holstein et al.: nucl-th/0407087
5. B. Desplanques, J. Donoghue, B. Holstein: Ann. of Phys. **124**, 449 (1980)
6. N. Kaiser, U.G. Meissner: Nucl. Phys. A **489**, 671 (1988); A **499**, 69 (1989); A **510**, 759 (1990)
7. U.G. Meissner, H. Weigel: Phys. Lett. B **447**, 1 (1999)
8. H.-J. Lee et al.: hep-ph/0405217
9. S.-L. Zhu et al.: Phys. Rev. D **63**, 033006 (2001)
10. J. Carlson et al.: Phys. Rev. C **65**, 035502 (2002)
11. W.C. Haxton et al.: Phys. Rev. C **65**, 045502 (2002)
12. B. Desplanques: Phys. Lett. B **47**, 212 (1973)
13. D.B. Kaplan et al.: Phys. Lett. B **449**, 1 (1999)
14. B. Desplanques: Phys. Lett. B **512**, 305 (2001)
15. C.H. Hyun et al.: Phys. Lett. B **516**, 321 (2001)
16. R. Schiavilla et al.: Phys. Rev. C **67**, 032501R (2003)
17. I.B. Khriplovich et al.: Nucl. Phys. A **690**, 610 (2001)
18. C.-P. Liu, C.H. Hyun, B. Desplanques: Phys. Rev. C **69**, 065502 (2004)
19. M. Fujiwara et al.: Phys. Rev. C **69**, 065503 (2004)
20. R. Schiavilla, J. Carlson, M. Paris: nucl-th/0404082
21. M.J. Savage, R.P. Springer: Nucl. Phys. A **644**, 235 (1998); A **657**, 457(E) (1999)
22. I.B. Khriplovich et al.: Nucl. Phys. A **665**, 365 (2000)
23. C.H. Hyun, B. Desplanques: Phys. Lett. B **552**, 41 (2003)
24. C.-P. Liu, C.H. Hyun, B. Desplanques: Phys. Rev. C **68**, 045501 (2003)
25. C.-P. Liu et al.: Phys. Rev. C **67**, 035501 (2003)
26. T. Oka: Phys. Rev. D **27**, 523 (1983)
27. M.J. Savage: Nucl. Phys. A **695**, 365 (2001)
28. S.-L. Zhu et al.: Phys. Rev. D **62**, 033008 (2000)

Eur Phys J A (2005) **24**, s2, 175–178
DOI: 10.1140/epjad/s2005-04-044-8

EPJ A direct
electronic only

Parity violation in nuclear systems

Experimental considerations in the deuteron photodisintegration with polarized photons

Efstathios Stiliaris[a]

Physics Department, University of Athens, GR-15771 Athens, Greece and
Institute of Accelerating Systems & Applications (IASA), P.O. Box 17214, GR-10024 Athens, Greece

Received: 15 December 2004 / Published Online: 8 February 2005

Abstract. Experimental measurements of Parity Non-Conserving (PNC) asymmetries in simple nuclear systems represent always a key-tool for the study of the weak nucleon-nucleon interaction and consequently an accurate experimental method for the determination of the meson-nucleon weak coupling constants of the underlying theory. Recent theoretical analysis on the deuteron photodisintegration with polarized photons, a few MeV above threshold, have drastically improved previous theoretical estimates. Based on that, the feasibility of measuring the photon asymmetry A_γ in the reaction $\vec{\gamma} + d \rightarrow n + p$ with the 10-MeV CW Linac at the Institute of Accelerating Systems and Applications (IASA) is considered here. A brief review on previous experimental results obtained in the deuteron photodisintegration and in the thermal-neutron radiative capture on protons (inverse reaction) is given. The most important parameters in the design of a nuclear parity experiment are presented and the crucial factors, such as beam intensity, beam polarization and neutron detection techniques with the required high accuracy are outlined.

PACS. 24.80.+y Nuclear tests of fundamental interactions and symmetries – 25.20.-x Photonuclear reactions

1 Introduction

The deuteron has played an important role in the study of the weak nucleon-nucleon interaction. Together with scattering experiments the studies of Parity Non-Conserving (PNC) transitions in the nucleon-nucleon system are very attractive because of the simplicity and the well understood structure of the system [1,2].

The first experiment which received a lot of attention during the 1970's was the "Lobashov" experiment. In this experimental study the Leningrad group has investigated the net polarization of the emitted photons in the radiative thermal neutron capture by proton, $n + p \rightarrow d + \gamma$. The experimental work was characterized by the novel techniques applied for measuring the integrated current. The nonzero polarization obtained in this reaction [3], $P_\gamma = -(1.3 \pm 0.45) \times 10^{-6}$, which is 30 times larger in magnitude than the theoretical prediction and, moreover, of opposite sign, motivated many theoretical calculations in the frame of strong and weak interaction models known at that time. Later experimental work for the circular polarization P_γ of the emitted 2.23 MeV photons reported values more consistent with theoretical estimates but with too poor accuracy to allow any definite conclusion about the strength of the PNC forces [4]. At present, a new PNC-asymmetry measurement for the radiative neutron-proton capture with polarized neutrons, $\vec{n} + p \rightarrow d + \gamma$, is in preparation at LANSCE in order to reduce the experimental error of the spatial asymmetry A_γ of the emitted photons [5,6].

Another tool to study PNC forces is the inverse reaction $\vec{\gamma} + d \rightarrow n + p$, where a deuteron is disintegrated by absorbing a circularly polarized photon. The asymmetry A_γ in this reaction near threshold is expected to be sensitive to the same components of the weak nucleon-nucleon interaction as the thermal neutron capture. A key measurement of this asymmetry has been also performed in the past at the Chalk River Nuclear Laboratories, established as a "Chalk River" experiment [7]. Unfortunately, the results of this measurement were characterized by poor control on the systematic errors, mainly due to beam instabilities. With the most recent advances in the beam instrumentation and in the detection techniques, taking also in account the new theoretical considerations on this subject, a measurement of A_γ becomes nowadays more realistic.

The feasibility of measuring the photon asymmetry A_γ in the deuteron photodisintegration at low energies with the 10-MeV CW Linac at the Institute of Accelerating Systems and Applications (IASA), Athens, is considered here. First, a brief review on the recent theoretical calculations and predictions of the expected asymmetry will be given. The most important parameters in the design of a nuclear parity experiment will be presented and the

[a] *e-mail address:* stiliaris@phys.uoa.gr

crucial factors, such as beam intensity, beam polarization and neutron detection techniques with the required high accuracy will be outlined. At the end, the IASA accelerator and the possibilities of using it as a dedicated machine for the nuclear parity study will also be discussed.

2 Theoretical calculations and predictions

The asymmetry in the deuteron photodisintegration $\vec{\gamma} + d \to n + p$ of an unpolarized target is defined as

$$A_\gamma = \frac{\sigma_+ - \sigma_-}{\sigma_+ + \sigma_-}$$

where σ_+ and σ_- denotes the total cross section using right- and left-handed polarized photons respectively. This asymmetry A_γ becomes asymptotically equal to the circular polarization P_γ of the emitted 2.23-MeV photons in the thermal neutron capture by protons (inverse reaction $n + p \to d + \gamma$) at threshold energy.

First calculations of A_γ done by Lee [8], up to photon energy 1 MeV above the disintegration threshold, show a reasonable result within the theoretical range of P_γ in this energy domain, where the regular transition M1 dominates. However, a later work by Oka [9], extended to higher photon energies $\omega_\gamma \simeq 35$ MeV, suggests that PNC effects are different compared to low photon energy, since the major contribution to the cross section comes now from the E1 transition (compare also Fig. 1). Due to the initiated parity-violating π-exchange contribution, it was predicted that A_γ shows a great enhancement at energies $\omega_\gamma \geq 5$ MeV. The experimental observation of this enhancement would provide an important and unambiguous determination of the weak πNN coupling constant h_π^1.

Unfortunately, a recent calculation of A_γ by Khriplovich and Korkin [10] showed critical contradictions to Oka's result, with a huge suppression of A_γ at energies $\omega_\gamma \geq 3$ MeV. It seems that in Oka's work, a non-vanishing value for the pion-exchange contribution to the asymmetry was obtained as a result of the incomplete account for P-odd mixing, since only the $^3P_1$ admixture to the deuteron ground state was included there. In order to examine the situation in a most accurate way, Liu, Hyun and Desplanques [11] did a careful calculation of this process by completing the missing parity-admixed components in the final state, in particular in the $^3P_1$ channel, and by including tensor and spin-orbit forces in the nucleon-nucleon interaction using the realistic Argonne AV18-potential. The result of this improved work confirmed the strong suppression of the π-exchange contribution to the asymmetry and is illustrated in Fig. 2.

At the same time, another independent theoretical work by Fujiwara and Titov [12] analyzed the energy dependence of the A_γ asymmetry in the deuteron photodisintegration. Although this second paper was based on the Paris potential with soft repulsion at short distances and the Hamada-Johnston potential, both studies come to consistent to each other results. The predicted theoretical values for the asymmetry A_γ of all mentioned models are summarized in Table 1.

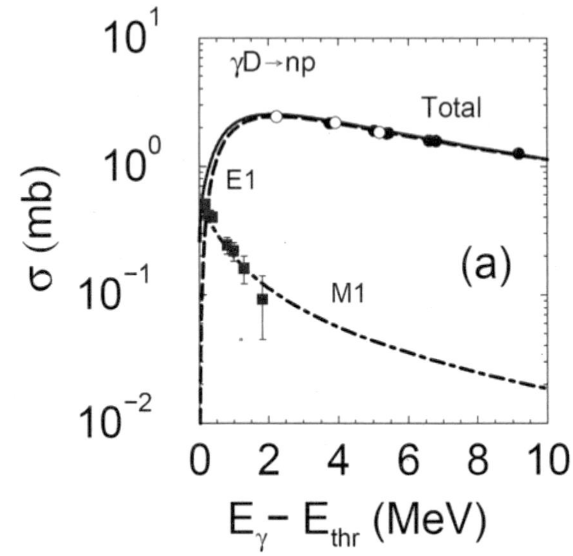

Fig. 1. Total cross section for the deuteron photodisintegration as a function of the energy excess above the disintegration threshold, calculated with the Paris potential (From [12]). The contributions of the M1 (*dot-dashed*) and E1 (*dashed*) transitions are shown together with experimental data

Table 1. Predicted values for the A_γ asymmetry in the $\vec{\gamma} + d \to n + p$ reaction

γ-Energy	A_γ	Dependence	Ref.
10 MeV	$\sim 2.4 \times 10^{-7}$	h_π^1	[9]
30 MeV	$\sim 5.7 \times 10^{-7}$	h_π^1	[9]
threshold	$\sim 1.0 \times 10^{-7}$	M1 dominance	[10]
3 MeV	$\sim 1.0 \times 10^{-8}$	M1 dominance	[10]
threshold	$\sim 2.5 \times 10^{-8}$	$h_\rho^0, h_\rho^2, h_\omega^2,$	[11],[12]

The experimental consideration following in the next section is based on the most realistic estimates by Liu, Hyun and Desplanques [11], which are in agreement with the results by Fujiwara and Titov [12]. The predicted value $A_\gamma \simeq 2.5 \times 10^{-8}$ near threshold will be the starting point for the discussion of possible future experiments measuring this asymmetry in the deuteron photodisintegration with polarized photons. From Fig. 2 it is clear that, when the photon energy gets larger, the asymmetry gets smaller; as the the photon energy reaches 1 MeV above the threshold, the asymmetry drops by an order of magnitude.

3 The future experiment

In the "Chalk River" experiment the obtained results for the A_γ asymmetry were mainly affected by the big systematic errors. The final values reported there $A_\gamma = (2.7 \pm 2.8) \times 10^{-6}$ at $E_\gamma = 4.1$ MeV and $A_\gamma = (7.7 \pm 5.3) \times 10^{-6}$ at $E_\gamma = 3.2$ MeV [7] put an experimental upper limit on the value of the PNC effect. But, as this experimental study

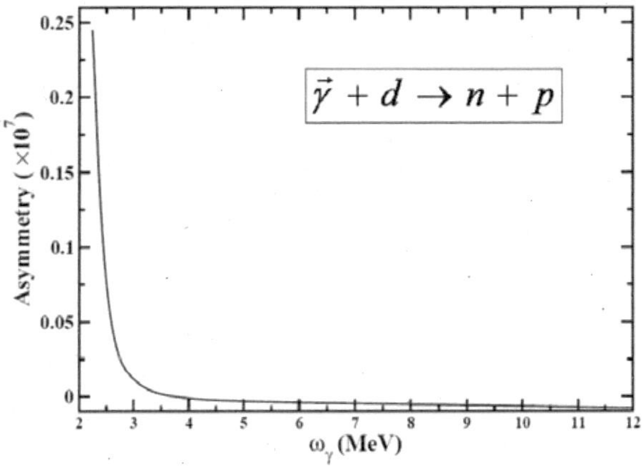

Fig. 2. Theoretical prediction for the asymmetry A_γ in the deuteron photodisintegration with polarized photons. The calculation is based on the DDH best values, taking in account tensor as well as spin orbit forces in the nucleon-nucleon interaction (From [11])

concluded, "... it is possible to consider future measurements with techniques similar to those used in the present measurement but having improved beam intensity, beam polarization and control of systematics" [7].

The experimental goal is therefore to reduce the systematic errors to a level better than 10^{-8} in measuring the neutron asymmetry of the reaction in the deuteron photodisintegration with polarized photons. The most important factors in the design of such an experiment are:

- Reasonable flow of polarized photons
- An improved neutron detection system
- Beam quality and stability with fast feedback systems

Starting with some conservative numbers all experimental requirements are presented in the following section.

3.1 Experimental considerations

The polarized photons are produced via Bremsstrahlung with a high density polarized electron beam and a gold radiator. Development of a highly polarized electron beam with energy stability on the level of a few 10^{-6} can be routinely achieved. Having in mind a recent Letter-Of-Intent presented at JLab for a similar project [13], following parameters are outlined here for the proposed apparatus:

- Beam energy in the range 2.3 – 8 MeV ($200\mu A$)
- Beam polarization $P_e \sim 80\%$
- Photon production target (radiator) made of 1mm Au plate
- 30 cm long Liquid Deuterium (LD_2) target
- Main detector consists of two components (slow neutrons & photons)
- Lead shielding to reduce the intensity of scattered photons on neutron detectors
- Heavy water moderator to slow down neutrons
- Additional Compton detectors for monitoring

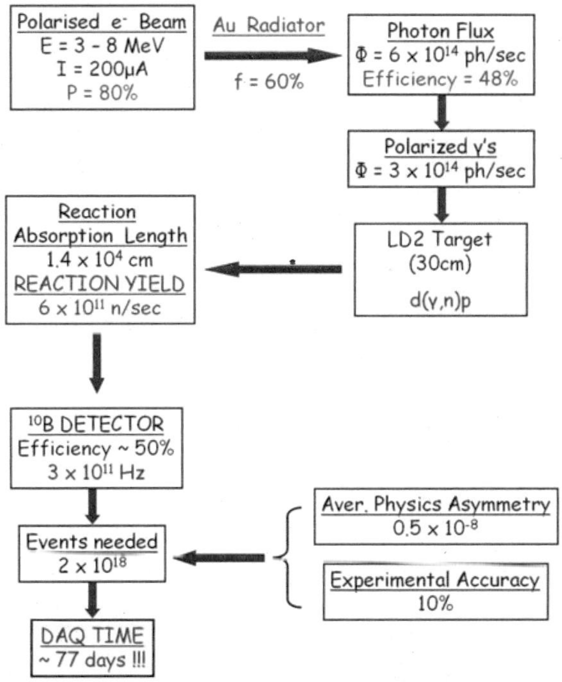

Fig. 3. Experimental statistics expected in a future experiment and the required acquisition time for data taking

Statistics obtained with the above mentioned parameters are presented in the diagram of Fig. 3. It is clear from this simple analysis, that the Data Acquisition (DAQ) time needed to successfully perform such an experiment and to reach the required accuracy comes to the order of several months!

The experimental apparatus is shown in Fig. 4. The main detector has segmentation on the forward and the back parts to get sensitivity to the directional asymmetry in addition to PNC asymmetry in the total cross section. The expected high counting rate of the neutron detectors ($\sim 10^{12} Hz$) will help to study tiny systematic effects. The choice of the spin flip frequency has to be optimized for minimum fluctuation of the beam energy and position.

3.2 The IASA electron accelerator

A 10-MeV CW linear electron accelerator has been already installed at the Institute of Accelerating Systems & Applications (IASA), which is currently under commissioning. This machine comprises a thermionic electron gun, a 100 keV-Line with a buncher-chopping system followed by a 5 MeV Linac with RF structures of the side-coupled type and a 4m-booster section of the same type [14]. Both are powered with a 500 kW multi-cavity CW klystron amplifier at 2380 MHz. The machine is hosted in the basement of the IASA building, which is extended with a new experimental hall (Fig. 5).

The IASA accelerator meets exactly the needs for a future parity-violation experiment as previously discussed. The energy range is optimally covered by this machine and

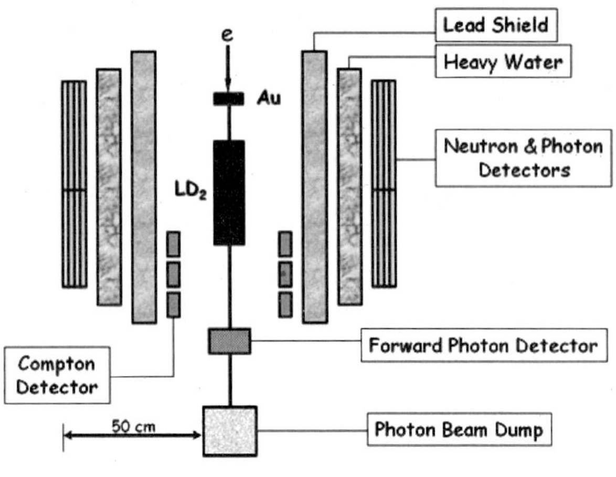

Fig. 4. Experimental setup for the study of the $\vec{\gamma} + d \to n + p$ reaction in a future experiment

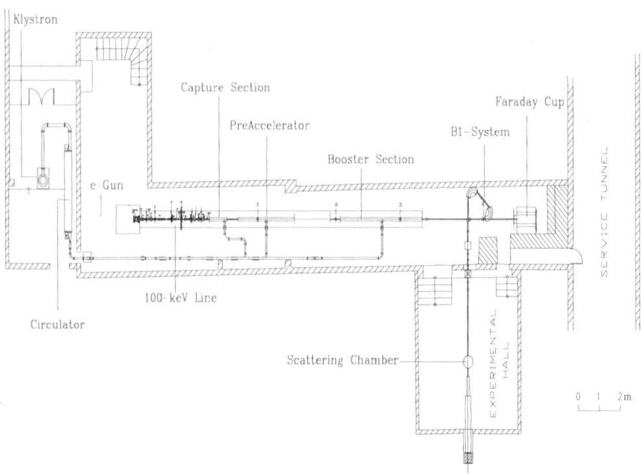

Fig. 5. The 10-MeV CW electron linear accelerator at IASA

the beam current and beam characteristics, already measured at 100 keV, could guarantee a high quality beam. At present, the beam is not polarized, but there are future plans for the installation of a high intensity polarized electron gun. Space has been already reserved in the accelerator vault for future parity experiments as indicated in Fig. 6. Taking into account the long acquisition time required by the parity-violation experiments, the IASA 10-MeV electron accelerator could ideally be devoted to the research of the PNC studies in the deuteron photodisintegration.

4 Conclusion

Parity Non-Conserving (PNC) experiments at low energy got large attention in the last years. With the recent theoretical improvements in the calculation of the asymmetry in the deuteron photodisintegration induced by polarized photons a few MeV above threshold and with the technical progress in the accelerator domain with polarized beams,

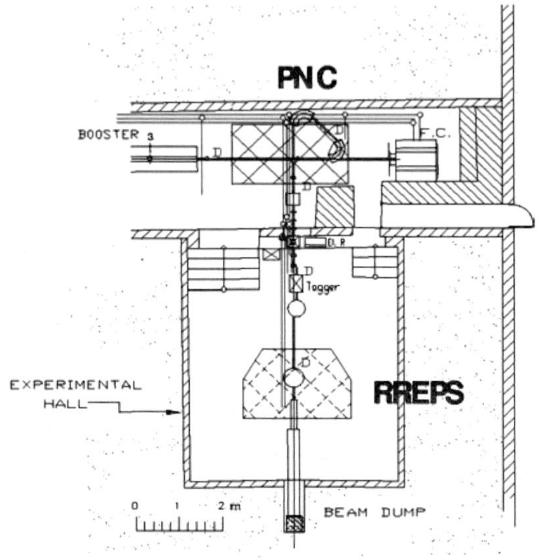

Fig. 6. Experimental area in the accelerator vault. Indicated are the area reserved for the future parity experiments (PNC) together with the area devoted to novel radiation sources (RREPS)

a precise measurement of parity-violating forces seems to be feasible. The 10-MeV IASA electron accelerator could serve as a machine dedicated to this kind of research.

Acknowledgements. I am indebted and would like to thank B. Desplanques from LPSC Grenoble, and B. Wojtsekhowski from JLab for the valuable suggestions and discussions on the deuteron parity-violation subject. I am also grateful to Prof. C.N. Papanicolas from the University of Athens for his continuous interest to this project.

References

1. B. Desplanques, J.F. Donoghue, B.R. Holstein: Ann. Phys. (N.Y.) **124**, 449 (1980)
2. B. Desplanques: Phys. Rep. **297**, 1 (1998)
3. V.M. Lobashov et al.: Nucl. Phys. A **197**, 241 (1972)
4. V.A. Knyazkov et al.: Nucl. Phys. A **417**, 209 (1984)
5. V.M. Snow et al.: Nucl. Instrum. Methods A **440**, 729 (2000)
6. V.M. Snow et al.: Nucl. Instrum. Methods A **515**, 563 (2003)
7. E.D. Earle et al.: Can. J. Phys. **66**, 534 (1988)
8. H.C. Lee: Phys. Rev. Lett. **41**, 843 (1978)
9. T. Oka: Phys. Rev. D **27**, 523 (1983)
10. I.B. Khriplovich, R.V. Korkin: Nucl. Phys A **690**, 610 (2001)
11. C.-P. Liu, C.H. Hyun, B. Desplanques: Phys. Rev. C **69**, 065502 (2004)
12. M. Fujiwara, A.I. Titov: Phys. Rev. C **69**, 065503 (2004)
13. B. Wojtsekhowski, W.T.H. van Oers: JLab Letter-Of-Intent 00-002 for PAC 17 (2000)
14. E. Stiliaris et al.: Proceedings of EPAC 2000, IoP, pp. 866–868

Eur Phys J A (2005) **24**, s2, 179–180
DOI: 10.1140/epjad/s2005-04-045-7

EPJ A direct
electronic only

Parity violating asymmetry in $\gamma + d \rightarrow n + p$ at low energy

C.H. Hyun[1], C.-P. Liu[2], and B. Desplanques[3]

[1] Institute of Basic Science, Sungkyunkwan University, Suwon 440-746, Korea
[2] Kernfysisch Versneller Instituut, Zernikelaan 25, Groningen 9747 AA, The Netherlands
[3] Laboratoire de Physique Subatomique et de Cosmologie (UMR CNRS/IN2P3-UJF-INPG), F-38026 Grenoble Cedex, France

Received: 15 October 2004 / Published Online: 8 February 2005

Abstract. We calculate the parity-violating asymmetry in the $\gamma + d \rightarrow n + p$ process where the deuteron is disintegrated by circular photons. The photon energy is considered up to 10 MeV above threshold, where the lowest electromagnetic transition modes $M1$ and $E1$ dominate. We employ the Argonne v18 potential for the strong interaction and the DDH potential for the parity-violating weak interaction of the two-nucleon systems. The asymmetry is about 2.5×10^{-8} at threshold, and decreases rapidly to have magnitude of order 10^{-9} or less for photon energies larger than 3 MeV. The exchange of vector mesons dominates the asymmetry, while the pion contribution is negligible.

PACS. 24.80.+y Nuclear tests of fundamental interactions and symmetries

1 Introduction

The interest to test the standard model in the realm of nuclear and atomic physics becomes more and more growing and popular nowadays. The activeness of the subject has been driven by recent experiments: measurement of the anapole moment of ^{133}Cs [1], parity-violating (PV) longitudinal asymmetry in $\mathbf{p}p$ scattering [2] and strange form factor of the nucleon from $\mathbf{e}p$ scattering [3]. In addition to these experiments, a PV asymmetry in $\mathbf{n}p \rightarrow d\gamma$ is being measured at LANSCE [4], and the possibility to measure a PV asymmetry in $\gamma d \rightarrow np$ is also being discussed. These experiments provide important information about the weak interaction of hadrons at low energy. From the theoretical viewpoint, weak as well as strong interactions can be described by means of the one-boson exchange, which introduces PV meson-nucleon coupling constants. Knowing precise values of the PV coupling constants is an important ingredient to understand the nuclear weak interaction. Unfortunately, for some of the PV coupling constants, recent experiments mentioned above give incompatible values with those from other experiments or theory calculations. We expect that future experiments will provide a way to resolve the present uncertainties, and shed light on the nuclear weak interaction problem.

In this work, we calculate the PV asymmetry A_γ in $\gamma d \rightarrow np$. Confronting the possibility of its experimentation, it is necessary to calculate A_γ with updated modern potentials, and compare it with old ones that show significant difference. In our calculation, the strong interaction is accounted for by the Argonne v18 (Av18) potential [5], and the PV potential given by Desplanques, Donoghue and Holstein (DDH) [6] is employed for the weak interaction.

In the next section, we briefly describe the basic formalisms. Numerical results are shown and discussions follow.

2 Result and discussion

We consider the photon energies up to 10 MeV above threshold, which leaves only a few low-orbital states in the final state sufficient for the result being without significant error. Opposite-parity states are admixed in the wave function by the PV meson-exchange potential. The DDH PV potential [6], adopted in our work, includes π-, ρ- and ω-exchanges with the vertices specified by isospin transfer $\Delta I = 0, 1, 2$. At the considered energies, $M1$ and $E1$ modes dominate the electromagnetic transition, and therefore we use the $M1$ and the $E1$ operators given as

$$\hat{O}_{M1} = \sum_i \frac{1}{2m_N}\left(e_i \mathbf{l}_i + \mu_i \sigma_i\right), \tag{1}$$

$$\hat{O}_{E1} = \sum_i e_i \mathbf{x}_i, \tag{2}$$

where e_i is the charge of the nucleon, μ_i the magnetic moment, $\mathbf{l}_i$ the angular momentum, σ_i the spin and $\mathbf{x}_i$ the position. Given the wave functions and the transition operators, the asymmetry A_γ can be calculated from the definition

$$A_\gamma \equiv \frac{\sigma_+ - \sigma_-}{\sigma_+ + \sigma_-} \tag{3}$$

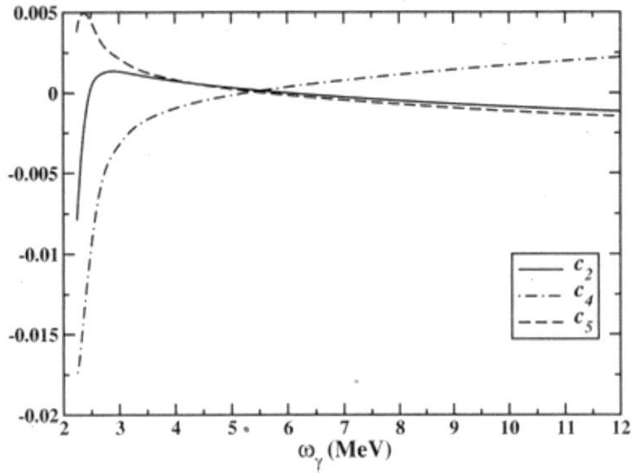

Fig. 1. Dominant coefficients

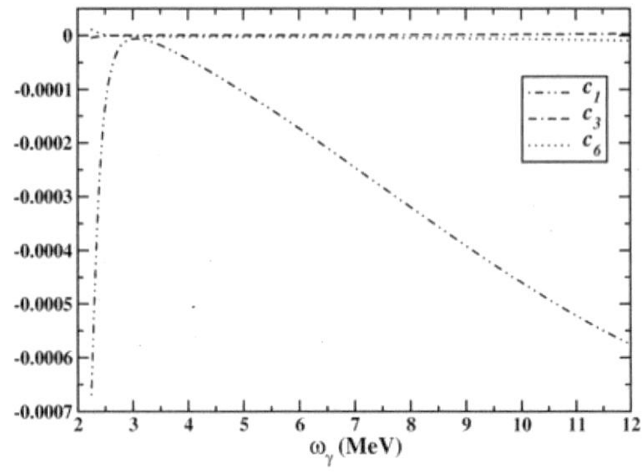

Fig. 2. Suppressed coefficients

where σ_+ (σ_-) denotes the total cross section for right- (left-) handed circular photons.

To first order in the weak coupling constants, A_γ can be written as

$$A_\gamma = c_1 h_\pi^1 + c_2 h_\rho^0 + c_3 h_\rho^1 + c_4 h_\rho^2 + c_5 h_\omega^0 + c^6 h_\omega^1. \quad (4)$$

Numerical results for the coefficients c_i's are shown in Fig. 1 and 2. The magnitude of c_i's for heavy-meson isoscalar and isotensor vertices (c_2, c_4, c_5) are larger than those of isovector ones (c_1, c_3, c_6) by an order or more. Since the weak coupling constants calculated from a quark model [6] or the Skyrme model [7,8] give a similar magnitude, A_γ will be dominated by isoscalar and isotensor PV vertices.

In Fig. 3, we show A_γ calculated with the DDH best values for the weak coupling constants. At threshold, $A_\gamma \simeq 2.5 \times 10^{-8}$, and it decreases to the order of 10^{-9} for $\omega_\gamma \geq 3$ MeV. The small magnitude of A_γ can be understood from the strong cancellation of c_4 with c_2 and c_5, and similar values of $h_\rho^0 (= -11.4 \times 10^{-7})$ with $h_\rho^2 (= -9.5 \times 10^{-7})$. The strong suppression of A_γ confirms the result of a recent schematic calculation [9], but does not support the earlier significant enhancement from the contribution of h_π^1 [10]. It is shown in [11] that the enhancement in [10] is due to the omission of the contribution from ${}^3P_1 \to {}^3\tilde{S}_1 - {}^3\tilde{D}_1$ transition, whose inclusion gives suppressed A_γ values.

The asymmetry A_γ was recently calculated with Av18 and CD-Bonn [12] potentials by Schiavilla *et al.* [13]. For Av18, they obtain a result similar to ours, but the CD-Bonn potential gives a larger result than Av18 by a factor of about 2. A_γ at threshold with r-space Bonn [14] and Bonn-A, B [15] is calculated, and, with the DDH best values, the results turn out to be similar with that of CD-Bonn [16]. The enhancement for the Bonn potentials is due to strong attraction at short distance in the ${}^1P_1$ channel, which leads to a bound state in ${}^1P_1$ state [17]. Therefore, this artifact of the ${}^1P_1$ Bonn potential should be treated properly, e.g. by introducing a cutoff, to obtain a sound result.

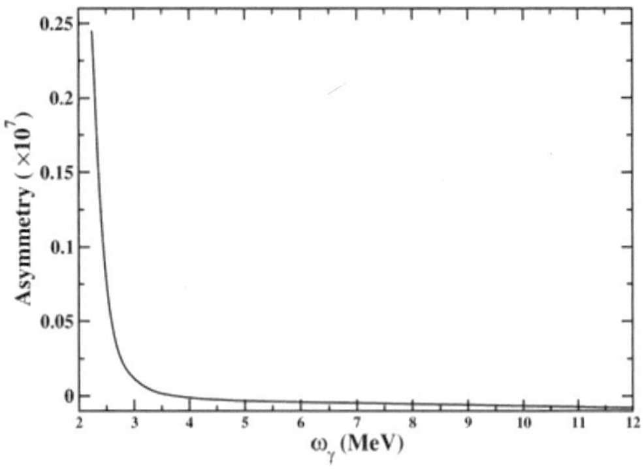

Fig. 3. A_γ with DDH best values

References

1. C.S. Wood et al.: Science **275**, 1759 (1997)
2. A.R. Berdoz et al.: Phys. Rev. Lett. **87**, 272301 (2001)
3. F.E. Maas et al.: Phys. Rev. Lett. **93**, 022002 (2004)
4. W.M. Snow et al.: Nucl. Instrum. Meth. A **440**, 729 (2000)
5. R.B. Wiringa, V.G.J. Stoks, R. Schiavilla: Phys. Rev. C **51**, 38 (1995)
6. B. Desplanques, J.F. Donoghue, B.R. Holstein: Ann. Phys. (N.Y.) **124**, 449 (1980)
7. N. Kaiser, U.G. Meissner: Nucl. Phys. A **489**, 671 (1988); A **499**, 69 (1989)
8. U.G. Meissner, H. Weigel: Phys. Lett. B **447**, 1 (1999)
9. I.B. Khriplovich, R.V. Korkin: Nucl. Phys. A **690**, 610 (2001)
10. T. Oka: Phys. Rev. D **27**, 523 (1983)
11. C.-P. Liu, C.H. Hyun, B. Desplanques: Phys. Rev. C **69**, 065502 (2004)
12. R. Machleidt: Phys. Rev. C **63**, 024001 (2001)
13. R. Schiavilla, J. Carlson, M. Paris: nucl-th/0404082
14. R. Machleidt, K. Holinde, Ch. Elster: Phys. Rep. **149**, 1 (1987)
15. R. Machleidt: Adv. Nucl. Phys. **19**, 189 (1989)
16. C.H. Hyun el al.: in preparation
17. J. Haidenbauer: private communication

V Hadronic structure. . . and more

V-3 Neutrino beam

Eur Phys J A (2005) **24**, s2, 183–186
DOI: 10.1140/epjad/s2005-04-046-6

EPJ A direct
electronic only

Precision physics at a neutrino factory

A. Blondel

Département de Physique Nucléaire et corpusculaire, Faculté des Sciences, Université de Genève, Quai Ansermet 24, Genève 4, CH1211 Switzerland

Received: 1 November 2004 / Published Online: 8 February 2005

Abstract. Neutrino beams of unprecedented flux could be produced in a Neutrino Factory from muon decays. In the vicinity of the storage ring, short baseline experiments would perform a new class of precise tests of the theory and original deep-inelastic-scattering (DIS) studies. Thanks to the availability of high energy ν_e and $\bar{\nu}_e$, the long baseline experiments will be capable of very precise measurements of neutrino oscillations, including ability to solve parameter ambiguities and study of leptonic CP violation, for any value of the mixing angle θ_{13} above a fraction of a degree. Finally, the Neutrino Factory is the first step towards muon colliders.

PACS. 1 3.88.+e Polarization in interactions and scattering – 1 2.15.Ff Quark and lepton masses and mixing – 1 4.60.Pq Neutrino mass and mixing – 2 9.25.-t Particle sources and targets – 2 9.27.-a Beams in particle accelerators – 2 9.20.-c Cyclic accelerators and storage rings

1 Introduction

Neutrinos have historically played an essential role in particle physics, with the discovery of Neutral Currents (NC), the first observations of open charm, precautious scaling and its violations and the early description of the structure of the nucleon. More recently LEP established that there are three species of light neutrinos, thus probably only three families of fermions. Finally, neutrinos have recently be in the limelight with the final demonstration that neutrinos have mass and mix. One of the main goals of the upcoming years will be the observation of leptonic CP violation, which is one of the leading explanations for the matter-antimatter asymmetry of the Universe. The importance of this search will justify substantial investments for the future. One of the leading ideas is that of a neutrino factory, in which neutrinos are produced in a controlled way by means of a stored muon beam.

This presentation and article summarize the work of several hundred of members of the CERN and ECFA studies of a Neutrino Factory Complex, which is contained in the very complete [1]. The Neutrino Factory has been proposed in 1998 [2] and studied extensively thanks in particular to the pioneering work of the Neutrino Factory and Muon Collider collaboration [3].

As we will see in the following sections, such a machine is very polyvalent, and offers opportunities in short baseline physics, neutrino oscillations and is also the first step towards muon colliders.

2 The neutrino factory

The basic layout of a Neutrino Factory is shown in Fig. 1. The principle [4] is to produce the largest possible intensity of low energy muons, accelerate them to an energy of 20 to 50 GeV and store them in a decay ring. A short (1-3 ns) high power (several MW) proton beam of energy in excess of a few GeV hits a renewable target (liquid mercury is the baseline design). The secondary pions are captured by a magnetic device (tapered solenoid or magnetic horn) and fed into a solenoidal magnetic channel where they decay into muons. The muons are then subject to phase rotation and ionization cooling after which the energy spread and transverse emittance are considerably reduced. The beam is by then small enough to be accelerated by means of recirculating linac or FFAG's, and stored in a decay ring equipped with long straight sections. Typically 10^{14} muons are injected per second, producing at the end of the straight sections a very intense, bunched, beam of neutrinos from the decay of muons, $\mu^+ \to e^+ \nu_e \bar{\nu}_\mu$ or $\mu^- \to e^- \bar{\nu}_e \nu_\mu$, the sign being determined by the polarity of a few magnetic elements in the system or by timing.

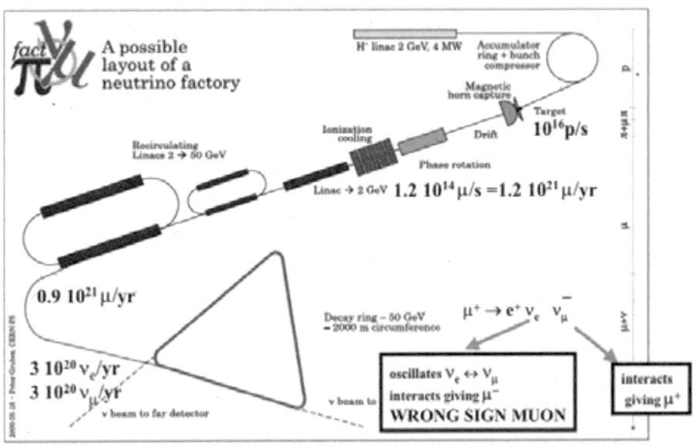

Fig. 1. Possible layout of a neutrino factory

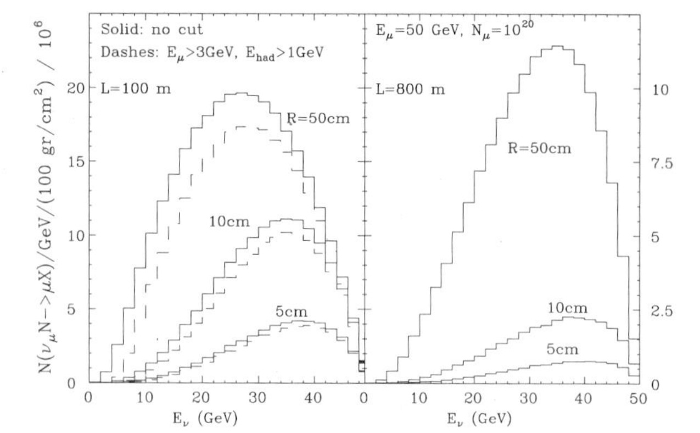

Fig. 2. Event rates at the near detector station of a neutrino factory. Note the vertical scale

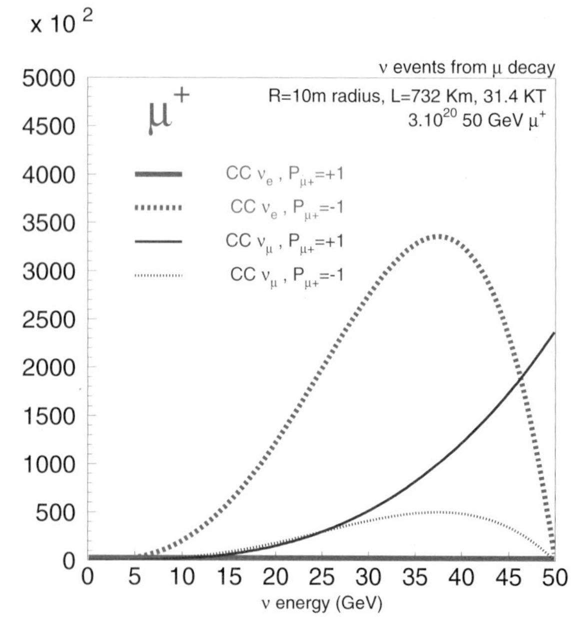

Fig. 3. Event rate at a far detector station of a neutrino factory for positive muon decay. The effect of muon beam polarization is shown. In the race-track or triangle geometry for the storage ring, the polarization can be preserved but averages out for each muon fill

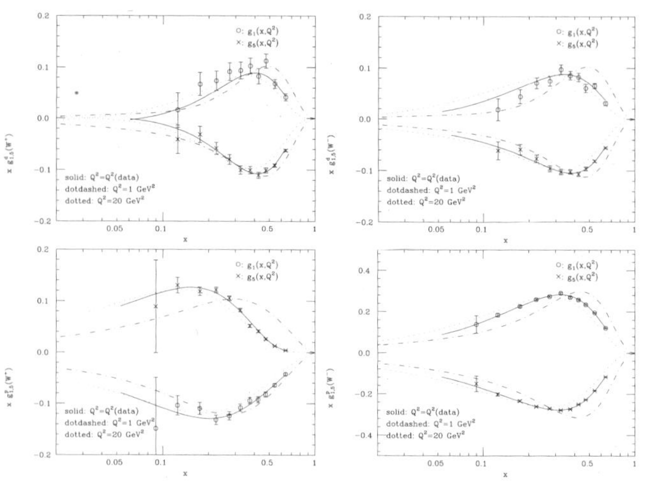

Fig. 4. Expected achievable precision on polarized structure functions g_1 and g_5 on proton (*lower part*) and deuteron (*upper part*) and for neutrino beam (*left part*) and anti-neutrino beam (*right part*) at the near detector of a neutrino factory

The resulting event rates are shown in Fig. 2 for the near detector station and in Fig. 3 for the far detectors. Originating from a stored and monitored beam, the flux of neutrinos should be known to a fraction of a percent [5].

3 Short baseline physics

In the near detector station, a kilogram of material placed on the beam axis would see typically 100 millions of interactions per year, more than in the old 1200 ton CDHS detector, but still only one event every few accelerator pulses. This, being obtained with a well defined flux of neutrinos (which are polarized by nature), opens a new realm of experimentation since the target material can be varied *ad infinitum* and the final state products can be measured in detail. The physics potential of high intensity short baseline physics has been discussed quantitatively in [6], although no experimental set up has been simulated so far.

The range of physics accessible to the near detector station is quite large. First, definitive measurements of unpolarized structure functions measurements and their flavour composition, in particular the strange sea, will be possible for the whole accessible kinematic range. This is, in particular, thanks to the assumed capability of the detectors to tag charm production. These ingredients should lead to an improved determination of the strong coupling constant from either global structure function fits or from the GLS sum rule.

Then, the use of small targets being possible, using polarized hydrogen or deuterium targets should allow a detailed decomposition of the spin structure functions, as shown in Fig. 4. Again, the use of charm tag should be determinant in the study of the polarized structure function of the strange quarks.

Maybe one of the most interesting topics shold be the possibility to use a variety of nuclear targets and map systematically the nuclear effects in structure functions.

The study of final states in neutrino scattering offers new possibilities. For instance the study of final state Λ and Λ_c polarization both in charged current and neutral current processes has been given as an example, allowing extraction of the newly introduced polarized fragmentation functions.

The high statistics available should allow precision measurements of standard model processes on electrons. This has a two-fold interest. First, the charged current inverse muon decay process, $\nu_\mu e^- \rightarrow \mu^- \nu_e$, although unfortunately applicable only to the muon neutrino component of the beam from μ^+ decay, should allow a very precise flux normalization (provided an adequate target and detector can be built), in a way similar to that provided by Bhabha scattering in e^+e^-experiments. Secondly, the electron final state, which is very rich due to the presence

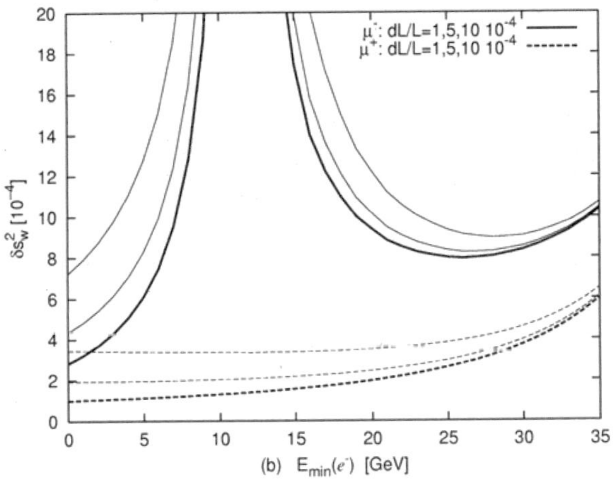

Fig. 5. Expected achievable precision on the weak mixing angle $\sin^2\theta_w^{eff}$ at small Q^2 from neutrino scattering off electrons. The precision is shown as a function of the cut on the final state electron energy. It is clear that the μ^+ exposure is more interesting for the mixing angle measurement, on the other hand the μ^- exposure is more sensitive to the interference between the NC and CC processes

of the electron neutrino component in the beam, is sensitive to $\sin^2\theta_w^{eff}$ at small Q^2 with a precision of the order of 2×10^{-4}, similar to that available at the Z pole, as shown in Fig. 5.

Finally, the high statistics allied with the improved knowledge of charm production should allow a complete revision of the measurement of the hadronic neutral current processes (NC/CC ratio) which have recently [7] exhibited a discrepancy with model expectations at the level of three standard deviations.

4 Neutrino oscillations

A large part of the present excitement for Neutrino Factories is, understandably, the long baseline oscillation physics. This has been extensively reviewed in the litterature [5],[8] and only the main results are summarized here.

Neutrino oscillations are now well established and demonstrate that neutrinos have mass. The most striking result of recent measurements are contained in our knowledge of the neutrino mixing matrix. For three flavours of neutrinos there are, naturally, three mixing angles, θ_{12}, θ_{13}, θ_{23}, and two mass differences Δm^2_{12}, Δm^2_{23}, that play a role in the oscillation process. The typical oscillation length is 500 km/GeV for the 'atmospheric oscillation' driven by Δm^2_{23}, and 18000 km/GeV for the 'solar oscillation' driven by Δm^2_{12}. In addition one expects the presence of a phase, yielding perhaps observable leptonic CP violation. The most interesting channel (so-called 'golden channel') to be studied is the $\nu_e \rightarrow \nu_\mu$ (and $\bar{\nu}_e \rightarrow \bar{\nu}_\mu$) oscillation which is suppressed at 'atmospheric' distances by the small value of the so far unknown θ_{13}. This suppression makes this channel particularly interesting since it makes it possibly sensitive to the interference between the solar and atmospheric oscillations, and thereby to the resulting CP violation, which would manifest itself by an

Table 1. Event rates in a 50 kton magnetized iron detector for one year running at a neutrino factory

Baseline	CC $\bar{\nu}_\mu$	CC ν_e	CC ν_μ Golden signal $\sin^2\theta_{13} = 0.01$
730 km	3.5 10^7	5.9 10^7	1.1 10^5
3500 km	1.2 10^6	2.4 10^6	1.0 10^5

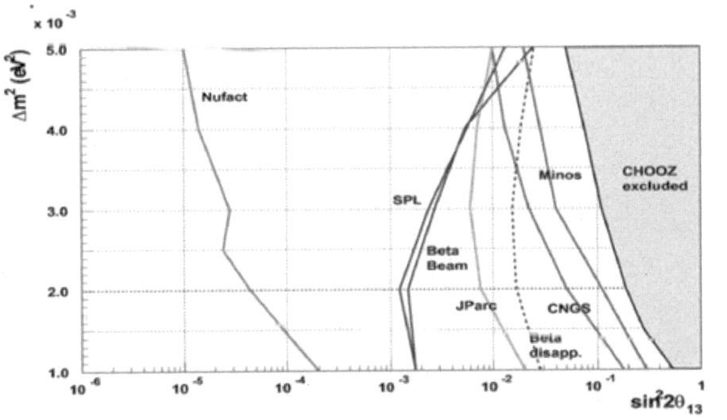

Fig. 6. Comparative sensitivity of various future measurements of θ_{13}

asymmetry between the neutrino and anti-neutrino oscillations. In addition this channel involving electron neutrinos is sensitive to matter effects and should allow a determination of the sign of the mass difference Δm^2_{23}, which is presently unknown.

The simulataneous presence of ν_e and $\bar{\nu}_\mu$ in the beam has for consequence that the detector has to be magnetic to separate the CC neutrino interactions generated by the $\bar{\nu}_\mu$ contained in the beam from those generated by ν_μ originating from $\nu_e \rightarrow \nu_\mu$ oscillations. A large 50 kton magnetic detector has been suggested, in extrapolation from the well known CDHS or MINOS experiments. The rates are astounding, as shown in Table 1, many times higher than in the case of more conventional neutrino beams. For this process the backgrounds are very small.

The other very strong point of the Neutrino Factory is the capability to study the $\nu_e \rightarrow \nu_\tau$ oscillation. This channel is particularly valuable since it allows to lift the unavoidable parameter ambiguities. The sensitivity or precision of the Neutrino Factory to the angle θ_{13} and to the CP violating phase δ are shown in Figs. 6 and 7, and are clearly superior to any other device imagined so far.

5 Muon collider

Finally, it is worth keeping in mind that the Neutrino Factory is the first step towards muon colliders. As shown in [9], the relevant characteristics of muons are that, compared to electrons, i) they have a much better defined energy, since they hardly undergo synchrotron radiation or beamstrahlung, ii) their coupling to the Higgs bosons is multiplied by the ratio $(m_\mu/me)^2$, thus allowing s-channel

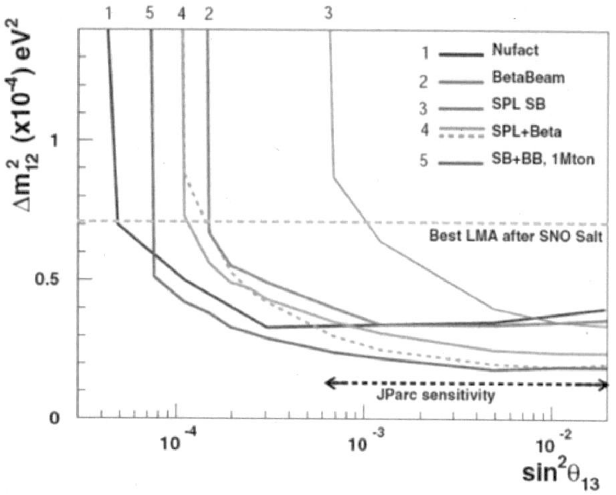

Fig. 7. Sensitivity of various future neutrino options to the CP-violating phase δ

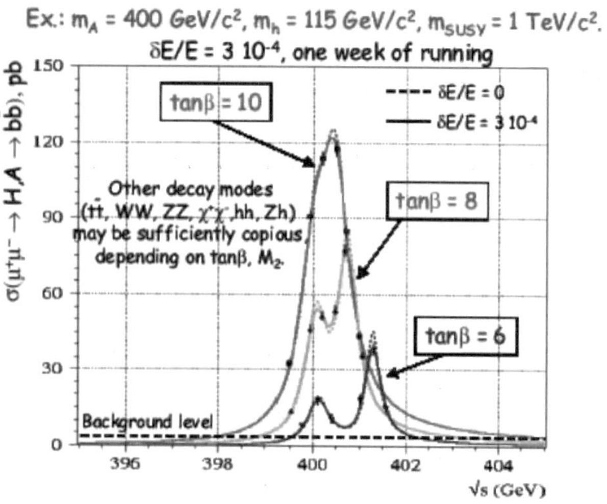

Fig. 8. Study of the supersymmetric H,A system at a muon collider

production with a useful rate. These remarkable properties make muon colliders superb tools for the study of Higgs resonances, especially if, as predicted in supersymmetry, there exist a pair H, A of opposite CP quantum numbers which are nearly degenerate in mass, as evidenced in Fig. 8. The study of this system is extremely difficult with any other machine and a unique investigation of the possible CP violation in the Higgs system would become possible.

6 Conclusions

We acknowledge the support of the European Community - Research Infrastructure Activity under the FP6 "Structuring the European Research Area" programme (CARE, contract number RII3-CT-2003-506395)

References

1. ECFA/CERN Studies of an European Neutrino Factory Complex, A. Blondel (ed.) et al.: CERN-2004-002 - ECFA-04-230, `http://preprints.cern.ch/cernrep/2004/2004-002/2004-002.html`
2. S. Geer: Phys. Rev. D **57**, 6989 (1998)
3. `http://www.cap.bnl.gov/mumu`
4. P. Gruber et al.: The Study of a European Neutrino Factory Complex, in [1] p. 7; Feasibility Study on a Neutrino Source Based on a Muon Storage Ring, D. Finley, N. Holtkamp, eds.: (2000), `http://www.fnal.gov/projects/muon_collider/reports.html`; Feasibility Study-II of a Muon-Based Neutrino Source, S. Ozaki, R. Palmer, M. Zisman, J. Gallardo, eds.: BNL-52623, June 2001, available at `http://www.cap.bnl.gov/mumu/studyii/FS2-report.html`; M.M. Alsharoá et al.: Phys. Rev. ST Accel. Beams **6**, 081001 (2003); Neutrino Factory and Beta Beam Experiments and Developments, (S. Geer, M. Zisman, eds.): Report of the Neutrino Factory and Beta Beam Working Group, APS Multi-Divisional Study of the Physics of Neutrinos, July 2004
5. M. Campanelli et al.: Oscillation Physics with a Neutrino Factory, arXiv:hep-ph/0210192 in [1] p. 138
6. M.L. Mangano et al.: Physics at the front-end of a Neutrino Factory: a quantitative appraisal, arXiv:hep-ph/0105155, in [1] p. 187
7. K. McFarland: these proceedings; G.P. Zeller et al.: CCFR coll., Phys. Rev. Lett. **88**, 091802 (2002); hep-ex/0110059
8. A. De Rùjula, M.B. Gavela, P. Hernàndez: Nucl. Phys. B **547**, 21 (1999); B. Autin, A. Blondel, J. Ellis (eds.): CERN yellow report CERN 99-02, ECFA 99–197; C.M. Ankenbrandt et al.: Phys. Rev. ST Accel. Beams **2**, 081001 (1999); A. Blondel et al.: Nucl. Instr. Meth. Phys. Res. A **451**, 102 (2000); C. Albright et al.: FERMILAB-FN-692, hep-ex/0008064; D. Harris et al.: Snowmass 2001 Summary, hep-ph/0111030; A. Cervera et al.: Nucl. Phys. B **579**, 17 (2000), Erratum-ibid. B **593**, 731–732 2001; M. Koike, J. Sato: Phys. Rev. D **62**, 073006 (2000)
9. See for instance S. Kraml et al: Physics opportunities at $\mu^+\mu^-$ Higgs factories, in [1], p. 337

Eur Phys J A (2005) **24**, s2, 187–187
DOI: 10.1140/epjad/s2005-04-047-5

EPJ A direct
electronic only

The MINERνA experiment at FNAL

Kevin S. McFarland[a]

University of Rochester, Rochester, NY 14610, USA

Received: 15 December 2004 / Published Online: 8 February 2005

Abstract. I present detector and physics capabilities of the MINERνA experiment.

PACS. 13.15.+g Neutrino interactions – 25.30.Pt Neutrino scattering from nuclei

MINERνA is a dedicated neutrino cross-section experiment planned for the NuMI beamline at Fermilab [1]. The detector (Fig. 1) consists of a low-mass active scintillator target surrounded by calorimetric detectors and upstream heavy nuclear targets. The low-mass target allows for separation of final state particles and therefore the identification of exclusive final states. The surrounding calorimeters ensure complete energy collection in the events, except for final state muons, which may be measured in the MINOS experiment's near detector located immediately downstream of MINERνA.

The physics goals of the experiment include measurements of the A-dependence of quasi-elastic ($\nu n \rightarrow \mu^- p$) scattering, measurement of the axial form factor of the nucleon at high Q^2 (shown in Fig. 2), tests of quark-hadron duality in the axial current and measurements of coherent single-pion prodution in the Coulomb field of a target nucleus (Fig. 3).

The physics of neutrino cross-sections is an exciting subject in its own right and explores physics in the axial current similar to that being probed at high precision in the vector current at high intensity electron scattering machines. These measurements are also important for future neutrino oscillation experiments planned with beams of energies 1–a few GeV, where neutrino cross-sections are difficult to predict theoretically and are poorly measured [2]. Results from the MINERνA experiment will significantly reduce errors from unknown neutrino cross-sections in the MINOS, T2K and NOνA experiments.

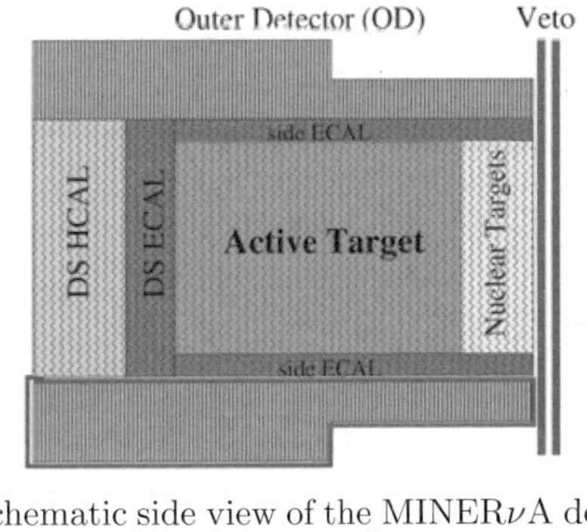

Fig. 1. A schematic side view of the MINERνA detector. Neutrinos enter from the *right*

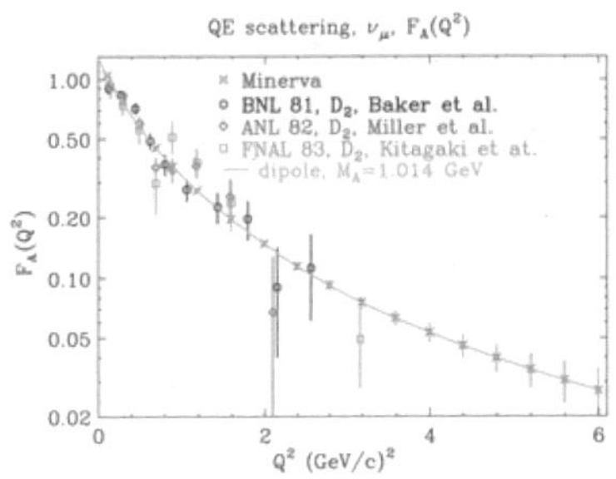

Fig. 2. Estimation of F_A from a sample of Monte Carlo neutrino quasi-elastic events recorded in the MINERνA active carbon target, assuming a dipole form with $M_A = 1.014$ GeV/c^2. Also shown is F_A from bubble chamber experiments

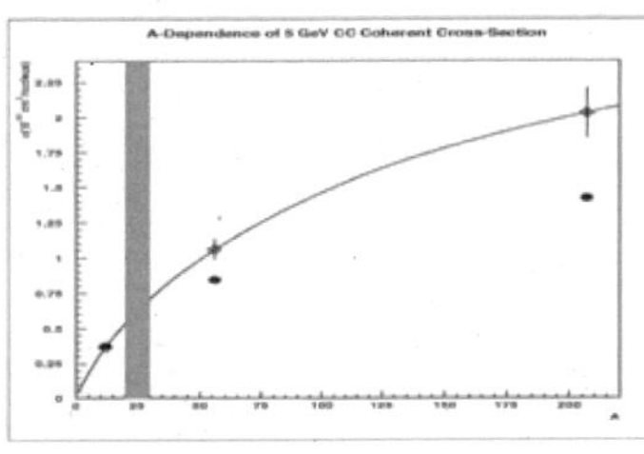

Fig. 3. Coherent cross-sections for 5 GeV neutrinos vs. atomic number. The *solid curve* and *circles* are two different models; crosses show expected MINERνA measurements. The *shaded band* indicates the A region of previous experiments

References

1. D. Drakoulakos et al. [Minerva Collaboration]: "Proposal to perform a high-statistics neutrino scattering experiment using a fine-grained detector in the NuMI beam," arXiv:hep-ex/0405002
2. D.A. Harris et al. [MINERvA Collaboration]: "Neutrino scattering uncertainties and their role in long baseline oscillation experiments," arXiv:hep-ex/0410005

[a] presenting on behalf of the MINERνA Collaboration

VI Concluding talks

Eur Phys J A (2005) **24**, s2, 191–195
DOI: 10.1140/epjad/s2005-04-048-4

EPJ A direct
electronic only

Frontiers of polarized electron scattering experiments

Krishna S. Kumar[a]

Department of Physics, University of Massachusetts, Amherst, MA 01003, USA

Received: 15 December 2004 / Published Online: 8 February 2005

Abstract. Parity-violating electron scattering has developed into a precise and sensitive tool to probe the structure of weak neutral current interactions at $Q^2 \ll M_Z^2$. Steady improvements in experimental techniques have made feasible asymmetry measurements with precision approaching 10 parts per billion and fractional accuracy of a few percent. In the future, upgrades of new facilities, new laboratories and further refinements of experimental techniques should allow us to explore new aspects of the strong and electroweak interactions. We describe some of these ideas in this article.

PACS. 12.15.Mm Neutral Currents – 11.30.Er C, P, T and other discrete symmetries

1 Introduction

The past couple of decades have seen the emergence of parity-violating electron scattering (PVES) experiments to the forefront of probing electroweak and strong interactions at low energy. Major experimental innovations have been made [1], such as the development of highly polarized electron beams, dense cryogenic targets, radiation-hard integrating detectors that effectively count at GHz rates, counting detectors that can count at MHz rates, precision beam monitoring and precision measurements of beam polarization with fractional accuracy approaching 1%.

When the accuracy of PVES measurements is combined with new accelerator capabilities, the investigation of novel aspects of weak neutral current (WNC) interactions becomes feasible. Further, the high rate capabilities and the ability to measure small asymmetries opens the window into measurements of beam-normal asymmetries, where the incident beam polarization is transverse to the scattering plane. In the following, we introduce the observables of interest and discuss possible new measurements.

1.1 Parity-violating asymmetries A_{PV}

In PVES, one measures the fractional difference A_{PV} of the scattered flux from an unpolarized target for incident longitudinally polarized electrons whose spins are aligned along or against the momentum. For $Q^2 \ll M_Z^2$, the leading contribution to the asymmetry is the ratio of the WNC amplitude (mediated by the Z boson) to the electromagnetic amplitude and thus A_{PV} is small and proportional to the four-momentum transfer Q^2:

$$A_{PV} \equiv \frac{\sigma_R - \sigma_L}{\sigma_R + \sigma_L} \approx \frac{|A_Z|}{|A_\gamma|} \approx \frac{G_F Q^2}{4\sqrt{2}\pi\alpha}. \qquad (1)$$

For typical Q^2 values of interest (between 0.01 and 1 GeV^2), A_{PV} ranges from 0.1 to 100 parts per million (ppm). The WNC amplitude is typically dominated by the piece of the low energy WNC interaction that arises from the product of the axial-vector electron coupling and the vector coupling of the target particle.

The discovery potential of PVES measurements arises in several different kinematic situations. In some cases, the weak vector coupling of the target provides access to a new linear combination of the underlying quark vector currents, leading to novel insights of nuclear and nucleon structure [2]. Experiments HAPPEX, SAMPLE, A4 and G0 (probing the strange structure of the nucleon) and the planned PREX experiment (probing the neutron skin of ^{208}Pb), fall under this category.

In special circumstances, if the vector coupling of the target is very well-known or if there are cancellations that make the ratio of weak and electromagnetic amplitudes insensitive to hadron structure, one can probe new particle physics at TeV scales [3]. The SLAC E122 (Deep Inelastic Scattering), SLAC E158 (Møller scattering) and Qweak (elastic electron-proton scattering) experiments fall under this category.

A third class of measurements access the small amplitudes that arise from the product of the electron vector coupling and the target axial-vector coupling. The A4, SAMPLE and G0 experiments probe the axial nucleon current in this manner.

[a] Research funded by US Department of Energy Grant No. DE-FG02-88R40415.A018

2 Beam-normal asymmetries A_T

If the incident beam is polarized transverse to the beam direction, one can construct an asymmetry:

$$A_T \equiv \frac{2\pi}{\sigma^\uparrow + \sigma^\downarrow}\frac{d(\sigma^\uparrow - \sigma^\downarrow)}{d\phi} \quad (2)$$

which is proportional to $\mathbf{S_e} \cdot (\mathbf{k_e} \times \mathbf{k'_e})$, where $\mathbf{S_e}$ is the incident electron spin vector and the incident and final electron 3-vectors $\mathbf{k_e}$ and $\mathbf{k'_e}$ define the scattering plane. The primary contribution to A_T comes from two photon-exchange. The transverse asymmetry is suppressed by the electron boost and thus an order of magnitude estimate is $A_T \sim \alpha m_e/\sqrt{s}$. However, even for elastic electron-proton scattering, the intermediate states between the two virtual photons have to be summed over the full virtuality kinematically available to each photon, resulting in enhancements that are quite sensitive to nucleon excited state structure [4,5,6].

Recently, analysis of data on proton elastic form factors have shown a disagreements between cross-section and asymmetry measurements. The real part of two-photon exchange amplitude is a good candidate to explain a signficant part of the discrepancy [7]. While A_T measurements are sensitive to the imaginary part of the same amplitude, detailed studies might be possible since the amplitude constitutes the leading contribution. Thus, A_T measurements have emerged as an important component of precision studies of nucleon structure.

3 PV deep inelastic scattering at high x

The upgrade of Jefferson Laboratory (Jlab) to 11 GeV incident energy will allow precision measurements in parity-violating deep inelastic scattering (PV DIS). For the first time, high statistics can be accumulated at high $x \sim 0.7$, where x is the fraction of the nucleon momentum carried by the struck quark. PV DIS provides access to novel aspects of nucleon structure, complementing and enhancing precision electromagnetic DIS studies.

A_{PV} in DIS can be written as

$$A_{PV} = Q^2 \frac{G_F}{2\sqrt{2}\pi\alpha}\left[a(x) + \frac{1-(1-y)^2}{1+(1-y)^2}b(x)\right], \quad (3)$$

$$a(x) \equiv \Sigma_i f_i(x) C_{1i} q_i / \Sigma_i f_i(x) q_i^2, \quad (4)$$

$$b(x) \equiv \Sigma_i f_i(x) C_{2i} q_i / \Sigma_i f_i(x) q_i^2. \quad (5)$$

Here, $C_{1i}(C_{2i})$ are the weak vector(axial-vector) weak charges for the ith quark flavor, $f_i(x)$ are parton distribution functions and q_i are the electromagnetic charges. The $a(x)$ term arises from the product of the electron axial-vector coupling and the quark vector coupling and is typically the dominant term. For an isoscalar target such as deuterium, the dependence on structure largely cancels out in the A_{PV} ratio of the weak and electromagnetic amplitudes:

$$a(x) = \frac{6}{5}\left[(C_{1u} - \frac{1}{2}C_{1d}) + \text{corrections}\right]; \quad (6)$$

$$b(x) = \frac{6}{5}\left[(C_{2u} - \frac{1}{2}C_{2d})\frac{q(x)-\bar{q}(x)}{q(x)+\bar{q}(x)} + \text{corrections}\right], \quad (7)$$

where $q(x) = u(x) + d(x)$. For scattering off the proton,

$$a(x) = \left[\frac{u(x) + 0.91d(x)}{u(x) + 0.25d(x)}\right]. \quad (8)$$

3.1 Charge symmetry violation

As can be seen from 6, $a(x) \approx 1.15$ for an isoscalar target, independent of x. This results from the assumption of charge symmetry, where the u-quark distribution in the proton is the same as the d-quark distribution in the neutron, with a similar assumption for the proton d-quark distribution: $u^p = d^n$ and $d^p = u^n$. If $a(x)$ can be measured with high precision over a range of x values, one is thus quite sensitive to charge symmetry violation (CSV). If one defines CSV parameters:

$$\delta u(x) = u^p(x) - d^n(x); \quad \delta d(x) = d^p(x) - u^n(x), \quad (9)$$

then the dependence on the parity-violating asymmetry for an isoscalar target is [8]:

$$\frac{\delta A_{PV}}{A_{PV}} = 0.28\frac{\delta u - \delta d}{u + d} \equiv 0.28 R_{CSV}. \quad (10)$$

While R_{CSV} is known to be less than 0.01 for $x < 0.4$ from neutrino DIS measurements [9], a bag model calculation suggests that $R_{CSV} \sim 0.01$ for $x \sim 0.4$ and rising to 0.02 for $x \sim 0.6$. At high x, knowledge of $u + d$ is limited. As $x \to 1$, if $u + d$ falls off more rapidly than $\delta u - \delta d$, then R_{CSV} might rise to 0.1 at $x \sim 0.7$, which would be observable with a 1% A_{PV} measurement. Further, R_{CSV} is quite unconstrained at large x. There is the tantalizing possiblity that R_{CSV} in the moderate and high x region is a factor of 3 bigger than above-mentioned values, which would be large enough to explain the 3σ discrepancy in the neutrino-nucleon DIS measurement (NuTeV anomaly) [9] and would be a very important discovery.

3.2 d/u at high x

As can be seen from 8 for PV DIS off the proton, $a(x)$ is quite sensitive to the ratio $d(x)/u(x)$. The value of d/u as $x \to 1$ is a very important parameter to pin down in DIS physics. It is required in order to properly constrain fits of parton distribution functions and impacts predictions for QCD processes at high energy colliders.

More importantly, d/u at high x provides new information on important pieces of the nucleon wave function. There is empirical evidence that the minority quark in the nucleon is suppressed at high x, an intuitive notion in terms of a hyperfine interaction. While the SU(6) wave function would predict $d/u \sim 0.5$, simple SU(6)-breaking arguments would predict $d/u \sim 0$. However, a perturbative QCD calculation predicts $d/u = 0.2$ as $x \to 1$ [10].

Currently, the best estimates of d/u comes from ^{2}H DIS structure function data, but uncertainties in the ^{2}H wave-function limits the ability to discriminate between various predictions for d/u [11]. There are plans to measure d/u via the ratio of ^{3}H and ^{3}He structure functions and also via measurements of deuteron structure functions with tagged slow recoiling protons.

A precise enough measurement of $a(x)$ for the proton at $x \sim 0.7$ would be able to distinguish between competing predictions for d/u as $x \to 1$. The advantage of A_{PV} measurements over other methods is that there are no nuclear corrections since the PV DIS measurement can be made on a proton target.

3.3 Higher twist effects

The topics discussed above can be accessed only if the physics of leading twist dominates the weak and electromagnetic amplitudes. However, it is well-known from unpolarized structure function studies that hadronic corrections from higher twist (HT) effects might add Q^2 dependence to the asymmetry:

$$A_{PV}(x, Q^2) = A_{PV}(x)(1 + C(x)/Q^2). \qquad (11)$$

It is not possible to calculate $C(x)$ from first principles, so ideally it must be extracted from data.

PV DIS has the potential to probe for interesting HT effects, since the A_{PV} weak-electromagnetic amplitude ratio would be sensitive to the amount of contributions from coherent quark combinations such as diquarks. For example, if spin-0 diquarks dominate, then the HT effects are expected to be small, with a $1/Q^4$ dependence [12]. Additionally, novel HT effects arising from the interference of HT effects involving γ exchange on one quark and Z exchange on a different quark might contribute [13].

Recently, an NNLO QCD analysis has shown that HT effects in the unpolarized structure functions are small for $x < 0.6$ but does not rule out 10 to 20% effects for $x \sim 0.7$ [14]. It would be interesting to search for HT effects in PV DIS. If they are large, they would point to dynamics specific to DIS processes involving $\gamma - Z$ interference. Determining whether HT effects are small or large is critical to ensure that the physics of CSV and the d/u ratio can be extracted cleanly. The size of HT effects can already be constrained or discovered with a 6 GeV beam; there is a proposal under consideration at Jlab to pursue this measurement [15].

3.4 Experimental equipment for PV DIS at high x

To comprehensively address the physics topics discussed above experimentally, a series of A_{PV} measurements in the range of 1 to 2% accuracy are required for the x range from 0.3 to 0.7, with a lever arm of a factor of 2 in Q^2 while keeping $W^2_{\text{min}} > 4$ and $Q^2_{\text{min}} > 1$. With the upgrade of Jlab, high luminosity with a beam energy of 11 GeV becomes possible for the first time. However, to achieve sufficient statistics at the highest possible Q^2, a spectrometer with at least 50% acceptance in the azimuth is required.

A conventional magnetic spectrometer with the requirements specified above would be prohibitively expensive. However, it might be possible to employ a calorimeter with the ability to identify and count clusters at rates of 10 to 20 MHz, such as the one employed by the Mainz A4 experiment. The scattered electron energies of interest would range from 2 to 4 GeV. A large acceptance toroid could serve as a sweeping magnet, to remove low energy pions, Møller electrons and to shield the calorimeter from line-of-sight photons. The conceptual design for such an apparatus is now under way, and might be potentially useful for other applications besides PV DIS.

3.5 Transverse asymmetries

As a bonus, the use of a device such as the one described in the previous section makes possible a precision study of an entirely new process. Beam-normal asymmetries in the DIS region would become measureable with high precision. While leading twist contributions to transverse asymmetries might be of the order of ppm, HT effects might be enhanced by one to two orders of magnitude [16] due to large logarithms in the sum over intermediate states in the 2-photon amplitude. Thus, HT effects could potentially be studied in detail.

4 Asymmetries at low Q^2 and forward angle

The unprecedented high luminosity available at Jlab as well as the upgrade of the energy provides new opportunities to measure asymmetries with sufficient precision at very forward angles, in a Q^2 range between 0.05 and 1 GeV^2, both in elastic scattering as well as in highly inelastic scattering ($W^2 > 4$ GeV^2).

4.1 Longitudinal asymmetries

The E158 experiment has carried out an auxiliary measurement of the PV asymmetry in electron-proton scattering at $Q^2 \sim 0.05$ GeV^2. The measurement is consistent with roughly $A_{PV} \approx -10^{-4}Q^2$ for the inelastic scattering component (mostly real and virtual photoproduction), which constituted about 30% of the flux. The remaining signal comes from elastic electron-proton scattering, with an asymmetry that is a factor of 5 smaller.

It would be interesting to measure A_{PV} in inelastic electron-proton scattering at forward angles, mapping out the asymmetry variation as function of the target recoiling mass W. This would provide new information on parity-violating real and virtual photoproduction, which is presumably related to electromagnetic photoproduction via an isospin rotation. Thus, prevailing parametrizations on leptoproduction, primarily from the HERA collider [17], would be tested in a new way.

4.2 Transverse asymmetries

As described in Sect. 2, A_T measurements have emerged as an important probe of nucleon structure. These asymmetries are enhanced as the incident beam energy is increased, since inelastic intermediate states with one quasi-real photon makes a significant contribution. At very low Q^2 (< 0.1 GeV^2) and forward angle, A_T can be predicted using the optical theorem [4,5]. At intermediate Q^2 ($0.1 < Q^2 < 1$ GeV^2), the inelastic amplitudes from single and multiple pion production are expected to dominate [4]. As Q^2 is increased, A_T receives increasing contributions from off-forward structure functions. At $Q^2 \sim 1$ GeV^2, A_T can be calculated in a perturbative QCD framework where it is related to Generalized Parton Distributions [18].

There are currently plans for A_T measurements at the G0 and A4 experiments. The kinematics are well-suited to test the regime of single pion production. On the other hand, it would interesting to measure A_T at forward angle at very low Q^2 (< 0.1 GeV^2) and at high $Q^2 \sim 1$ GeV^2 in order to connect to the optical theorem at one extreme and parton distributions at the other extreme.

4.3 Experimental program

The longitudinal and transverse asymmetry measurements discussed above require the ability to measure very high flux rates ($\sim$ 100 MHz) and small asymmetries ($\sim$ 1 ppm). A spectrometer/detector package that provides this along with azimuthal coverage to measure A_T with high efficiency would have to be similar in concept to the E158 design, where the entire primary and scattered beam were enclosed in a set of quadrupole doublets. This allows the separation of elastic and inelastic electron-proton scattering events from background while providing acceptance in the full range if the azimuth. A compact radiation-hard calorimeter would be placed downstream of the quadrupole spectrometer to integrate the scattered flux.

Alternatively, a more conventional forward angle spectrometer setup perhaps enhanced with septum magnets could be contemplated, although the solid angle would be significantly smaller. Indeed, these studies can be launched already with a 6 GeV beam and a proposal is under consideration at Jlab to pursue this measurement [19].

5 Electroweak physics

As mentioned in the introductory paragraphs, with judicious choice of target and kinematics, it is possible to probe the structure of the WNC interaction itself, with little uncertainty from hadron structure. High precision measurements of such amplitudes at $Q^2 \ll M_Z^2$ continue to be important and complementary to measurements at high energy colliders [20]. The latter are typically sensitive to parity-conserving observables that probe contact interaction scales better than 10 TeV. They are typically less sensitive to PV contact interaction scales by an order of magnitude.

There are only a select few reactions that can probe the 10 TeV scale in fixed target WNC interactions [3]. Successful measurements have been reported in atomic parity violation [21], neutrino DIS (NuTeV) [22] and fixed target Møller scattering (E158) [23]. One legacy of such precision measurements is that they are sensitive enough to observe electroweak radiative corrections (higher-diagrams that involve W and Z bosons in quantum loops). A convenient manifestation of this sensitivity is the "running", as a function of momentum transfer Q^2, of the weak-mixing angle $\sin^2\hat{\theta}_{\mathrm{W}}(Q^2)_{\overline{\mathrm{MS}}}$ [24].

It is important to measure $\sin^2\theta_W$ with sufficient precision to observe the running in as many different reactions as possible in order to comprehensively probe for physics beyond the standard model at the TeV scale. The E158 measurement has established the running of $\sin^2\theta_W$ at the 7σ level. The NuTeV measurement has a 3σ descrepency with the standard model prediction, the origin of which is a subject of active theoretical and experimental debate. In the following, we describe plans and new ideas to make more precision measurements of $\sin^2\theta_W$ at low Q^2.

5.1 Elastic electron-proton scattering

At sufficiently forward angles and low Q^2, the hadronic structure undertainty in the WNC elastic electron-proton amplitude becomes small enough such that one can measure the underlying coherent $2u + d$ e-q amplitude combination to high precision. This combination is proportional to $1 - 4\sin^2\theta_W$, so that a 4% measurement of A_{PV} would achieve a precision of $\delta(\sin^2\theta_W) = 0.0007$

The experiment would use a 1 GeV 180μA electron beam incident on a 35 cm liquid hydrogen target. Scattered electrons would be focused by a toroidal field in the full range of the azimuth on to Čerencov detectors. The predicted asymmetry is 0.3 ppm. The experiment, named Qweak [25], has been approved for initial construction at Jlab and the first data collection is projected for 2008.

It turns out to be quite complementary to compare A_{PV} measurements from elastic electron-proton scattering and Møller scattering. Various models for new physics at the TeV scale can induce constructive or destructive interferences in one or both reactions. In particular, one might be able to distinguish between conventional and R-Parity breaking SUSY models, which is relevant to the viability of a SUSY dark matter candidate [26].

5.2 Deep inelastic scattering

If the PV DIS asymmetry can be measured to an accuracy of 1% at $Q^2 \sim 5$ GeV^2 and $x \sim 0.35$ for an isoscalar target such as deuterium, then the parameter $a(x)$ can be measured with high precision, free from hadron structure uncertainties [27]. This measurement would be robust only if the program described in Sect. 3 is carried out. Indeed, as can be seen from 6, $a(x)$ is independent of x and simply a function of $\sin^2 \theta_W$ under the assumption of charge symmetry and assuming that HT effects are either directly measured or constrained.

The measurement is interesting for several reasons. Firstly, it would test the WNC amplitude in the lepton-quark sector, where there is currently a 3 σ discrepancy in the NuTeV result. Secondly, combined with other measurements in elastic electron-proton scattering, precise constraints would be possible on the lesser known axial-vector quark couplings C_{2i}. This would, among other things, provide complementary constraints on various models with new heavy Z' bosons [3].

5.3 Møller scattering

The PV asymmetry in fixed target Møller scattering is unique among WNC processes at $Q^2 \ll M_Z^2$ in being able to probe for new physics at the 10 TeV scale in a purely leptonic reaction. The recently completed E158 experiment has provided the most precise measurement of $\sin^2 \theta_W$ at low energy [23]. Nevertheless, it is useful to consider future ideas to measure the asymmetry with higher precision.

One way to achieve higher precision is to consider a high luminosity measurement with the 12 GeV beam at the upgraded Jlab [27]. If the 40 ppb asymmetry can be measured with a precision of 3%, then the weak mixing angle could be measured with a precision of 0.0004. At this level, the asymmetry is sensitive to SUSY loops in the radiative corrections. In particular, the higher precision facilitates further stringent constraints on R-parity violating terms, with important implications for SUSY dark matter candidates [26].

The figure of merit for Møller scattering (the precision with which the asymmetry can be measured for a given luminosity) rises linearly with incident beam energy. Thus, at a future linear collider operating at 250 to 500 GeV, Møller scattering offers a way to measure the weak mixing angle to a precision of 0.00007 [28]. At this level, the asymmetry becomes sensitive to radiative corrections involving fundamental scalars such as the Higgs boson, providing an important crosscheck of the symmetry breaking sector of the electroweak theory. Such a measurement would be competitive with precision electroweak measurements such as the W boson mass and left-right asymmetries at future lepton and hadron colliders.

6 Outlook

Parity-violating electron scattering will continue to remain in the forefront of studies of low energy electroweak interactions in the future. Over the next few years, there are exciting possibilities for novel experimental programs as new facilities are developed, addressing physics topics from nucleon structure to physics beyond the standard model at the TeV scale.

Acknowledgements. It is a pleasure to thank the organizers for a stimulating workshop. Input and discussions with A. Afanasev, B. Holstein, T. Londergan, K. McFarland, W. Melnitchouk, B. Pasquini, M.J. Ramsey-Musolf, P.A. Souder, A.W. Thomas and M. Vanderhaeghen are gratefully acknowledged.

References

1. K.S. Kumar, P.A. Souder: Prog. Part. Nucl. Phys. **45**, S333–S395 (2000), and references therein; D.H. Beck, B.R. Holstein: Int. J. Mod. Phys. E **10**, 1–41 (2001)
2. M.J. Musolf et al: Phys. Rept. **239**, 1 (1994)
3. M.J. Ramsey-Musolf: Phys. Rev. C **60**, 015501 (1999)
4. B. Pasquini, M. Vanderhaeghen: Phys. Rev. C **70**, 045206 (2004)
5. A. Afanasev, N.P. Merenkov: Phys. Rev. D **70**, 073002 (2004)
6. F. Maas et al: nucl-ex/0410013 (2004)
7. P.A.M. Guichon, M. Vanderhaeghen: Phys. Rev. Lett. **91**, 142303 (2003); P.G. Blunden et al: Phys. Rev. Lett. **91**, 142304 (2003)
8. P.A. Souder: to be published in the proceedings of *HiX2004*, Marseilles, France, (2004)
9. J.T. Londergan, A.W. Thomas: hep-ph/0407247 (2004), and references therein
10. G.R. Farrar, D.R. Jackson: Phys. Rev. Lett. **43**, 246 (1979)
11. W. Melnitchouk et al: Phys. Rev. Lett. **84**, 5455 (2000); W. Melnitchouk, A.W. Thomas: Phys. Lett. B **377**, 11 (1996)
12. R.L. Jaffe: private communication
13. S. Brodsky: In *Proceedings from Jlab/Temple University HiX2000 Workshop*, Philadelphia, USA, (2000)
14. A.D. Martin et al: Eur. Phys. J. C **35**, 325 (2004)
15. Jlab Proposal P-05-007, X. Zheng: contactperson
16. A. Afanasev: private communication
17. A.C. Caldwell: Acta Phys. Polon. B **33**, 3599–3608 (2002)
18. M. Gorshteyn et al: Nucl. Phys. A **741**, 234–248 (2004)
19. Jlab Proposal P-05-005, P. Bosted: contactperson
20. K.S. Kumar et al: Mod. Phys. Lett. A **10**, 2979 (1995)
21. S.C. Bennett, C.E. Wieman: Phys. Rev. Lett. **82**, 2484–2487 (1999)
22. G.P. Zeller et al: Phys. Rev. Lett. **88**, 091802 (2002)
23. P.L. Anthony et al: Phys. Rev. Lett. **92**, 181602 (2004)
24. A. Czarnecki, W.J. Marciano: Phys. Rev. D **53**, 1066 (1996); A. Czarnecki, W.J. Marciano: Int. J. Mod. Phys. A **15**, 2365 (2000); J. Erler, M.J. Ramsey-Musolf: hep-ph/0409169; A. Ferroglia, G. Ossola, A. Sirlin: Eur. Phys. J. C **34**, 165 (2004)
25. D.S. Armstrong et al: Qweak Collaboration, Published in *AIP Conf. Proc.* **698**, 172–175 (2004)
26. A. Kurylov, M.J. Ramsey-Musolf, S. Su: Phys. Rev. D **68**, 035008 (2003)
27. D. Mack et al: in *Pre-Conceptual Design Report for The Science and Experimental Equipment for the 12 GeV Upgrade of CEBAF*, (2004) unpublished
28. K.S. Kumar: in *DPF/DPB Summer Study on New Directions in High-Energy Physics*, econf **C960625**, (1996) NEW168

Eur Phys J A (2005) **24**, s2, 197–204
DOI: 10.1140/epjad/s2005-04-049-3

EPJ A direct
electronic only

Parity-violation with electrons: Theoretical perspectives

M.J. Ramsey-Musolf

Kellogg Radiation Laboratory, California Institute of Technology, Pasadena, CA 91125, USA, e-mail: mjrm@caltech.edu

Received: 1 January 2005 / Published Online: 8 February 2005

Abstract. I review recent progress and developments in parity-violating electron scattering as it bears on three topics: strange quarks and hadron structure, electroweak radiative corrections, and physics beyond the Standard Model. I also discuss related developments in parity-conserving scattering with transversely polarized electrons as a probe of two-photon processes. I conclude with a perspective on the future of the field.

PACS. 11.30.Er Parity symmetry – 25.30.-c Lepton-induced reactions – 12.15.Mm Neutral currents

1 Introduction

Parity-violating electron scattering (PVES) was once considered something of a specialized – almost exotic – subfield of nuclear physics. In the past decade, however, the field has made substantial advances and has become something of a mainstream area of research. This maturation of the field has been nicely summarized in the two PAVI meetings: Mainz (2002) and Grenoble (2004). My own involvement has now spanned a decade and a half, so I feel somewhat justified in providing a theoretical perspective on the status of the field. A rather extensive review can be found in my contribution to the Mainz PAVI proceedings [1], and this one will be somewhat abbreviated. Here, I will try to highlight what I think we have learned since the early 1990's and what I think may be the important directions for the future.

In reviewing the evolution of PVES, one can identify three rough eras: (1) the 1970's and 1980's, which Wieman and Masterson have called "ancient history" [2]; (2) the 1990's and first half of this decade, the "modern era"; and (3) the future, lasting perhaps into the 2020's. The focus in the early era was on testing the neutral current structure of the Standard Model (SM). Pioneering experiments were carried out in both atomic PV (see [2] for a review) as well as in the SLAC deep inelastic scattering (DIS) experiment [3,4] that were followed by the Mainz ^{8}Be quasielastic [5] and MIT-Bates ^{12}C elastic [6] PVES measurements. In the 1990's, the structure of the SM neutral currents had been well established, so the emphasis shifted to using the SM neutral current as a probe of nucleon and nuclear structure. Here, the basic idea was that the weak neutral current depends on a different linear combination of the light quark currents than enters the electromagnetic current. Thus, by making a judicious choice of targets and kinematics, one can use PVES – in combination with ordinary electron scattering – to perform a flavor decomposition of the nucleon's vector current response. More recently, there has been a resurgence of interest in exploiting PVES to study the electroweak interaction itself. In this regard, the pioneering experiment has been the SLAC Møller experiment [7] that will be followed by the equally demanding JLab Q-Weak experiment [8]. Looking down the road, one may see follow-ups to these measurements, possibly including more precise versions of the original SLAC deep inelastic experiment at JLab or the Møller experiment at either JLab or the linear collider.

With this context in mind, let me try to summarize the recent developments in three fields, focusing primarily on the "modern era": strange quarks and nucleon structure; radiative corrections, including both electroweak and QED; and the search for new physics.

2 Strange quarks

The heyday of PVES has been dominated by strange quarks in the nucleon. Coming amidst the "spin crisis" of the early 1990's, wherein there were indications that strange quarks carried a more substantial fraction of the nucleon's spin (and mass) than one might naïvely think (for a review, see [9]), the question arose as to whether strange quarks might also play a substantial role in the nucleon's electromagnetic structure. Three developments – two theoretical and one experimental – catalyzed the use of PVES to address this question. Theoretically, Kaplan and Manohar observed how the weak neutral current (WNC) of the SM – in conjunction with the electromagnetic (EM) current – provided the tool needed to carry out the flavor decomposition of the nucleon's vector current structure [10]. Subsequently, Jaffe noted that dispersion theory analyses of the nucleon's isoscalar form factors implied large couplings of the nucleon to the $\phi(1020)$ and that this OZI-violation would imply sizeable strange quark vector current form factors as a consequence [11]. On the experimental side, McKeown showed that by measuring the backward angle PV asymmetry under conditions that were experimentally feasible, one would be sensitive to strange magnetic form factors of the magnitude implied by Jaffe's analysis [12]. (For other important, early work on this topic, see [13,14,15,16]).

The field was shortly off and running, resulting in the SAMPLE program at MIT-Bates [17,18,19,20,21,22], the HAPPEX [23,24] and G0 [25] experiments at JLab, and the A4 program at Mainz [26,27,28]. A related development that I will not focus on here is the development of the JLab experiment designed to study the neutron distribution of ^{208}Pb (see [29] and references therein). The idea – for which Donnelly, Dubach, and Sick deserve the credit [30] – is again to exploit the complementarity of the WNC and EM charge operators: the EM charge operator is sensitive primarily to protons, while the WNC primarily sees neutrons. The details of each of these experiments has been discussed elsewhere in this meeting, so I will not comment on any further on the specifics.

While the G0, HAPPEX, and A4 experiments are not yet completed, the results obtained to date have taught us that the role of strange quarks in the nucleon vector current form factors is likely to be relatively less important than in the case of the nucleon mass and spin. In particular, the benchmark Jaffe predictions appear to be too large compared to experiment. At least, it is not consistent to have both sizable strange magnetism and strange electricity in the nucleon. The completion of the experimental program will definitively tell us whether there is room for either one to be relatively important.

Theoretically, we have learned that strange quark dynamics are more subtle than the simplest pictures might suggest and that – because the effects are not so large – obtaining a quantitatively reliable description and realistic physical picture remains a challenging task. In terms of the simplest physical pictures that one might use to guide intuition, we have learned that neither vector meson dominance nor kaon cloud dominance is right. Vector meson dominance works well in describing the dynamics of pseudoscalar mesons; it has long been known, for example, that the *a priori* unknown low energy constants (LECs) in the $\mathcal{O}(p^4)$ chiral Lagrangian for the octet of pseudoscalar mesons are well-described by vector meson dominance [32]. Although a similar *ansatz* allows one to obtain a reasonable fit to the nucleon isoscalar vector form factor in dispersion theory, Jaffe's generalization of it to the strange quark sector is now ruled out by experiment (for an up-date of the analysis in [11]).

The other picture that used to guide one's intuition is that of a kaon cloud around the nucleon. Its field theoretic description involves fluctuations of the nucleon into kaon-hyperon intermediate states whose effects are computed using loop graphs in perturbation theory. Now, one might rightfully object that given the large [$\mathcal{O}(1)$] meson nucleon coupling constants (*e.g.*, $g_A \approx 1.26$), one has no basis for believing perturbation theory. Nevertheless, many theorists (myself included) proceeded to carry out such calculations anyway, using various models for the hadronic vertex form factors (for a list of references, see [18]). The motivation behind this mini-industry was an old computation of the nucleon's isovector form factors by Bethe and DeHoffman using pion loops [33]. Despite the presence of the $\mathcal{O}(1)$ couplings, it worked, and so, based purely on this phenomenological success, people proceeded with the same approach for the strange form factors.

What most people did not realize, however, is that in the late 1950's, Federbush, Goldberger, and Treiman (FGT) showed – using dispersion theory – that the success of the Bethe and DeHoffman pion loop calculation was a big accident [34]. In fact, the pion loop calculation as naïvely performed does not even respect unitarity. When the latter requirement is imposed, the pion loop calculation falls short of the nucleon's isovector magnetic moment by half. FGT showed how to compute the pion cloud contribution correctly in dispersion theory by using measured strong interaction πN scattering amplitudes and π form factor data, effectively summing the loop graphs to all orders in the strong couplings. Remarkably, they obtained stunning agreement with the measured nucleon form factors.

In the 1990's, my collaborators and I applied the FGT dispersion theory approach to the strange quark form factors, focusing on the kaon cloud contribution that undergirded so many model computations [35,36,37,38]. Using experimentally obtained KN scattering amplitudes and e^+e^- data, we performed the all-orders summation of the kaon loop graphs analogous to the one carried out by FGT. We found that the resulting spectral functions for the strangeness vector form factors have a peak at invariant mass $t \approx m_\phi^2$. In short, the kaon cloud contribution is dominated by the $\phi(1020)$ resonance – consistent with the earlier pole-dominance *ansatz*.

This result has two implications. First, the plethora of kaon cloud model calculations are all wrong; only an all-orders summation is credible. Second, the kaon cloud cannot be the whole story, because the large effect implied by the ϕ-resonance is ruled out by experiment. In principle, this situation is not incompatible with dispersion theory, which implies contributions from a tower of intermediate states. In practice, however, we have reached the end of the road, because there simply is not enough high-quality e^+e^- and strong interaction scattering data involving these other states[1] . In the case of the isovector form factors, FGT got lucky because the pion cloud contribution – when computed correctly – saturated the experimental form factor result. There was no need to analyze other, higher mass intermediate states. The strange quark, in contrast, is more elusive. The PVES experiments suggest that contributions from other intermediate states cancel against the kaon cloud, but theoretically, we simply cannot compute these canceling effects. Parenthetically, my experience with this dispersion theory analysis leads me to view simple meson cloud model computations of other nucleon structure effects with a degree of skepticism.

[1] For a quark model treatment of part of this tower of states, see [39]

It is reasonable to ask, however, why one has to rely on dispersion theory in the "modern era". Indeed, the effects of pseudoscalar mesons in various aspects of nucleon structure – such as the isovector form factors or polarizablities – have been treated with considerable success in chiral perturbation theory (ChPT). In ChPT, any observable is given by the sum of a loop contribution and an LEC, or counterterm[2]. Even though one never carries out an all-orders loop calculation as dispersion theory suggests one should, this approach is consistent with dispersion theory because of the presence of the LEC contribution and because for low-energy processes, higher-order effects are suppressed by p/Λ, where Λ is either the nucleon mass or chiral symmetry-breaking scale. The actual value of the LEC, however, cannot be predicted theoretically; it has to be taken from experiment. To the extent that one has enough independent experiments to determine all the LECs at a given order in p/Λ, one can then make predictions for other observables.

The problem for strange quarks is that we don't have the required set of independent experiments. As Ito and I pointed out some time ago, the leading-order LEC's contain an SU(3)-singlet component that cannot be taken from any existent experiments [40]. In order to determine these constants, one has to measure the very strange quark form factors that one would like to predict – a situation of circular logic. Formally, an exception occurs in the case of the strange magnetic radius, $\langle r_s^2 \rangle_M$. As shown by Meissner, Hemmert, and Steininger, the leading-order contribution is entirely non-analytic, so that a counterterm-free prediction can be obtained from a one-loop computation [41]. Unfortunately, the chiral expansion for SU(3) is slowly-converging, since the relevant expansion parameter is $m_k/\Lambda \sim 1/2$. Thus, one might worry that higher-order effects are not negligible. In fact, my collaborators and I showed that – in the case of the strange magnetic radius – the next-to-leading order (NLO) loop contribution cancels most of the LO loop effect, exposing one to a dependence on the NLO LEC, b_s^r [42]. The resulting expression for $\langle r_s^2 \rangle_M$ is

$$\langle r_s^2 \rangle_M = -\left[0.04 + 0.3\, b_s^r\right]\ \text{fm}^2 \quad , \tag{1}$$

where the first term on the RHS gives the loop contribution evaluated at a renormalization scale $\mu = 1$ GeV and the second term is the NLO counterterm contribution. Naturalness considerations suggest that b_s^r should have magnitude of order unity, implying that the second term on the RHS of (1) dominates over the first. Since the precise value of b_s^r cannot be determined except by measuring the strange magnetic radius itself, one is back to the original problem. It seems there is no free lunch with strange quarks. The situation is not entirely bleak, as one can get some indications of the size of the LEC from either dispersion theory or lattice QCD, and a reasonable (though not rigorous) range of predictions can be obtained in this case. Indeed, the final SAMPLE result for the strange magnetic moment – which requires extrapolation to $Q^2 = 0$ by using the magnetic radius – includes this range in its quoted error [18].

It goes without saying that one would like to carry out reliable, microscopic calculations of the strangeness form factors in QCD, and the lattice is our only tool for doing so. Some time ago, the Kentucky-Adelaide collaboration obtained a non-zero signal for the form factors in a quenched calculation, the results of which suggested a negative strangeness magnetic moment [43]. More recently, the authors of [44] performed an quenched computation and found no evidence for non-zero strangeness form factors. The two computations differed in the number of gauge configurations and Z_2 noise vectors employed. To my knowledge, the two groups have not yet sorted out the reasons for the difference in their two results. As we also heard in this meeting, D. Leinweber and collaborators have exploited quenched lattice results and the assumption of charge symmetry to predict the strangeness magnetic moment. The details may be found in his talk, but the approach has successfully reproduced the measured octet baryon magnetic moments. The predicted strangeness magnetic moment is negative: $\mu_s = -0.051 \pm 0.021$ [45]. The conundrum for this prediction as well as for the previously reported lattice results is that the PVES experiments suggest a *positive* value for μ_s (see, *e.g.*, [26]). If these indications are solidified with the final results from the experiments, more work will be needed to understand the sign of μ_s from first principles in QCD (for an observations regarding the possible role of C transformation properties, see [46]) .

Finally, let me note that hadron models may provide some insights into the strangeness form factors. At present, one model – the chiral quark soliton model – gives the only prediction for a positive μ_s that I am aware of [47]. In light of all other theoretical work that seems to suggest a negative sign, this prediction must be taken seriously. Since I am not an expert on this model, I cannot comment in detail on what it may mean for the QCD dynamics of strange quarks. It is suggestive, however, that the topology of the QCD vacuum has a nontrivial impact on sea quark dynamics.

3 Radiative corrections

A significant consideration in the theoretical interpretation of the PV asymmetries and their implications for strange quarks has been contributions from electroweak radiative corrections. In high energy processes, one can compute these corrections with a high degree of reliability. At the low energies relevant to the PVES experiments, the situation is quite different due to the interplay of the strong interaction and higher-order electroweak effects. The electroweak Ward Identities protect the weak neutral vector current from strong interaction renormalization, but the the axial vector current can experience substantial effects. Since the axial vector response contributes to a generic PV asymmetry, one has to take into

[2] Usually, one takes for the loop contribution only the part of a loop result that is non-analytic in external momenta or masses that cannot be written down in a Lagrangian.

account the *uncertainties* associated with these axial vector radiative correction effects. Uncertainties also arise in the vector channel *via* box diagrams in which two electroweak gauge bosons are exchanged; no symmetry protects one from QCD effects in this case. Fortunately, however, the potentially largest uncertainties, which would appear in the $Z-\gamma$ box graphs, are fortuitously suppressed by $1-4\sin^2\theta_W \sim 0.1$ [48,76]. Consequently, one's primary concern involves the axial vector channel.

As part of my Ph.D. research – carried out in collaboration with Barry Holstein – I showed that one class of axial vector corrections – the so-called "anapole moment" terms – can be both surprisingly large and theoretically uncertain [49]. These corrections involve the exchange of a virtual γ between the lepton and hadron target, so their effect in neutrino-hadron scattering is suppressed by an additional power of $\alpha/4\pi$ compared to the effect in PVES. Consequently, it is useful to distinguish between the axial vector form factor as measured in electron scattering, G_A^e, from the corresponding form factor probed by neutrinos, G_A^ν. As Holstein and I pointed out in [49], the uncertainties associated with the anapole contributions to G_A^e are at least as large as the contribution expected from strange quarks (as extrapolated from polarized deep inelastic scattering), so PVES provides a rather poor probe of axial vector strangeness. Neutrino scattering is theoretically much cleaner, since the radiative correction uncertainties are insignificant.

Subsequent to this work, Donnelly and I observed that the radiative correction uncertainties in G_A^e could mimic the effect of strange magnetism on the elastic, **e***p* PV asymmetry [50]. Consequently, without a better handle on G_A^e, one faced an intrinsic, theoretical uncertainty in the value of G_M^s that could be extracted from PVES experiments. More generally, this kind of delicate interplay between various theoretical inputs – including radiative corrections – and the extracted values of the strange quark form factors was laid out in [51]. Fortunately, Hadjimichael, Poulis, and Donnelly observed that by measuring the PV asymmetry in quasieslastic (QE) **e***D* scattering, one could obtain a different linear combination of G_M^s and G_A^e than enters elastic **e***p* scattering, thereby facilitating an *experimental* separation of both [52]. With this goal in mind, the SAMPLE Collaboration measured both asymmetries. As initially reported, the results implied a value for G_A^e consistent with zero and a value of G_M^s that was somewhat positive [20]. The result for G_A^e was particularly surprising. Originally, Holstein and I predicted that radiative corrections would reduce G_A^e by roughly 30%, but the SAMPLE results suggested a 100% reduction. This surprise set off a flurry of theoretical efforts to explain the large effect. My collaborators and I up-dated the earlier work with Holstein using heavy baryon chiral perturbation theory and found little room for a larger effect than originally predicted [53]. Others explored the Q^2-dependence of the anapole form factor [54,55], quark model estimates of the anapole contribution [56], and nuclear PV effects in the **e***D* reaction [57,58]. Despite this hard work, no one could explain the SAMPLE result. One remaining suspect remained to be analyzed: the $Z-\gamma$ box contribution. In contrast to its effect in the vector channel, its contribution to the axial response is not $1-4\sin^2\theta_W$ suppressed, so it was thought that a large effect might be present here. Subsequently, however, the SAMPLE collaboration reanalyzed the deuterium data and found three previously underestimated corrections [19]. The largest involved neutral pion backgrounds. The resulting shifts in the PV QE asymmetry was just enough to change the extracted value of G_A^e, bringing it into agreement with the original Musolf and Holstein prediction.

What we have learned from this experience is that nonperturbative QCD effects on electroweak radiative corrections in the axial vector channel can be significant, and one must take them into account somehow. Doing so has become important not only for the strange quark program, but also for interpretation of the G0 measurement of the PV $N \to \Delta$ asymmetry. Here, the same kind of anapole effects that enter the elastic asymmetry can be as significant [59]. Consequently, the normalization of the axial vector transition form factor probed by PVES, $G_A^{\Delta,e}$ at $Q^2=0$, will contain a theoretical radiative correction uncertainty of order 10-20%. As Shi-lin Zhu and I recently showed, a determination of these corrections could be achieved by combining an experimental value for $G_A^{\Delta,e}(Q^2=0)$ with the off-diagonal Goldberger-Treiman relation (ODGTR), as the latter predicts the value of $G_A^{\Delta}(Q^2=0)$ in the absence of these corrections [60]:

$$G_A^{\Delta}(0) = \sqrt{\frac{2}{3}}\frac{g_{\pi N\Delta}F_\pi}{m_N}(1-\Delta_\pi) \quad , \tag{2}$$

where $g_{\pi N\Delta}$ is the strong $\pi N\Delta$ coupling constant that is known with 5% accuracy from πN scattering in the resonance region, F_π is the pion-decay constant, and Δ_π is a chiral correction that is of order a few percent. Since $G_A^{\Delta,e}(0)$ differs from $G_A^{\Delta}(0)$ by the large electroweak radiative corrections, a measurement of the former with PVES will primarily probe these radiative corrections.

In addition, a new effect arises in the inelastic asymmetry that does not occur for elastic scattering. Specifically, the asymmetry no longer vanishes at $Q^2=0$. This result is a consequence of Siegert's theorem [61], which implies that elastic matrix elements of the $E1$ operator vanish (even in the presence of parity-mixing in the initial and/or final states), but that they need not do so for inelastic reactions. The resulting, low-Q^2 $N \to \Delta$ asymmetry has the form

$$A_{LR}^{N\to\Delta}(Q^2=0) = -2\frac{d_\Delta}{C_3^V}\frac{m_N}{\Lambda_\chi} + \cdots \tag{3}$$

where C_3^V is the transition magnetic form factor, d_Δ parameterizes the $N \to \Delta$ $E1$ amplitude, and the "$+\cdots$ indicate higher order chiral corrections.

This effect for the $N \to \Delta$ asymmetry was first pointed out by in [62], where we argued that measuring d_Δ could be important for two reasons. First, without knowing its value, any attempt to determine the Q^2-dependence of $G_A^{\Delta,e}$ with PVES could be confused by this term. Second, d_Δ is the neutral current analog of the $E1$ amplitudes for

PV, electromagnetic hyperon decays, such as $\Sigma^+ \to p\gamma$. The asymmetries associated with the latter are of order four times larger in magnitude than one would expect based on simple symmetry considerations – a puzzle that has largely eluded explanation. Studying the neutral current analog (*i.e.*, d_Δ) could, in principle, shed new light on this old problem. New efforts are underway to develop experiments that would determine d_Δ.

More recently, another aspect of radiative corrections – those involving the exchange of two photons – has grabbed the attention of the PVES community. While this effect is pure QED and conserves parity, one manifestation of it can be measured using similar methods to those used in the PV experiments. Specifically, by scattering transversely polarized electrons from a given target and measuring the transverse-spin asymmetry (or vector analyzing power), A_n, one probes the imaginary part of the two-photon exchange amplitude, $\mathcal{M}_{\gamma\gamma}$. The vector analyzing power (VAP) has been measured by the SAMPLE Collaboration at relatively low energy [65] and by the Mainz A4 Collaboration at higher energy [66]. The SAMPLE result differs substantially from an old, potential scattering prediction carried out by Mott [67], leading to considerable theoretical interest in this observable. Although the VAP has been discussed by several others in this meeting, I would like to comment on why I think it is an important topic to pursue experimentally and theoretically.

By now it is well-known that a proper treatment of the real part of $\mathcal{M}_{\gamma\gamma}$ may help resolve the apparent differences in the Q^2-dependence of the proton electric form factor as determined polarization transfer experiments *vs.* Rosenbluth separation (for a discussion, see, *e.g.*, [63,64] and references therein). Clearly, having an experimental test of theoretical calculations of $\mathcal{M}_{\gamma\gamma}$ is important from this standpoint, and the VAP provides such a test. It is also important for our understanding of electroweak radiative corrections. Indeed, the kinds of two-boson exchange box graphs mentioned above have to be computed theoretically in order to arrive at Standard Model predictions for various electroweak observables. In particular, the dominant theoretical uncertainty associated with the interpretation of neutron and nuclear β-decay involves such graphs, where one of the exchanged bosons is a $W^\pm$ and the other is a γ. In order to extract a value for the CKM matrix element V_{ud} from these measurements, one must compute the $W\gamma$ box contribution. The theoretical machinery needed for this computation is the same as required for $\mathcal{M}_{\gamma\gamma}$, but there is no hope of ever measuring $\mathcal{M}_{W\gamma}$ directly. Hence, the VAP may provide our only experimental test of the theoretical framework used in computing this important electroweak radiative correction.

The β-decay correction applies to a very low-energy process, for which the use of effective field theory (EFT) might be particularly suitable. Similarly, the recent SAMPLE measurement of the VAP was performed in the low-energy/EFT domain. With this context in mind, my student L. Diaconescu and I recently computed the VAP for low-energy, elastic *ep* scattering using an EFT approach. Our primary objective was to determine if one could use EFT to resolve the discrepancy between the SAMPLE result and the old potential scattering calculation performed by Mott. If successful, we would have some confidence in applying the same framework to the electroweak box correction. The details of our calculation can be found in [68]. Let me comment, however, on a few aspects of the calculation.

First, given the technical challenge in carrying out analytic computations, we decided to work initially with an EFT involving only electrons, photons, and nucleons. In the heavy baryon formalism, the nucleon is essentially a "static" source, and there is no large momentum associated with this degree of freedom. In this EFT, the pion is integrated out, being treated as "heavy". While this assumption would certainly hold for the β-decay correction, it is admittedly questionable for the VAP at the SAMPLE kinematics. Nonetheless, we explored how much of the experimentally observed VAP could be accounted for by this simplest EFT. In this case, A_n has a power series expansion in p/M, where M is the nucleon mass and p is either the incident electron energy or electron mass. The Mott computation corresponds to the $\mathcal{O}(p/M)^0$ contribution – the one that survives for an infinitely heavy target. We showed that, working to $\mathcal{O}(p/M)^2$, one can make a parameter-free prediction for the VAP. To this order, there are no unknown LEC's, and the VAP arises entirely from a one-loop effect. The first unknown constants appear at $\mathcal{O}(p/M)^4$. To obtain a consistent computation to $\mathcal{O}(p/M)^2$, one must include both the nucleon magnetic moment and charge radius at the γNN vertices in the loop, as well as "kinetic" terms in the nucleon propagator.

To our surprise, we found that the $\mathcal{O}(p/M)^2$ VAP agrees beautifully with the SAMPLE result. Given that the momentum transfer is of order m_π, one would have expected that inclusion of pions would be necessary to produce agreement. What the result suggestions, however, is that for this particular observable, pions do not appear to play a significant role at these kinematics. As a corollary, it seems reasonable that using the same EFT (without dynamical pions) at the lower-energies relevant for β-decay will produce a reliable result for that process. Carrying out the latter computation is clearly a task for the future. On the other hand, taking the EFT expression for A_n to the higher energies relevant to the Mainz VAP measurement leads to significant disagreement – not a surprising outcome given that the Mainz energies are well beyond the limit of this EFT. For this domain, one must likely include dynamical pions and the Δ-resonance. I don't know how to do that in a model-independent way that retains the systematic power counting of EFT, so some other approach may be necessary. B. Pasquini has reported on one such attempt using the MAID program [69]. As I understand the results, the MAID calculation comes closer to the Mainz results than the EFT calculation, but fails to reproduce the lower energy SAMPLE result that can be explained using EFT (for other model computations, see, *e.g.*, [70,71]). It is clearly of interest to understand the reasons behind these differing theoretical results, and having more experimental data in the low- to medium-energy domain would be quite helpful.

4 New physics

A subject that is close to my heart – but one that I will spend less time on here – is the use of PVES to search for physics beyond the Standard Model. Two experiments are currently on the books: the E158 Møller experiment at SLAC [7] that has finished its data taking and the Q-Weak experiment that will take place in the future at JLab [8]. The beauty of these two experiments is that they are highly precise, that they are being carried out at essentially the same Q^2, and that they are theoretically "clean" probes of the Standard Model and its extensions. Indeed, these two experiments, when taken in tandem with the results of the Cesium atomic PV experiment, provide a unique diagnostic of new physics. My collaborators and I have recently illustrated this point in several papers, focusing particularly intently on supersymmetric extensions of the Standard Model [72,73,74,75,76]. I refer the reader to those papers for details. Even in the absence of new physics effects, the two experiments will provide the most precise determinations of the running $\sin^2\theta_W$ at a scale below the Z^0-pole (see, *e.g.*, [77,72,78,79]).

Looking to the future, the diagnostic power of the PVES experiments could be amplified with additional measurements. A more precise version of the Møller experiment is under consideration for JLab in its post-12 GeV upgrade phase. As I understand it, this experiment would be quite challenging, but several talented people are working on ways to meet this challenge. A second possibility would be to carry out a more precise version of the SLAC deep inelastic experiment. Recently, a proposal to do so was given high marks by the SLAC E-PAC, but the funds to carry out this experiment do not seem to be available. There is also considerable interest in performing a lower-energy version of the PV deep inelastic, or "DIS-Parity" experiment at the up-graded JLab. In this case, sorting out the theoretical implications of a precision measurement would be more involved than for the SLAC kinematics, primarily because of potential higher-twist (HT) contributions. The SLAC Q^2 would be sufficiently large that one could neglect these HT effects and perform a precision electroweak test. On the other hand, "twist pollution" will creep in at the JLab kinematics, and we do not have enough information on HT contributions to make definitive statements about their size at these kinematics. It may be that carrying out multiple measurements at different momentum transfers would allow one to simultaneously constrain the degree of twist pollution and deviations from the Standard Model electroweak contribution. Given that the figure of merit for the deep inelastic measurements is relatively large, such a program may be realistic.

Theoretically, several questions pertaining to the deep inelastic asymmetry remain to be studied extensively. In the case of HT, for example, one does not know what QCD predicts for the evolution of the corresponding structure functions. Similarly, the degree to which isospin violation in the leading twist parton distribution functions would affect the asymmetry is an interesting question. As we heard in K. McFarland's talk, the currently favored explanation of the NuTeV anomaly [80] is just such isospin violation. To be a viable one, this violation would have to be large, so that one might expect similarly large effects on the deep inelastic asymmetry. Finally, there is the question as to what combination of measurements would yield the most useful information on the electroweak sector of the Standard Model as well as on the aforementioned aspects of nucleon structure. Clearly, there is considerable room for future theoretical work on this topic.

5 Conclusions

It seems to me that the "modern era" of PVES experiments will be coming to an end in the next few years. With the conclusion of the SAMPLE, HAPPEX, G0, and A4 programs, the experimental questions about the size of the strange quark form factors and axial vector radiative corrections will have been settled; the ^{208}Pb measurement will have provided new information about the distribution of neutrons in heavy nuclei; and the first generation of precision electroweak tests (E158 and Q-Weak) will have produced results. I believe that those who have been involved in this field will be able to look back with considerable satisfaction at the unique and varied physics that this subfield of nuclear physics has illuminated. At the same time, there will be ample grist for the theoretical mill. If the present indications of a positive strangeness magnetic moment persist, it will be a challenge to provide a QCD explanation for this result and to understand the dynamics behind it. Similarly, if G0 and possible new experiments are able to cleanly separate the $G_A^{\Delta,\ e}$ and d_Δ contributions to the inelastic asymmetry, then new confrontations with hadron structure theory will be available. Finally, comparisons of first results from the LHC with information obtained from precision measurements – such as Cesium atomic PV, E158, Q-Weak, and possible second generation progeny – may help us determine which of the currently popular extensions of the Standard Model – if any – are most viable. In short, it appears that we have hardly heard the last word from parity violation.

Acknowledgements. I am indebted to my many theoretical and experimental collaborators and colleagues who are too numerous to list and who, over the years, have helped me understand various aspects of this field with greater clarity and insight. This work was supported in part by U.S. Department of Energy DE-FG03-02ER41215 and by a National Science Foundation Grant PHY00-71856.

References

1. M.J. Ramsey-Musolf: arXiv:nucl-th/0302049

2. B.P. Masterson, C.E. Wieman: in *Precision Tests of the Standard Model*, P. Langacker ed. (World Scientific, Singapore, 1995), Ch. IV. D
3. C.Y. Prescott et al.: Phys. Lett. B **77**, 347 (1978)
4. C.Y. Prescott et al.: Phys. Lett.B **84**, 524 (1979)
5. W. Heil et al.: Nucl. Phys. B **327**, 1 (1989)
6. P.A. Souder et al.: Phys. Rev. Lett. **65**, 694 (1990)
7. P.L. Anthony et al. [SLAC E158 Collaboration]: Phys. Rev. Lett. **92**, 181602 (2004), [arXiv:hep-ex/0312035]
8. D.S. Armstrong et al. [Qweak Collaboration]: AIP Conf. Proc. **698**, 172 (2004), [arXiv:hep-ex/0308049]
9. B.W. Filippone, X.D. Ji: Adv. Nucl. Phys. **26**, 1 (2001), [arXiv:hep-ph/0101224]
10. D.B. Kaplan, A. Manohar: Nucl. Phys. B **310**, 527 (1988)
11. R.L. Jaffe: Phys. Lett. B **229**, 275 (1989)
12. R.D. Mckeown: Phys. Lett. B **219**, 140 (1989)
13. G. Feinberg: Phys. Rev. D **12**, 3575 (1975), [Erratum-ibid. D **13**, 2164 (1976)]
14. J.D. Walecka: Nucl. Phys. A **285**, 349 (1977)
15. D.H. Beck: Phys. Rev. D **39**, (1989) 3248
16. J. Napolitano: Phys. Rev. C **43**, 1473 (1991)
17. E.J. Beise, M.L. Pitt, D.T. Spayde: Prog. Part. Nucl. Phys. **54**, 289 (2005)
18. D.T. Spayde et al. [SAMPLE Collaboration]: Phys. Lett. B **583**, 79 (2004), [arXiv:nucl-ex/0312016]
19. T.M. Ito et al. [SAMPLE Collaboration]: Phys. Rev. Lett. **92**, 102003 (2004), [arXiv:nucl-ex/0310001]
20. R. Hasty et al. [SAMPLE Collaboration]: Science **290**, 2117 (2000), [arXiv:nucl-ex/0102001]
21. D.T. Spayde et al. [SAMPLE Collaboration]: Phys. Rev. Lett. **84**, 1106 (2000), [arXiv:nucl-ex/9909010]
22. B. Mueller et al. [SAMPLE Collaboration]: Phys. Rev. Lett. **78**, 3824 (1997), [arXiv:nucl-ex/9702004]
23. K.A. Aniol et al. [HAPPEX Collaboration]: Phys. Rev. C **69**, 065501 (2004), [arXiv:nucl-ex/0402004]
24. K.A. Aniol et al. [HAPPEX Collaboration] Phys. Lett. B **509**, 211 (2001), [arXiv:nucl-ex/0006002]
25. G. Batigne [G0 Collaboration]: Eur. Phys. J. A **19**, SUPPL1206 (2004)
26. F.E. Maas et al.: arXiv:nucl-ex/0412030
27. F.E. Maas et al. [A4 Collaboration]: Phys. Rev. Lett. **93**, 022002 (2004), [arXiv:nucl-ex/0401019]
28. S. Baunack [A4 Collaboration]: Eur. Phys. J. A **18**, 159 (2003)
29. C.J. Horowitz, S.J. Pollock, P.A. Souder, R. Michaels: Phys. Rev. C **63**, 025501 (2001), [arXiv:nucl-th/9912038]
30. T.W. Donnelly, J. Dubach, I. Sick: Nucl. Phys. A **503**, 589 (1989)
31. H.W. Hammer, U.G. Meissner, D. Drechsel: Phys. Lett. B **385**, 343 (1996), [arXiv:hep-ph/9604294]
32. G. Ecker, J. Gasser, A. Pich, E. de Rafael: Nucl. Phys. B **321**, 311 (1989)
33. H. Bethe, F. deHoffman: *Mesons and Fields* (Row and Peterson, Evanston, 1955), Vol. II
34. P. Federbush, M.L. Goldberger, S.B. Treiman: Phys. Rev. **112**, 642 (1958)
35. H.W. Hammer, M.J. Ramsey-Musolf: Phys. Rev. C **60**, 045204 (1999), [Erratum-ibid. C **62**, 049902 (2000)], [arXiv:hep-ph/9903367]
36. H.W. Hammer, M.J. Ramsey-Musolf: Phys. Rev. C **60**, 045205 (1999), [Erratum-ibid. C **62**, 049903 (2000)], [arXiv:hep-ph/9812261]
37. M.J. Ramsey-Musolf, H.W. Hammer: Phys. Rev. Lett. **80**, 2539 (1998), [arXiv:hep-ph/9705409]
38. M.J. Musolf, H.W. Hammer, D. Drechsel: Phys. Rev. D **55**, 2741 (1997), [Erratum-ibid. D **62**, 079901 (2000)], [arXiv:hep-ph/9610402]
39. P. Geiger, N. Isgur: Phys. Rev. D **55**, 299 (1997), [arXiv:hep-ph/9610445]
40. M.J. Musolf, H. Ito: Phys. Rev. C **55**, 3066 (1997), [arXiv:nucl-th/9607021]
41. T.R. Hemmert, U.G. Meissner, S. Steininger: Phys. Lett. B **437**, 184 (1998), [arXiv:hep-ph/9806226]
42. H.W. Hammer, S.J. Puglia, M.J. Ramsey-Musolf, S.L. Zhu: Phys. Lett. B **562**, 208 (2003), [arXiv:hep-ph/0206301]
43. S.J. Dong, K.F. Liu, A.G. Williams: Phys. Rev. D **58**, 074504 (1998), [arXiv:hep-ph/9712483]
44. R. Lewis, W. Wilcox, R.M. Woloshyn: Phys. Rev. D **67**, 013003 (2003), [arXiv:hep-ph/0210064]
45. D.B. Leinweber et al.: arXiv:hep-lat/0406002
46. X.D. Ji, J. Tang: Phys. Lett. B **362**, 182 (1995), [arXiv:hep-ph/9507465]
47. A. Silva, H.C. Kim, K. Goeke: Eur. Phys. J. A **22**, 481 (2004), [arXiv:hep-ph/0210189]
48. W.J. Marciano, A. Sirlin: Phys. Rev. D **29**, 75 (1984), [Erratum-ibid. D **31**, 213 (1985)]
49. M.J. Musolf, B.R. Holstein: Phys. Lett. B **242**, 461 (1990)
50. M.J. Musolf, T.W. Donnelly: Nucl. Phys. A **546**, 509 (1992), [Erratum-ibid. A **550**, 564 (1992)]
51. M.J. Musolf, T.W. Donnelly, J. Dubach, S.J. Pollock, S. Kowalski, E.J. Beise: Phys. Rept. **239**, 1 (1994)
52. E. Hadjmichael, G.I. Poulis, T.W. Donnelly: Phys. Rev. C **45**, 2666 (1992)
53. S.L. Zhu, S.J. Puglia, B.R. Holstein, M.J. Ramsey-Musolf: Phys. Rev. D **62**, 033008 (2000), [arXiv:hep-ph/0002252]
54. C.M. Maekawa, U. van Kolck: Phys. Lett. B **478**, 73 (2000), [arXiv:hep-ph/0006161]
55. C.M. Maekawa, J.S. Veiga, U. van Kolck: Phys. Lett. B **488**, 167 (2000), [arXiv:hep-ph/0006181]
56. D.O. Riska: Nucl. Phys. A **678**, 79 (2000), [arXiv:hep-ph/0003132]
57. R. Schiavilla, J. Carlson, M. Paris: Phys. Rev. C **67**, 032501 (2003), [arXiv:nucl-th/0212038]
58. C.P. Liu, G. Prezeau, M.J. Ramsey-Musolf: Phys. Rev. C **67**, 035501 (2003), [arXiv:nucl-th/0212041]
59. S.L. Zhu, C.M. Maekawa, G. Sacco, B.R. Holstein, M.J. Ramsey-Musolf: Phys. Rev. D **65**, 033001 (2002), [arXiv:hep-ph/0107076]
60. S.L. Zhu, M.J. Ramsey-Musolf: Phys. Rev. D **66**, 076008 (2002), [arXiv:hep-ph/0207304]
61. A.J.F. Siegert: Phys. Rev. **52**, 787 (1937)
62. S.L. Zhu, C.M. Maekawa, B.R. Holstein, M.J. Ramsey-Musolf: Phys. Rev. Lett. **87**, 201802 (2001), [arXiv:hep-ph/0106216]
63. J. Arrington: Phys. Rev. C **69**, 022201 (2004), [arXiv:nucl-ex/0309011]
64. P.A.M. Guichon, M. Vanderhaeghen: Phys. Rev. Lett. **91**, 142303 (2003), [arXiv:hep-ph/0306007]
65. S.P. Wells et al. [SAMPLE collaboration]: Phys. Rev.C **63**, 064001 (2001), [arXiv:nucl-ex/0002010]
66. F.E. Maas et al.: arXiv:nucl-ex/0410013
67. N.F. Mott: Proc. R. Soc. London, Ser. A **135**, 429 (1932)
68. L. Diaconescu, M.J. Ramsey-Musolf: Phys. Rev. C **70**, 054003 (2004), [arXiv:nucl-th/0405044]

69. B. Pasquini, M. Vanderhaeghen: Phys. Rev. C **70**, 045206 (2004), [arXiv:hep-ph/0405303]
70. A.V. Afanasev, N.P. Merenkov: Phys. Lett. B **599**, 48 (2004), [arXiv:hep-ph/0407167]
71. A.V. Afanasev, N.P. Merenkov: Phys. Rev. D **70**, 073002 (2004), [arXiv:hep-ph/0406127]
72. J. Erler, A. Kurylov, M.J. Ramsey-Musolf: Phys. Rev. D **68**, 016006 (2003), [arXiv:hep-ph/0302149]
73. A. Kurylov, M.J. Ramsey-Musolf, S. Su: Phys. Rev. D **68**, 035008 (2003), [arXiv:hep-ph/0303026]
74. A. Kurylov, M.J. Ramsey-Musolf, S. Su: Phys. Lett. B **582**, 222 (2004), [arXiv:hep-ph/0307270]
75. M.J. Ramsey-Musolf: Phys. Rev. D **62**, 056009 (2000), [arXiv:hep-ph/0004062]
76. M.J. Ramsey-Musolf: Phys. Rev. C **60**, 015501 (1999), [arXiv:hep-ph/9903264] [77,78,79]
77. A. Czarnecki, W.J. Marciano: Phys. Rev. D **53**, 1066 (1996), [arXiv:hep-ph/9507420]
78. A. Ferroglia, G. Ossola, A. Sirlin: Eur. Phys. J. C **34**, 165 (2004), [arXiv:hep-ph/0307200]
79. J. Erler, M.J. Ramsey-Musolf: arXiv:hep-ph/0409169
80. G.P. Zeller et al. [NuTeV Collaboration]: Phys. Rev. Lett. **88**, 091802 (2002), [Erratum-ibid. **90**, 239902 (2003)], [arXiv:hep-ex/0110059]

Eur Phys J A (2005) **24**, s2, 205–208
DOI: 10.1140/epjad/s2005-04-050-x

EPJ A direct
electronic only

Workshop summary

R.D. McKeown

California Institute of Technology, Pasadena, CA 91125, USA

Received: 15 October 2004 / Published Online: 8 February 2005

Abstract. This latest in a series of workshops on parity-violating electron scattering comes at a momentous time in the history of this subject. The first experiments to determine strange form factors of the nucleon have produced intriguing final results, and several powerful new experiments are now producing data. In addition, the precision of the technique has been improving and new experiments testing the electroweak theory have reported remarkably precise data. There has also been a great deal of progress on both the theory of strange form factors and interpretation of electroweak symmetry tests.

PACS. 01.30.Cc Conference proceedings – 25.30.Bf Elastic electron scattering – 11.30.Er Charge conjugation, parity, time reversal, and other discrete symmetries

1 Introduction

It has been 15 years since the first papers [1,2,3] proposing to study strange quark contributions to nucleon electroweak form factors gave new impetus to the field of parity-violating electron scattering. Indeed shortly after those papers in February of 1990, the first of these workshops [4] was held in Pasadena, California. The warm and sunny winter weather seemed to portend a bright future (we didn't know the workshop should have been called PAVI-90!) and provided additional motivation for participants from colder climes in the US, Canada, and Germany to attend. That workshop certainly reaffirmed the motivation for planning a program of parity-violating electron scattering experiments to explore the strangeness in the nucleon. But it also served to highlight many concerns on both the theoretical and experimental sides, which have fortunately all been mitigated through diligent and clever efforts in the subsequent years.

Since that time, we have seen truly remarkable progress in this field with great strides made by theorists as well as experimentalists. This workshop was an excellent opportunity to take stock and chart a new course for the future. The organizers should be commended for providing a stimulating program and delightful environment, and I certainly hope that their plans to continue this tradition are fruitful.

2 Strange form factors

The first proposed experiment to study the strangeness in the nucleon using parity-violating electron scattering was the SAMPLE [5] experiment at MIT/Bates. As reported at this workshop by D. Spayde, the SAMPLE experiment has measured the strange magnetic form factor for the first time, with the result [6]

$$G_M^s(Q^2 = 0.1(\mathrm{GeV/c})^2) = 0.37 \pm 0.20 \pm 0.26 \pm 0.07\,. \quad (1)$$

Although this result is consistent with zero strangeness, it is not consistent with the prevailing theoretical view as presented in Fig. 1. Most theoretical predictions favor a substantially negative strange magnetic moment, whereas the SAMPLE result indicates $\mu_s \geq 0$.

Figure 1 also indicates the level of interest in the quantity μ_s among theorists in this field. There have been a great number of studies, and this has made for very pleasant and productive interactions between theorists and experimentalists. What is difficult to tell from Fig. 1 is that we have actually learned something from this discourse. Early models with simplistic treatments of vector meson dominance or lowest order $K - \Lambda$ loops are known to be inadequate to reproduce the the experimental result, have well-studied theoretical shortcomings, and so are no longer used in modern calculations. Unfortunately, as reported in this conference by Kubis and Ramsey-Musolf, effective field theory is not very effective for these observables as there are too many unknown counterterms and the convergence of the series is, at best, very slow. It now appears that the remaining theoretical treatments that are consistent with the data are the chiral soliton model [9] (as discussed by Silva at this workshop) and the lattice-based treatment [10] as discussed by Leinweber. The prediction of the chiral soliton model for the SAMPLE result is shown as the black diamond at entry 25 in Fig. 1, one of the few results with $G_M^s > 0$. The Leinweber *et. al.* result is shown as the last magenta star (entry 27), and represents a bold and precise prediction

$$\mu_s = -0.051 \pm 0.021 \quad (2)$$

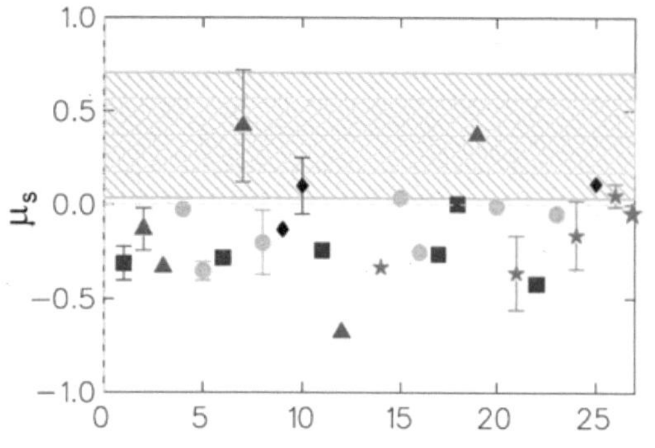

Fig. 1. Theoretical predictions for μ_s are shown as symbols, with the SAMPLE experimental result indicated by the hatched region, statistical uncertainty (*inner*) and total uncertainty (*outer*). The theoretical results are compiled and discussed in [7,8]

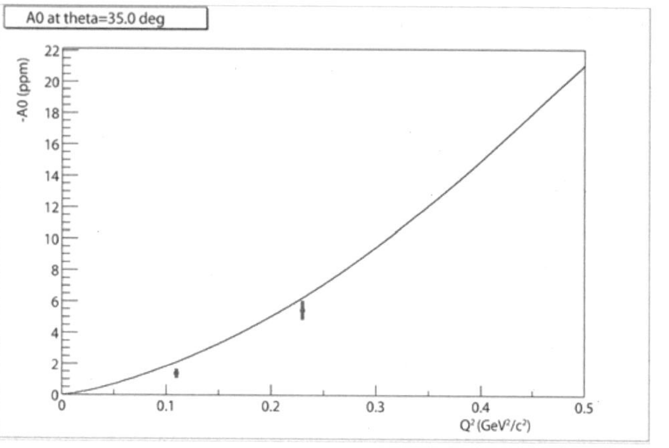

Fig. 2. Results reported at this workshop by the A4 collaboration. The *solid line* is the prediction for the case where both strange form factors vanish

that lies just outside the 1σ error bar of the experiment. There was much discussion about the justification for the very small quoted theoretical error, and I am sure that this discussion will continue at many additional conferences and workshops.

Of course during the last few years we have also seen the remarkably precise data [11] from HAPPEX. This experiment has the notable distinction of being the first parity-violation experiment to use a high polarization beam from a strained GaAs crystal. They obtain a result for the combination of form factors at $Q^2 = 0.477$ $(\mathrm{GeV/c})^2$:

$$\frac{G_E^s + 0.392 G_M^s}{G_M^p/\mu_p} = 0.091 \pm 0.054 \pm 0.039 \qquad (3)$$

which also can be used to rule out many models. Future planned running of HAPPEX involves lower $Q^2 = 0.1$ $(\mathrm{GeV/c})^2$ measurements of elastic asymmetries for Helium as well as Hydrogen targets.

And the major news at this conference are the new results from the Mainz A4 collaboration. They report asymmetries at $\theta = 35°$ at two Q^2 values, as shown in Fig. 2. These are very interesting results that may indicate a non-vanishing strange quark contribution, and the A4 collaboration has plans for many more measurements including backward angles.

In addition, we heard that the G^0 experiment has completed its production run at forward angles. The data are under analysis and will hopefully be available in the very near future. Clearly, many years of planning, systematic studies, and equipment construction are now yielding an impressive data set that will surely provide us with a clearer picture of strangeness in the nucleon.

The broader context of strangeness in the nucleon was also discussed at the workshop. M. Sainio updated the situation regarding the sigma term. And J. Ellis presented an overview discussing the relation to the spin of the nucleon, the possible connection with exotic baryon states, and ideas for evading the OZI-rule. It is intriguing to note that the chiral soliton model, which is uniquely successful in producing a result for G_M^s that quantitatively agrees with the SAMPLE experimental data, is also achieving significant notoriety for it's recent predictions of pentaquark states (for which there is increasingly substantial, but also controversial, experimental evidence).

3 Transverse spin asymmetries

The subject of transverse spin asymmetries is receiving increased attention in the last year or so. Due to the fact that it is higher order in the fine structure constant (i.e., 2 photon exchange) and it is further suppressed by the lorentz factor $1/\gamma$, there was not much previous interest in this subject. The existence of recent data from the SAMPLE experiment [12] demonstrated the feasibility of the method for measuring the small asymmetries (although larger than parity-violation asymmetries). In addition, there has been increased recent interest due to the apparently significant 2 photon exchange effects in the interpretation of proton form factor measurements at high Q^2 (see Sect. 4). At this workshop, we heard presentations of the SAMPLE data by D. Spayde as well as new data from A4 by S. Baunack. In addition, high energy data from SLAC E158 will be available soon.

The A4 collaboration measures transverse asymmetries with their luminosity monitors, or "Lumis" (mostly Møller scattering), as well as with their detectors for elastic scattering off protons. The Lumi data are quite precise (5%), but disagree with the NLO calculations for Møller scattering by 2-5 σ. This is not understood.

The theoretical interpretation of the transverse asymmetries in elastic *e-p* scattering was discussed by Pasquini and by Ramsey-Musolf. Pasquini *et al.* use the MAID description of photon couplings to the nucleon and the continuum. This prescription results in qualitatively correct behavior, but misses all the data by about 2 σ (they *underpredict* the magnitude of the SAMPLE asymmetry but *overpredict* the magnitude of the A4 asymmetries). Ramsey-Musolf presented calculations [13] in the framework of effective field theory. They obtain reasonable agreement with the SAMPLE data, but it seems that the A4 data are at too high a Q^2 for this treatment to be valid.

4 Other form factors

K. DeJager presented a thorough review of the situation with the elastic electromagnetic form factors of the nucleon. There are recent G_E^n data that are beginning to map out the Q^2 dependence of this elusive form factor with impressive precision. This form factor is an important constraint on nucleon structure as well as an important input in the interpretation of parity-violation experiments. Another major topic of current interest relates to the controversy regarding the G_E^p/G_M^p data from recoil polarization measurements [14] at $Q^2 > 1$ GeV^2. The form factor ratio extracted from these measurements disagrees with the results determined previously by Rosenbluth separation. More recently, new Rosenbluth separation data [15] taken by detection of the recoil proton (to reduce systematic errors and radiative corrections) give strong support for the earlier Rosenbluth separation data.

This conundrum appears to be resolvable by consideration of 2 photon exchange effects [16]. Guichon presented their analysis at this workshop, and it appears that quantitative agreement with the two data sets is possible although additional free parameters must be employed. Their analysis indicates that $2-3\%$ two photon exchange contributions with reasonable magnitudes can distort the Rosenbluth plots by about the amount necessary to resolve the discrepancy. Clearly further measurements are desirable, including comparison of positron and electron scattering cross sections to test the model quantitatively. If this explanation is correct, then it would seem prudent to revisit all cases where delicate Rosenbluth separations have been performed to extract small amplitudes (e.g. the famous $R = \sigma_L/\sigma_T$ in deep inelastic scattering).

E. Beise discussed a recent reanalysis [17] of the Q^2 dependence of the axial form factor of the nucleon as determined in quasielastic neutrino charged current interactions. That work produced a slightly smaller value of the axial mass as compared to previous studies: $M_A = 1.001 \pm 0.020$ GeV. This quantity is relevant to the interpretation of backward angle parity-violation experiments such as future planned measurements by G^0 and A4. These collaborations also plan to run with deuterium, which allows separation of the axial term (as was done for SAMPLE [18]).

5 Electroweak tests

The birth of parity-violating electron scattering was marked by the famous Prescott experiment at SLAC [19], which had the distinction to provide the first quantitative evidence for violation of parity in the neutral current as predicted by the standard model. Other early experiments in parity-violating electron scattering also were attempts at testing the standard electroweak theory, notably the Mainz ^{9}Be experiment and the Bates ^{12}C experiment. The advances in experimental techniques that have been achieved in the pursuit of strange quark form factors during the last decade have also motivated new higher precision experiments to test the standard electroweak model. At this workshop we heard a report by A. Vacheret on the SLAC E158 experiment, which measures parity-violating Møller scattering and a presentation of the plans for Qweak at JLab by G. Smith.

In a fundamental sense, the parity-violating effect in electron scattering relates to the weak neutral couplings to the electrons and quarks. At tree-level, these couplings depend on one parameter, the weak mixing angle θ_W. Thus the various precision parity-violating electron scattering experiments aim to constrain $\sin^2\theta_W$ and compare with the very precise value measured at the Z-pole in e^+-e^- scattering. Due to radiative corrections, the value of $\sin^2\theta_W$ "runs" with Q^2 in a predictable fashion according to the standard model. Thus these experiments test for new particles in loops and exchanges associated with these radiative corrections to search for evidence of new physics beyond the standard model.

Another method to precisely determine $\sin^2\theta_W$ at low Q^2 is via neutral current neutrino scattering. Recently, the NuTeV experiment at Fermilab has reported a measurement [20] of ratios of neutral current to charged current cross sections for both neutrinos and antineutrinos. They observe a substantial effect in the neutrino ratio, which they interpret as an anomalous value of ("on-shell" scheme)

$$\sin^2\theta_W = 0.2277 \pm 0.0013 \pm 0.0009 \qquad (4)$$

which is 3σ from the standard model prediction of 0.2227 ± 0.00037. K. McFarland presented these results and discussed many alternate explanations for the discrepancy. It appears that relatively large isospin violation in the parton distribution functions could generate the observed effect and would be consistent with all other experiments. The cause of the NuTeV anomaly remains a subject of much active study.

The recently published E158 results [21] are based on the data from the first of three data runs. The results reported at this workshop included the first and second data sets:

$$\sin^2\theta_W = 0.2379 \pm 0.0016 \pm 0.0013 \qquad (5)$$

which agrees well with the standard model prediction of 0.2386 ± 0.0006. (Note that E158 uses a different renormalization scheme to quote $\sin^2\theta_W$). The situation is shown graphically in Fig. 3.

The Qweak proposal [23] to JLab is to measure parity-violating elastic electron-proton scattering at very forward angles ($\theta \sim 8°$) to achieve a low $Q^2 = 0.028$ GeV^2. This will ensure that the strange quark effects in the electric form factor are small enough to be manageable. The high statistics will be achieved by using 180μA of electron beam from strained GaAs with high polarization. The 35 cm long liquid hydrogen target must absorb 2.5kW of beam power without boiling effects. (Previous high power targets at SAMPLE and E158 achieved 500W and maintained excellent thermal stability.) The goal is to measure

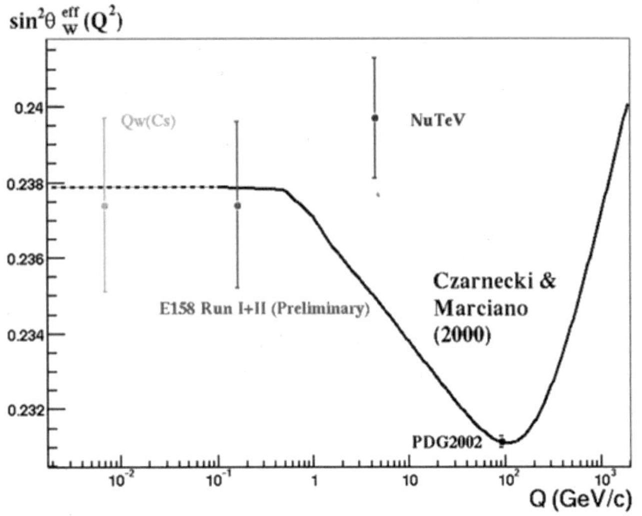

Fig. 3. Plot of effective $\sin^2\theta_W$ vs Q^2 showing the E158 result, NuTeV result, Z-pole value, and Cs atomic parity-violation. The theoretical curve is from [22]

$\sin^2\theta_W$ to a precision of ± 0.0007 (compared to the ultimate precision of E158 of $\sim \pm 0.0012$). It is hoped that this ambitious experiment could be mounted in 2007 and begin commissioning studies shortly thereafter. One should keep in mind that this is the time scale for LHC to start, and the future of such experiments in the post-LHC era is far from clear.

6 Outlook

The field of parity-violating electron scattering has entered an extremely productive phase. Over the next 3-5 years we should have in hand a definitive dataset mapping out the role of strange quarks in the electroweak form factors of the nucleon. It is difficult to judge how that will turn out and how the theoretical interpretation will develop in response to the data. Perhaps a round of higher precision experiments will be indicated. Already we see that if one takes the Leinweber, *et al.* prediction seriously, one could justify building a super-SAMPLE experiment with high polarization CW beam to achieve precision on μ_s comparable to this theoretical prediction.

Perhaps we will even have new data from Qweak in this time frame to further test the standard electroweak theory. And it might even be worth revisiting the method of elastic scattering from a spinless nucleus like ^{12}C to perform other high precision tests.

In the shorter term, it seems that the methods developed for parity-violation experiments can be used to obtain more data on transverse asymmetries to explore the 2 photon exchange process in more detail. At high Q^2 there could be a useful connection with the Generalized Parton Distributions (see the talk by M. Gorshteyn) and at low Q^2 there could be a connection with the nucleon polarizabilities. Clearly there is a great deal of experimental and theoretical work to do in this area, and we are just seeing the beginning of that endeavor.

So it seems that at least one more workshop on this topic is well-justified, and if the organizers can find a location that is of comparable quality to the Grenoble workshop I am sure we (and many new younger people) will be there to hear of the exciting developments that are sure to come.

References

1. D. Kaplan, A. Manohar: Nucl. Phys. B **310**, 527 (1988)
2. R.D. McKeown: Phys. Lett. B **219**, 140 (1989)
3. D.H Beck: Phys. Rev. D **39**, 3248 (1989)
4. E.J. Beise, R.D. McKeown: *Parity Violating Electron Scattering*, (World Scientific, Singapore, 1990)
5. Bates proposal 89-06, R.D. McKeown, D.H. Beck: spokespersons
6. D. Spayde et al.: Phys. Lett. B **583**, 79 (2004)
7. D.H. Beck, B.R. Holstein: Int. Journ. Mod. Phys. E **10**, 1 (2001)
8. D.H. Beck, R.D. McKeown: Ann. Rev. Nucl. **51**, 189 (2001)
9. A. Silva, H.-C. Kim, K. Goeke: Nucl. Phys. A **721**, 417 (2003)
10. D.B. Leinweber et al.: hep-lat/0406002
11. K.A. Aniol et al.: Phys. Lett. B **509**, 211 (2001)
12. S.P. Wells et al.: Phys. Rev. C **63**, 064001 (2001)
13. L. Diaconescu, M.J. Ramsey-Musolf: nucl-th/0405044
14. M.K. Jones et al.: Phys. Rev. Lett. **84**, 1398 (2000); O. Gayou et al.: Phys. Rev. Lett. **88**, 092301 (2002)
15. JLab experiment E01-001, J. Arrington, R. Segel: spokespersons
16. P.A.M. Guichon, M. Vanderhaeghen: Phys. Rev. Lett. **91**, 142303 (2003)
17. H. Budd, A. Bodek, J. Arrington: hep-ex/0308005
18. T. Ito el al.: Phys. Rev. Lett. **92**, 102003 (2004)
19. C.Y. Prescott et al.: Phys. Lett. B **77**, 347 (1978)
20. G.P. Zeller et al.: Phys. Rev. Lett. **88**, 091802-1 (2002)
21. P.L. Anthony et al.: Phys. Rev. Lett. **92**, 181602 (2004)
22. A. Czarnecki, W.J. Marciano: Int. J. Mod. Phy. A **15**, 2365 (2000)
23. JLab experiment E02-020: J. Bowman, R. Carlini, J. Finn, S. Kowalski, S. Page, spokespersons

Finito di stampare nel mese
di marzo 2005 da
Compositori Industrie Grafiche